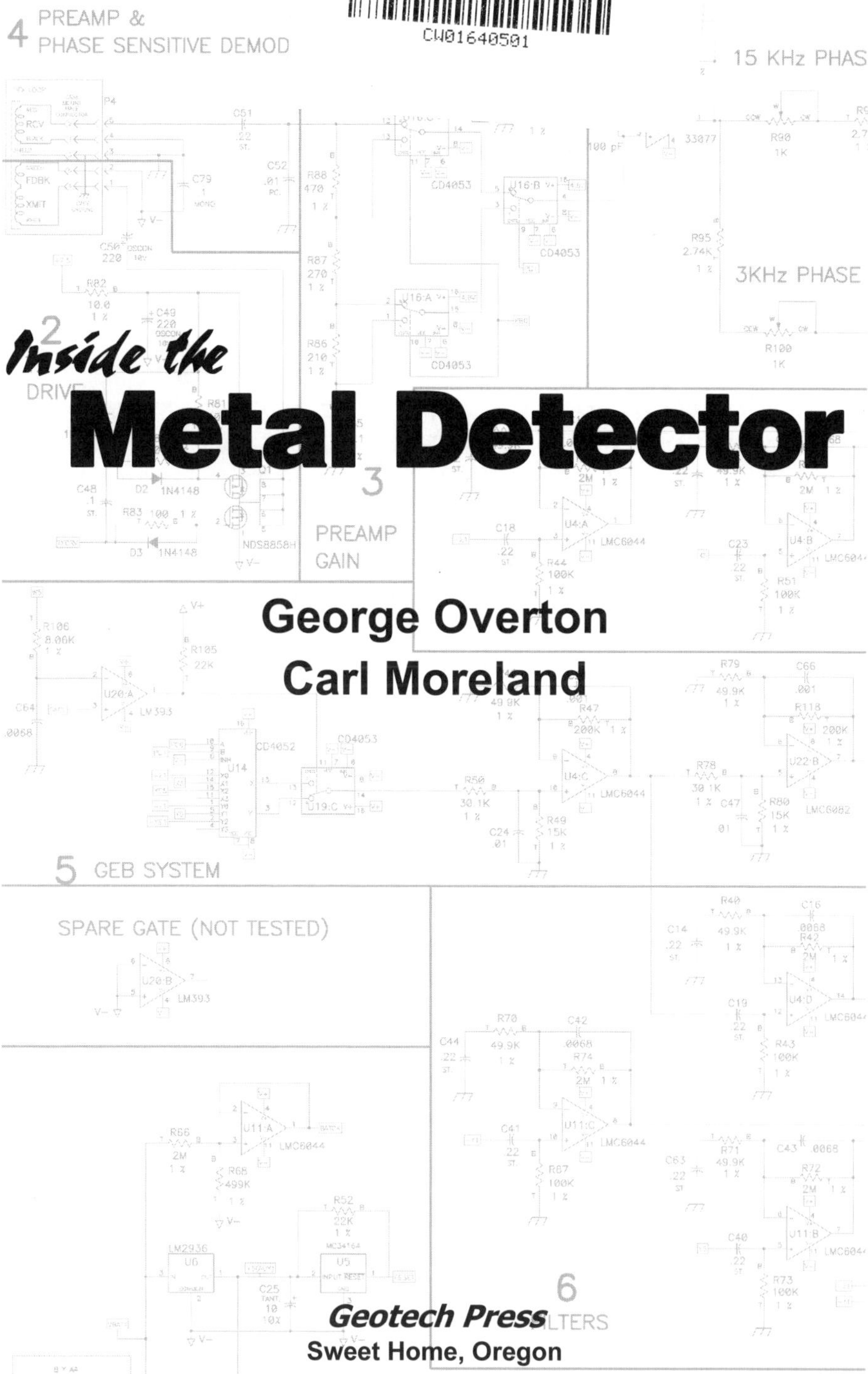

Inside the Metal Detector

George Overton
Carl Moreland

Geotech Press
Sweet Home, Oregon

Published by:
Geotech Press
PO Box 666
Sweet Home, OR 97386
www.geotech1.com

Cover: Inside the Bounty Hunter Red Baron; background schematic: White's XLT

The circuits presented in this book are for experimental purposes only. Some designs may be covered under active patents and may not be commercially developed without permission from the patent holder. Some of the projects might also fall under government regulation for radio emission. It is the responsibility of the user to ensure adherence of these regulations.

Some studies have indicated that devices which emit radio frequencies may be linked to certain health problems, including disruption of pacemakers. Although there has been no indication that metal detectors are among these devices, health safety is the responsibility of the user.

Neither the authors nor the publisher of this book shall be held responsible, whether legal, financial, or any other way, for any damages incurred by the use of the contents of this book.

This book was written in Adobe FrameMaker in New Times Roman 10pt. Line art was drawn in Microsoft Visio.

ISBN 978-0-9858342-0-3
[redacted]

Table of Contents

Introduction

"In science one tries to tell people, in such a way as to be understood by everyone, something that no one ever knew before. But in poetry, it's exactly the opposite."
— *Paul Dirac* (1902–1984)

It's hard to imagine a single instrument that has transformed treasure hunting more than the metal detector. The hobby of metal detecting barely existed in the 1950's, when surplus military mine detectors were just about the only metal locator available. Twenty years later, inexpensive transistorized metal detectors could be found everywhere. And the magnetometer, a close cousin of metal detectors, has been responsible for the vast majority of shipwreck discoveries since the 1960's.

The metal detecting hobby experienced its "boom" period a number of years ago, but even nowadays it remains a popular hobby amongst all ages. It is relatively inexpensive to get started; an entry level detector can be purchased for around $100, although you won't get much at that price, either in performance or quality. Many newcomers have made the mistake of buying a cheap discount-store detector only to get discouraged and quit. Although some high-end models can cost well over $1000, a decent quality entry-level detector can be found for $200-$300.

In the early years, metal detectors had to be tuned, balanced, adjusted, and tweaked continuously during use. On some models, the user really had to know exactly what all the knobs did, and how they interacted. But many of the newest models feature turn-on-and-go operation, automatic ground tracking, and "target identification," so today's user can get reasonable results with no knowledge about how the instrument actually works.

Most models now have such an array of features that it can be extremely confusing for the beginner deciding which machine to purchase. Regardless of whether a particular design is all-analog or microprocessor-controlled, has a simple tonal response or an LCD display, or has a single control knob or twenty, it must abide by the same basic electromagnetic principles as any other detector. The addition of more complexity, for the most part, provides more bells and whistles with little substantial gain in depth.

Even though metal detectors are not especially expensive, and many detectorists possess more than one model, the engineers amongst us experience an irresistible urge to develop our own homebrew detectors, to both understand the basic principles and to try and match the achievements of the commercial models.

Hence the purpose of this book.

This book is aimed at anyone with an interest in metal detectors, who would like to understand in a clear and easy way how a metal detector works, irrespective of whether it is of the BFO, IB or PI variety. Before we go any further, we should probably take a look at basic metal detector types, and what "BFO," "IB," and "PI" stand for.

Basic Types and the Contents of this Book

There are several ways to categorize metal detector types, but we will use three that are fairly popular with practically all detector engineers:

1. Frequency shift
2. Induction Balance
3. Pulse Induction

Frequency shift detectors are usually characterized by having a single loop in the search head that oscillates at a particular frequency, and in which targets cause a measurable shift in that frequency. By far, the most popular frequency shift design is the beat-frequency oscillator, or BFO. Since the BFO played such a prominent role in hobby detecting, Chapters 2 and 3 are dedicated to this technique.

There are other ways to design frequency shift detectors. One is called "off-resonance" and was popularized in the 1970's in an early discriminating detector called the "A.H. Pro." Chapter 4 covers off-resonance. Another method is to use a phase-locked loop (PLL) to detect the frequency shift, and these detectors are called, naturally, PLL designs.

The second major category is called induction balance, or IB, and is covered in Chapters 6-10. IB designs use a search head that contains both a transmit (TX) coil for transmitting a continuous waveform (usually a sinusoid) and a receive (RX) coil for receiving target signals. The coils are arranged in such a way that there in very little mutual inductance between them, and therefore very little transmit signal gets directly coupled to the receive coil. In this way, the coils are said to be *inductively balanced*, hence the name. There are many, many ways to balance coils, and Chapter 5 explores some of these.

Induction balance includes some other descriptive terms that have been popular. The so-called "transmit-receive" (T-R, T/R, or TR) metal detectors of the 1960's and 70's were really IB detectors. The term VLF (very low frequency) is used for IB detectors that operate in the VLF radio band, which is 3kHz-30kHz. This includes most IB detectors being sold today. In the early days of detectors, the "two-box" designs were often referred to as "RF" (radio frequency) designs but they, too, were really IB detectors.

The final category is pulse induction, or PI, covered in Chapters 11 and 12. Like BFOs, these detectors can operate with only one coil which acts as both the transmit and receive coil. It does so by transmitting a rapid impulse, then switching to receive mode and listening to the impulse response. However, it is possible to use separate TX and RX coils for this and they can be induction balanced, though it's not necessary. Even when a PI detector uses an IB coil arrangement, we still refer to it as a PI detector. A major difference is that the PI detector transmits an impulse train, not a continuous-time waveform.

Besides metal detectors, treasure hunters use a variety of other instruments. One popular device is the magnetometer, including the "proton precession" design whose operation is somewhat similar to pulse induction. Other methods include ground-penetrating radar and side-scan sonar. These devices are beyond the scope of this book. But another device that shows up quite often is the so-called *long-range locator* (LRL). It is covered in Chapter 13 of this book because the LRL (and the basis of most LRLs, dowsing) doesn't work[1] and are often the realm of unscrupulous "wallet miners." It is important to see where the limits of technology lie, and why LRL claims are based on nonsense.

Finding Technical Information

In the realm of consumer electronics, it's remarkable just how little information has been published on metal detector technology. With other electronic technologies—radio, television, computers, remote control, etc.—there not only exist vast amounts of information about the theory, but often full schematics of consumer products are easily obtainable. Not so with metal detectors. Companies rarely will provide schematics, and often take steps to prevent reverse-engineering, such as potting circuits in epoxy, grinding off IC part numbers, or using code-protected microcontrollers.

But there have been a few books on detector technology. In 1927, a book with the strange title "The Modern Divining Rod" by R. J. Santschi offered what was probably the first technical book on metal detectors. Most detectors of that era were the orthogonal two-box locators, such as those sold by Fisher Labs, and all had tube-based circuits. Santschi not only covered the basic types of locators of his era, but did a masterful job with technical analyses and included a number of full schematics. Ironically, the book does not cover divining rods.

A book in 1969 by Dr. Arnold Kortejarvi, "Official Handbook of Metal Detectors," included a chapter on "How Detectors Function." Also in 1969, treasure hunter E. S. "Rocky" LeGaye produced "The Electronic Metal Detector Handbook," of which a majority is dedicated to technical descriptions of detectors. He included both BFO and induction-balance, though not PI which was in its infancy and not well-known.

1. Do you doubt this statement? Read Chapter 13!

LeGaye provided little in the way of actual schematics. Most other books on detector technology are more focused on do-it-yourself projects, such as "Metal Detector Projects" by Charles Rakes, "Metal Locators" by Traister and Traister, "Proximity Sensors and Metal Locators" by John Potter, and "How to Build Your Own Metal & Treasure Locators" by F.G. Rayer. Although these books are chocked full of project circuits, they offer little in the way of theory and detailed technical description.

Besides these few books, there have been a fair number of articles published in various electronics hobby magazines, describing metal detector projects. Although a few of these articles made decent attempts at explaining theory, most of the projects have been variations of basic designs, and most do not cover operational theory very well. Appendix B includes a list of many of the better designs.

Perhaps the most valuable source of technical information are the patents that have been filed by detector manufacturers. A patent is usually very thorough in describing all the details of the invention and, in the case of metal detectors, often include schematics. However, patents tend to be written rather obscurely, often in intentionally obfuscated legalese, perhaps to confuse the reader as to exactly what new circuit they have come up with. Appendix B contains a list of several key patents, and also what to watch out for when walking through a patent minefield.

Finally, the Internet is slowly accumulating technical information on metal detectors. Several web sites and forums focus largely on the technological aspects, and many schematics and articles are now available if you dig around a little. Appendix A lists some of these web sites and forums.

Why bother?

So why would anyone want a book on how metal detectors work? With very capable detectors available at low prices, relatively few people are going to be motivated to build their own. But there are those who like to experiment, and this book will provide an excellent starting point in detector technology. For those who are inclined to build circuits, the projects offered in this book range from a simple and seemingly useless detector, up to some quite powerful designs that rival the performance of many commercial models. Some of the projects are:

- Multi-purpose BFO detector
- Off-resonance pinpointing probe
- Basic TR detector
- VLF motion discriminator
- Saltwater pulse-induction detector

In addition, some engineering hints are included that might allow a more experienced hobbyist to wring out a little more performance from a given design. In the end you can wind up with several different detectors, each optimized for a specific task, all for about the same money as a mid-level general-purpose commercial unit.

For those who are not inclined to build circuits, there is something that is of value to any detectorist: knowledge of how a detector works. The most successful detectorists are those who have a good understanding of what all those knobs are doing; they know how to get the most out of their machine. How does discrimination work? What, exactly, is happening when you ground balance a detector? What are the advantages and disadvantages of different coils? Even if you never build a detector circuit, this book will help answer those questions, and more. But this book does not cover detecting techniques or the need to do proper research; many other books are available that attempt to do that.

Building Detectors

For those interested in building circuits, unless you are an experienced electronics hobbyist, it is not advised that you jump straight to an advanced project. You might end up with something that does not work, and you will not have a clue as to why. One of the things this book does not attempt to do is teach basic electronics. It is assumed you know Ohm's Law, those of Kirchhoff, transistor basics, opamps, filtering, and some trigonometry. There are many other books that teach these concepts.

As in using a metal detector, building one requires patience. Some chapters include small side experiments that are useful in demonstrating key concepts. Some projects are slowly developed from a simple circuit to a more complex and complete version. Don't try to jump ahead and build the complete version if you don't have a thorough understanding of the concepts; use the simpler versions as stepping stones.

What will you need to build these circuits? There are two areas to consider: the hardware needed to build the circuit itself, and the test equipment needed to debug it when things go wrong[2]. In the old days, Veroboard or perf board was used with point-to-point soldered wiring. It was a slow process. Today, there are two easy methods: plug-in breadboards, and custom PC boards. Figure 1 shows a plug-in breadboard, which happens to be hosting a metal detector circuit. Breadboards offer a wonderful method of building quick circuits that are easy to modify. When done, you can rip everything up and re-use the parts in another design. Breadboards are not good for high-frequency circuits, but work just fine for most metal detectors.

Fig. 1: **Breadboard**

For permanent solutions (something you can stuff in a box and actually use), Veroboard or perfboard is still an option, but custom PC boards are even better. With the use of iron-on transfers, these have become very easy to make at home and have a professional appearance. Even small-quantity manufactured boards have come down in price, and are a reasonable solution[3]. A big advantage to using a PC board is the

2. Oh, yes, things *will* go wrong.
3. At the time this was written, small-quantity PCBs could be bought for ~$25 each.

ability to use surface-mount components and shrink the size of the board dramatically. Figure 2 shows the same pulse induction design implemented using through-hole components and surface-mount components. Some designs in this book include a PC board layout, limited to singled-sided foil and through-hole components.

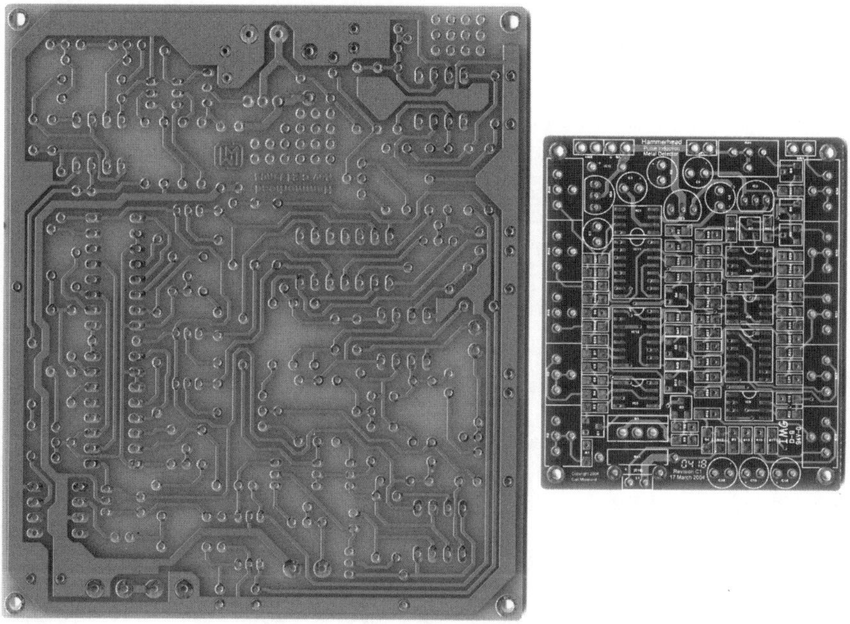

Fig. 2: **PC board comparisons**

Experimentation and testing requires test equipment. It is usually not reasonable to expect that you can build circuits without encountering occasional problems that need debugging, and some of the circuits in this book require adjustment settings that can only be done with the proper equipment.

First and foremost is a good digital multimeter (DMM or DVM). Besides the normal voltage, current, and resistance, a DMM that can measure capacitance and frequency will be of great value, and generally costs less than $50. The ability to measure inductance is even better, but the cost of this feature is often expensive. Secondly is an oscilloscope. These used to be fairly expensive, but have come way down in price, especially in the used market[4]. For metal detectors, you can get by with a basic 2-channel model with a bandwidth as low as 20MHz. These can be had for as little as $50 used[5], generally without probes. New probes can be found on the Internet for as little as $15 each.

4. Dominated by eBay.
5. Be very very very careful in buying used test equipment. Many units are sold "as-is" with no warranty and no returns; try to make sure it is listed as working, with a return policy for problems.

Finally, you will of course need components. It's getting more difficult to find local parts houses, but far easier to find everything on the Internet. See Appendix A for some sources.

So you want to build a metal detector? Let's get started!

Acknowledgements

This book began as two independent efforts that turned into a collaboration after meeting on the *Geotech* forums. Both authors have greatly benefitted from the online forum discussions that have occurred over the years. It is because of this that we partially dedicate this book to the many contributors to the *Geotech* forums.

But the *Geotech* web site itself was created largely as an extension of the efforts of Eric Foster, who was generously helping people understand pulse induction techniques. Eric was therefore not only a pioneer in pulse induction technology, but also a pioneer in the on-line proliferation of metal detector knowledge. It is because of this that we also dedicate this book to Eric.

Both authors also thank their families for their extreme patience while we built, tested, wrote, and re-wrote, again and again and again.

CHAPTER 1 *History of the Metal Detector*

"The most exciting phrase to hear in science, the one that heralds new discoveries, is not *Eureka! (I found it)* but *That's funny …*"
— Isaac Asimov (1920 – 1992)

The nineteenth century saw rapid development in the new frontier of electrical science. One area in particular — the relationship between electrical currents and magnetic fields — became the foundation for the development of the metal detector. Although we could dive right in to metal detector circuits, it is interesting to look back just a little bit further, to the scientists who developed some of the underlying principles that enabled this invention, and to the applications that have resulted. Also, the hobby of metal detecting has seen many companies and technologies come and go — we'll take a quick look at some of the ones that have made a lasting impact.

The People

The information on our historical heros will be brief. The reader is urged to seek out more detailed biographies on these fascinating people.

Hans Christian Øersted

In 1820, Øersted (1777-1851), quite accidentally[1], noticed that the current through a wire caused a nearby compass needle to deflect. He subsequently investigated, and determined that the current was producing a magnetic field. His findings sparked widespread interest in electrodynamics.

Michael Faraday

One person in particular, Michael Faraday (1791-1867), was instrumental in discoveries involving electricity and magnetism. With little more than a primary education, Faraday worked his way up to be the director of the laboratory of the Royal Institute of London. Although not a theorist like many of his peers, Faraday had a keen insight and his experimental methods won him acclaim. It is in his honor that the unit of capacitance is called the *farad.*

One of Faraday's greatest discoveries was the principle of *induction*. It had already been shown that an electric current through a wire produced a magnetic field, and Faraday reasoned that, therefore, a magnetic field should produce an electric cur-

1. See the quotation for this chapter!

rent. He showed this to be true, and his discovery directly led to the inventions of the electric generator and the electric motor.

What Faraday found was that when a wire moves through a magnetic field, an electric current is produced in the wire. It is also true that when a wire is brought near a *changing* magnetic field a current is produced. The electric current that is developed in the wire will also result in a counter-magnetic field around the wire. This is the very principle used in metal detectors.

David Hughes

David Hughes (1831-1900) was primarily a professor of music, but also worked in experimental physics, winning numerous awards for his work. He likely invented wireless radio transmission long before Guglielmo Marconi, but Hughes' weakness was his inability to analyze and describe the science and math behind his experiments. In the realm of metal detecting, Hughes invented the *induction balance* system which is used in practically all T/R and VLF style metal detectors. His induction balance was initially used to investigate the conductive properties of ore samples, likely making it the first applied metal detector.

Alexander Graham Bell

Without a doubt, David Hughes experimented with the effects that metal targets had on his induction-balanced systems. It's very possible that other scientists, perhaps someone we haven't mentioned, did similar experiments. It's even possible that someone intentionally built a crude portable metal detector using this method. But it was a medical emergency that spurred what is considered to be the first known hand-held metal detector.

In 1881, United States President James Garfield lay dying from an assassin's bullet lodged in his back. Using the horribly unsanitary technique of bare fingers, doctors attempted to remove the bullet but were unable to locate it. The bullet was thought to be critically close to the liver, and doctors were personally unwilling to probe too deeply and risk killing the President in order to find the bullet.

Inventor Alexander Graham Bell (1847-1922) decided to volunteer his efforts. On his initial visit to the White House, he found that a Hughes Induction Balance had been sent by a Mr. George Hopkins of New York. Bell took Hughes' idea and merged it with some of the same techniques he developed for the telephone. Bell was successful in building a metal detector, and it could easily detect a bullet held in a clenched fist. He even tested the device on at least one Civil War veteran, and showed that it could successfully locate an embedded bullet.

The drawing in Fig. 1 (from Harper's Weekly) shows Bell and an assistant using the detector on Garfield. Notice in the lower right corner there is a wire going out the door. The (incomplete) inset in the upper right corner shows the clapper circuit that functioned as the oscillator. The clapper made such a loud racket that it had to be placed in another room so the assistant could listen for faint signals.

At first, when they tried to use it on Garfield, they got readings everywhere. They soon realized Garfield was laying on one of the earliest spring-coil mattresses. More attempts failed because, it turned out, the bullet lodged in Garfield was too deep, and

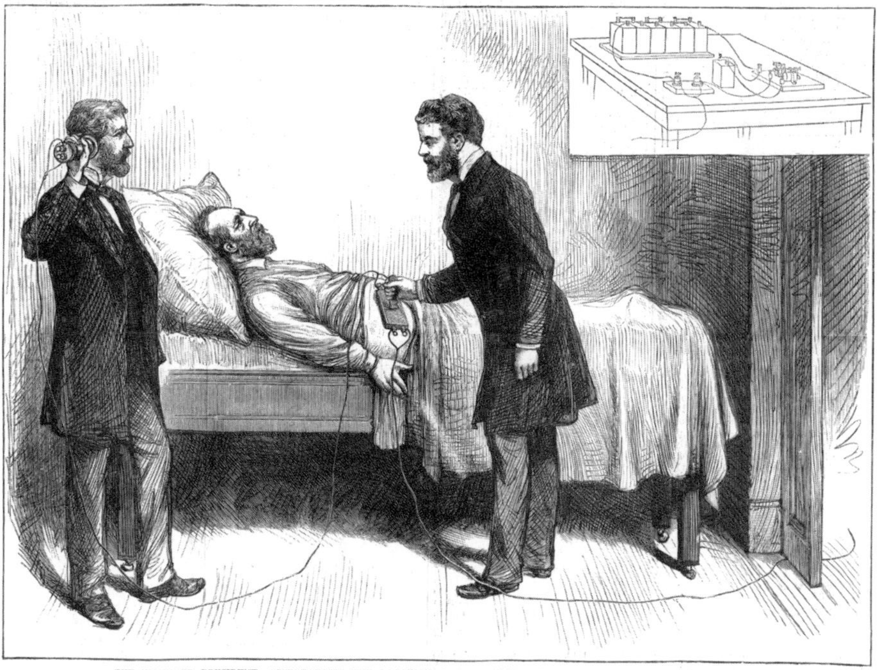

THE WOUNDED PRESIDENT—ASCERTAINING THE LOCATION OF THE BULLET.—From a Sketch by W. Shinkle.—[See Page 555.]

Fig. 1-1: **Alexander Graham Bell using his metal detector on U.S. President James Garfield (1881)**

he died shortly afterward from an infection caused by the efforts of the doctors. It also turned out that the bullet was not in a critical location, and Garfield would have survived if they had simply dressed the bullet wound. Bell was distraught over his failure to locate the bullet, and wrote to his wife:

> *I feel much disturbed by the result of the Autopsy of the President. It is now rendered quite certain why it was that the result of the experiment with the Induction Balance was "not satisfactory" as I stated in my report -- for the bullet was not in any part of the area explored. This is most mortifying to me and I can hardly bear to think of it -- for I feel that now the finger of scorn will be pointed at the Induction Balance and at me -- and all the hard work I have gone through -- seems thrown away. I feel all the more mortified -- because I feel that I have really accomplished a great work -- and have devised an apparatus that will be of inestimable use in surgery -- but this mistake will re-act against its introduction.*

Bell's metal detector design work included experiments with coils which achieve induction balance by partially overlapping them. This same exact method is still used in DD-style (a.k.a. *widescan*) coils today. Figure 1-2 shows his original drawings; notice the ability to slide the two coils until balance is achieved.

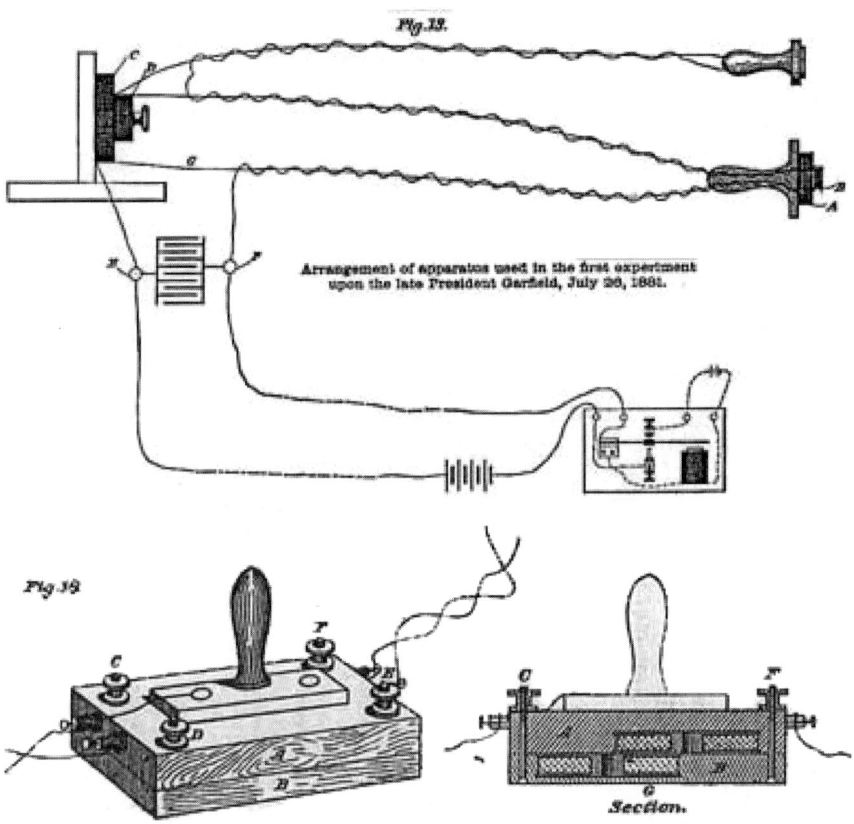

Fig. 1-2: **Drawings by Alexander Graham Bell (1881)**

Other Early Inventors

Bell's work with metal detectors is known only because it revolved around a high-profile case. Bell never received — and probably never filed for — a patent for his detector, but we can look at the patent records to see the work of other people who might otherwise be unknown.

The first U.S. patent we find (269,439) is from 1882, *Apparatus for Finding Torpedoes*, by Charles Ambrose McEvoy. Interestingly, the patent states that the invention was also patented in England (5,581) in 1881. This puts McEvoy's work coincident with Bell's, so he may have had a working detector prior to Bell. McEvoy's patent illustrates a coil assembly that is lowered into the water (Figure 1-2), so this would probably make it the first underwater detector.

The next patent we run across (1,126,027) is from 1915, *Apparatus for Detecting Pipe-leads or Other Metallic Masses Embedded in Masonry*, by Max Jullig. The interesting things to note from this patent are two illustrations; one shows the coils arranged in an orthogonal configuration, and another shows the schematic equiva-

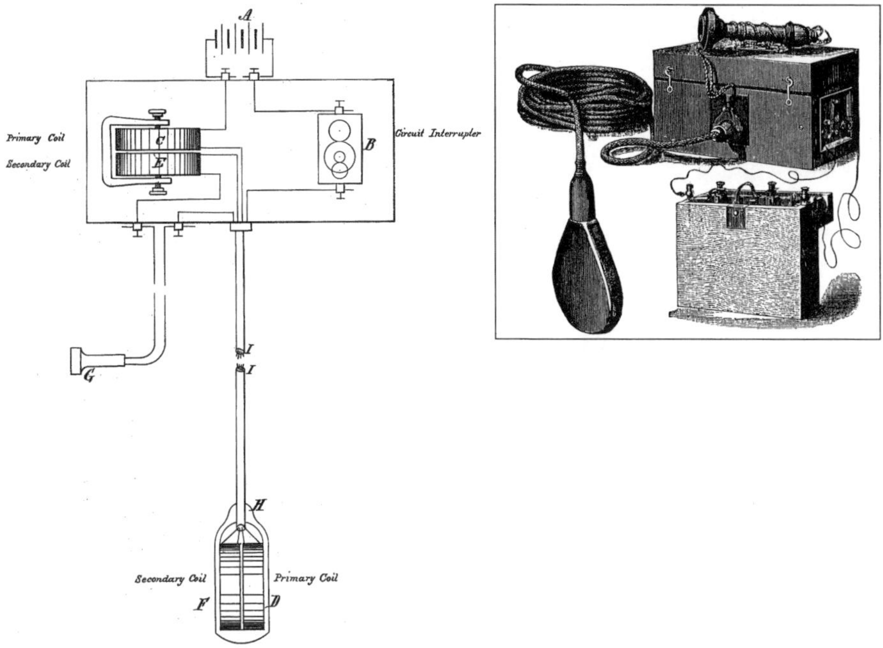

Fig. 1-3: **McEvoy's underwater metal detector (1882)**

lence of those coils drawn in a figure-8 style.

A patent from 1924 (1,492,300) is *Means For Electro Aviatic Proof and Measuring of the Distance of Electric Conductive Bodies* by Heinrich Lowy of Austria. This patent describes a detector for ore bodies carried by an airship, such as a Zeppelin. It is interesting in that the design is a type of off-resonance detector, and it also describes the possibility of using a secondary oscillator to form a BFO detector.

Shirl Herr was issued a patent (1,679,339) in 1928 for a *Hidden-Metal Detector.* This detector uses an orthogonal coil configuration but, unlike other orthogonal designs of the time, the transmit coil was fully contained within the receive coil.

Two other early patents are *Method Of and Apparatus For Locating Terrestrial Conductive Bodies* (1,812,392 in 1931) by Theodor Zuschlag, and *Electrical Apparatus for Locating Bodies Having Anomalous Electrical Admittances* (1936, 2,066,135) by William Barret and Randolph Mayer. The former has a rather bizarre receive coil setup, and the latter shows the detector as a traditional two-box design.

Gerhard Fisher

It was apparently not until 1931 when the first portable metal detectors were widely marketed, based on an original design invented in 1925 by Gerhard Fisher[2] (1899-1988). In the 1920's he was working with radio navigation equipment for Naval aircraft when he noticed that the equipment would occasionally register small

2. Fischer was his German name; he adopted Fisher as an American.

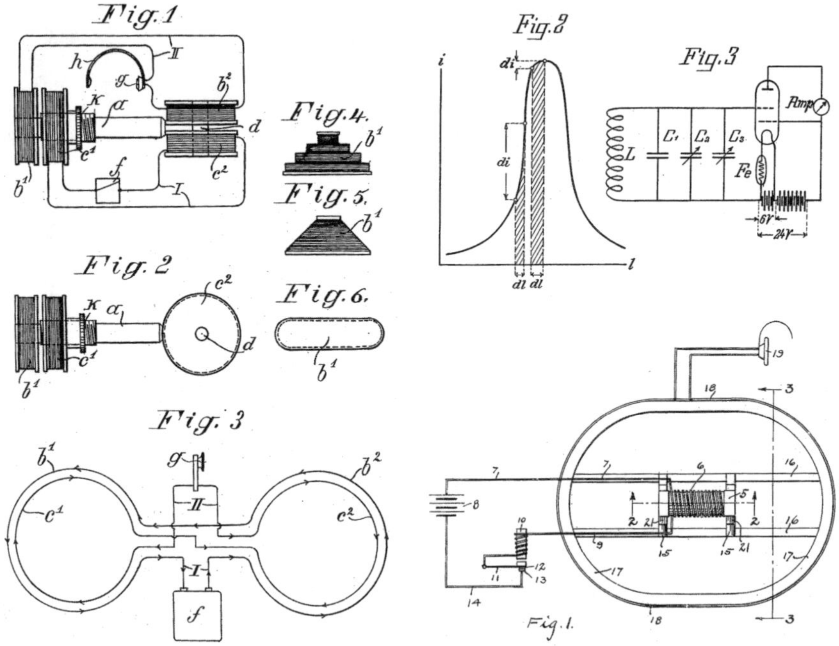

Fig. 1-4: **Max Jullig patent; shows orthogonal coils, plus a Figure-8**

errors in the directional reading. Subsequent investigation led Fisher to the discovery that it was due to large metal objects, such as buildings.

He received U.S. Patent 2,066,561 for his "Metalloscope" in 1937. Though this was a year after the Barret and Mayer two-box patent above, Fisher actually filed his patent in 1933, a year prior to Barret and Mayer. It is often stated that Fisher received the first metal detector patent, but even discounting the Barret/Mayer patent this is an erroneous claim.

In 1931 Fisher left his Naval work and founded Fisher Research Laboratory for the purpose of developing his own commercial detectors. His earliest designs were two-box locators, in which the coils are inductively balanced in an orthogonal configuration. Figure 1-5 shows an early Fisher, the Model 47B, which is constructed entirely of wood. In the 1950's and 60's, Fisher T10 and T20 models were very popular in the emerging hobby of metal detecting. Fisher is considered by many to be the father of the modern metal detector and, today, the Fisher brand lives on as a subsidiary of First Texas Holdings.

Later Contributors

Many other people have been instrumental either in the development of detectors, or in the promotion of the industry. Two of the earliest promoters in the hobby market were Robert Gardiner and Kenneth White, both of whom began in the 1950's. Gardiner developed and sold a number of innovative designs, including an early discriminating detector. White started out producing the Oremaster Geiger counter dur-

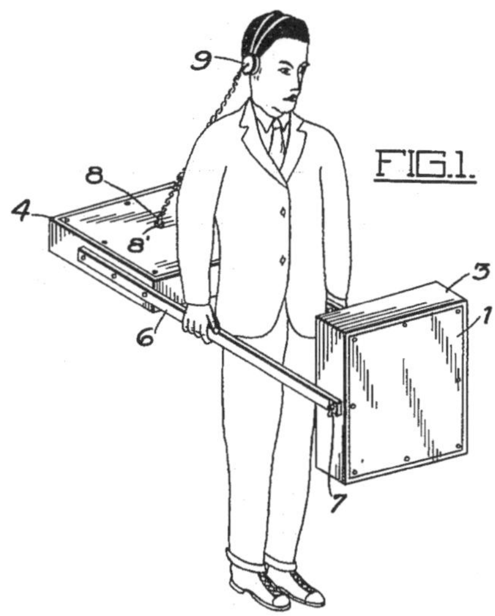

Fig. 1-5: **Fisher patent drawing (1937); Fisher Model 29**

ing the uranium craze of the 50's, but switched to metal detectors in 1959 when demand for uranium fizzled. White's Electronics continues to be a leading manufacturer of hobby detectors.

Bill Mahan, a developer of BFO detectors and founder of D-Tex, was a prominent figure in the early days of treasure hunting. Charles Garrett likewise began with BFOs in the early 60's, and Garrett Electronics is now one of the largest detector companies in the world, especially in the security sector. Both men were strong ambassadors for the hobby.

It is uncertain who invented basic target discrimination in metal detectors. The method widely used for discrimination — determining phase using synchronous demodulation — is easily found in patents back to the early 1960's and even before, though largely applied to other types of instruments. Basic TR discrimination first showed up in hobby detectors in the early 1970's, in the Technos *Phase Readout Gradiometer* and in models produced by White's and others.

George Payne is widely recognized as one of the most significant contributors to metal detector technology, having invented the ground canceling motion discriminator, tonal target ID, and visual target ID. In pulse induction, Eric Foster has been a leading developer since the 1960's, and some of his designs were instrumental in the discovery and recovery of Spanish shipwreck treasure, such as with Mel Fisher and the *Atocha*. Bruce Candy has made numerous contributions to both VLF and PI technologies which are found in Minelab detectors. Other noted contributors to metal detector design include Dave Johnson (Fisher, Tesoro, White's), Robert Podhrasky (Garrett), Jack Gifford (C&G, Tesoro), Rick Maulding (White's), Mark Rowan (White's), John Earle (Compass, White's), Alan Hametta (A.H. Pro), Dick Hirschi (Compass, Teknetics), and Jim Karbowski (loop design), among many, many others.

The Applications

The first known application of a metal detector, as described in the section on Alexander Bell, was medical in nature. Let's look at some other areas where metal detectors are used.

Military Use

Even before Fisher began building metal detectors, they were being used by military forces searching for unexploded ordinance. A Popular Mechanics article from 1917 shows two operators using a detector, described as an induction-balance design, to search a French field for unexploded ordinance during World War I.

POPULAR MECHANICS 205

BURIED SHELLS FOUND BY INDUCTION BALANCE

This Instrument Detects the Presence of Shells or Other Metal Buried in the Soil, Thereby Enabling the Farmer to Remove Them before Tilling a Field

Fig. 1-6: **UXO detection in WW1 (Popular Mechanics, 1917)**

By the second World War, mine detectors were in widespread use, with the SCR-625 being the dominant model. During Vietnam, the AN/PSS-11 mine detector was popular. It used an unusual 4-over-1 induction balanced coil scheme. Before the widespread availability of commercial detectors, treasure hunters used these kinds of surplus mine detectors which are still widely available on the used-surplus market, though they are now only useful as collector items.

Commercial & Industrial Use

In manufacturing, metal detectors can be used in robotic assembly as position sensors for accurately placing objects. In the food industry, detectors are used to ensure no foreign metal objects get into the final product. In logging, detectors look for embedded metal objects such as nails that could damage cutting tools.

Security is a rapidly growing market due to theft and threats to safety. Walk-through metal detectors are widely employed in high-risk security areas such as airports, government buildings, and prisons. Most walk-throughs are pulse induction designs, but newer technologies such as terrahertz imagers that can "see" non-metallic weapons are gaining popularity. Businesses are using walk-through detectors in an effort to stem theft, both from customers and employees.

Treasure Hunting

With the advent of transistorized circuitry in the 1960's, low-cost metal detectors began to emerge specifically for the consumer market. Unlike Fisher's original design 30 years prior, most of these units were simple BFO detectors which could only discern between ferrous and non-ferrous targets. There were some T-R machines but they were generally more expensive and, while exhibiting some impressive air test depths, had severe problems with mineralized ground. With some of the cheaper detectors selling well below $100 treasure hunting became a hobby for the masses.

Initially, the major players in the consumer market were D-Tex, Relco, White's, and Garrett. Because the BFO was such a simple design, it was not long before new start-ups entered the market. Most of these companies did not have the research and development teams needed to produce new innovations that competition ultimately demands, and so many of them were quickly overtaken by the more established manufacturers.

The introduction of integrated circuitry, in particular the operational amplifier (opamp), made the TR design much more attractive to develop. Like the BFO, early TR detectors could only differentiate between ferrous and non-ferrous targets. But in 1974 another breakthrough technology was introduced by Technos called the *Phase Readout Gradiometer*[3], which was a TR discriminator. This new circuitry looked at the phase shift of the received signal and, using a threshold level set by the user, either signaled the target as good or ignored it as undesirable. The discovery that led to this new feature was that different metals — aluminum, nickel, copper, gold, silver, etc. — exhibited slightly different phase shifts in the return signal due to differing conductivities. By measuring this phase shift the detector could discriminate between targets. This new feature was a two-edged sword, however, as some desirable targets — rings and nickels — appeared to have a similar response as aluminum pull tabs. Thus discriminating out the "trash" could also mean losing some of the treasure. However, in

3. The PRG wasn't really the first development of a TR discriminator. The patent for the PRG is US3826973 which was filed in 1973. Several other patents predate this one, notably US3020470 was filed in 1943 and awarded in 1962, so the idea of phase discrimination clearly dates back to the 1940's. However, the PRG was the probably the first discriminator in the hobby metal detecting market.

areas with heavy concentrations of buried bottle caps and pull tabs, the trade-off was well worth it, especially in the days when there were so many silver coins to find.

TR detectors normally operate at a few 10's of kilohertz, called the low frequency range (LF), while BFO's operate at 100kHz or more. The lower frequencies yielded better soil penetration. By lowering the frequency even further to just a few kilohertz, called the very low frequency range (VLF), designers found that they could achieve one more desired result, that of ground cancelation. With both BFO and TR detectors, ground mineralization creates a false signal. However, ground mineralization is fairly consistent over a given search area so the user can retune the detector once the search coil is lowered near the ground. However, any variation in the height of the search coil above the ground will result in a change in signal, a very annoying phenomenon.

Some detectors up to the mid-70's included both a ground-cancel mode and a discrimination mode, but the two modes could not operate at the same time. The next technological breakthrough occurred in 1977 when Bounty Hunter introduced the world's first VLF motion discriminator, the Red Baron. Analog filter techniques allowed users to both ground cancel and discriminate at the same time, as long as the search loop remained in motion.

The Companies

As mentioned before, the transition of technology from BFO to IB resulted in the failure of many early companies. Subsequent technological developments have also left other companies unable to compete, and the brief history of the commercial detector is littered with names that have come and gone.

Of the current major players, the longest-surviving company is Fisher Labs, which was created in 1931 by Gerhard Fisher and sold to First Texas in 2006. White's Electronics (1950) and Garrett Electronics (1962) are both still family-owned and largely unchanged. Tesoro was created in 1984 after founder Jack Gifford broke off from C&G Technologies, a late 1970's detector company. England's C-Scope was founded in 1975, and Minelab of Australia was founded in 1985. One of the latest major entries in the hobby detector market is XP Detectors of France, begun in 1998.

Bounty Hunter was an early player back the 60's when it resided in Klamath Falls, Oregon (known then as Pacific Northwest Instruments). Around 1970 it was sold to Space Data Corporation and moved to Tempe, Arizona. In the early 80's it was sold to Teknetics and moved back to Oregon (Lebanon) until Teknetics went bankrupt. Teknetics was founded by George Payne when he left White's in 1983 and produced some innovative detectors, but poor management lead to its demise. Teknetics and Bounty Hunter were bought by Techna of El Paso, Texas.

First Texas, current producer of Bounty Hunter, Teknetics, and Fisher, is an incarnation of an earlier company, Techna, which was begun by John E. Turner and produced inexpensive entry-level detectors in the 1980's. Before Techna, Turner had also run a similar detector company, Jetco, of which the "Jet" came from his initials. Turner now owns Ranger Security, which builds security walk-through detectors.

The last company we'll comment on is Compass Electronics. Compass (of Forest

Grove, Oregon) was an early innovator in TR and TR discrimination, and did fairly well transitioning to VLF and motion discrimination. It had some very competitive and well-respected detectors. But, as with so many other detector companies, mismanagement caused its downfall. In 1994, the factory mysteriously burned to the ground, and brought Compass to an abrupt end. Interestingly, Compass, Bounty Hunter, Teknetics, and another detector company, Discovery Electronics — all Oregon-based — were started by ex-employees of White's Electronics.

CHAPTER 2

The BFO Principle

"A child of five would understand this.
Send someone to fetch a child of five."
— Groucho Marx

The term BFO stands for *Beat Frequency Oscillator.* Machines using the BFO principle are nowadays relegated to the lower end of the metal detector market, and are mostly sold as toys or as introductory detectors for those with a fledgling interest in the hobby.

BFO designs consist of two radio frequency (RF) oscillators, usually oscillating at a frequency of around 100 kHz. One of the oscillators incorporates the search coil, and the second oscillator uses a much smaller coil that is hidden inside the control box, and is known as the reference oscillator. The two oscillators are adjusted so that their frequencies are nearly the same, and their outputs are summed (or mixed) to produce sum and difference frequencies. For example:

f1 = 100.0 kHz (fixed oscillator)
f2 = 100.5 kHz (search coil oscillator)

Sum and difference frequencies produced are: 200.5 kHz and 500 Hz.

The sum of the two frequencies is filtered out by the electronics leaving the difference signal which conveniently lies within the audio band of 20 Hz to 20 kHz. This signal is amplified and used to drive either an internal speaker or a set of headphones. In radio terms, a heterodyne is the tone or note created by mixing together the outputs from the two oscillators, where the frequency difference lies in the audible range, and the beat frequency is the frequency of the heterodyne. Hence the origin of the term Beat Frequency Oscillator.

When a metal object is brought close to the search coil it affects the inductance value of the coil and the oscillator frequency is altered slightly, and this in turn affects the difference frequency. If (for example) a metal object causes the search oscillator to alter frequency from 100.5 kHz to100.6 kHz, then the user would hear a change in audio tone from 500 Hz to 600 Hz, signifying the presence of a metal target.

A BFO detector is ideal for the beginner, as it is easy to become proficient very quickly. Most units possess only one control knob to tune the fixed oscillator, and possibly a second knob for the audio volume. BFOs are also sensitive over almost the entire search coil area, are good at pinpointing the exact location of the metal target, and have a good battery life. In addition, they have the capability of rudimentary discrimination between ferrous (iron-containing) and non-ferrous metals. Very useful if you are searching for coins, but wishing to ignore junk, such as nails or other iron fragments. BFO detectors will give a rise in pitch for non-ferrous (good) targets and a fall for ferrous (unwanted iron) items – assuming the search oscillator frequency is set above the reference oscillator frequency. Also, from a construction point of view, the BFO is one of the simplest designs to implement.

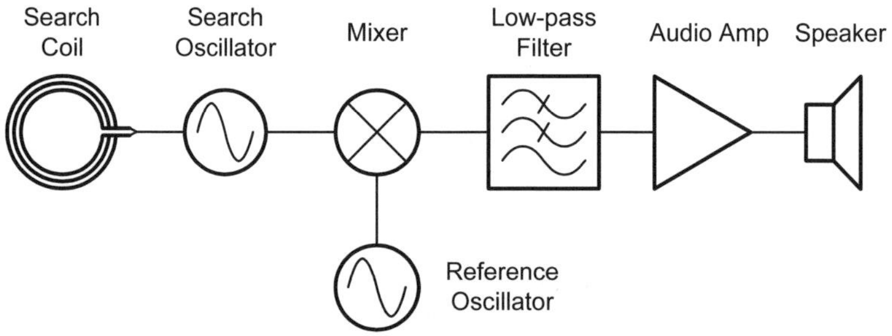

Fig. 2-1: **BFO Metal Detector Block Diagram**

You may be excused for thinking that the BFO sounds like the ideal metal detector with its ease-of-use and discriminating ability, but as usual there is always a downside.

Unfortunately they are prone to drift (small changes in component values) as the temperature changes, which requires the user to frequently retune the fixed oscillator. Since most BFOs operate at a relatively high frequency of 100 kHz (the old mine detectors used to operate at 150 kHz) they tend to be affected by ground capacitance. This means that simply bringing the search coil close to the ground can cause a decrease in the difference frequency. The higher the frequency of the search oscillator the more susceptible it will be to ground effects. However, there is a solution to the problem. It is possible to create a Faraday shield around the search coil by tightly binding it with a foil screen. The screen actually adds more capacitance, and the search frequency is lowered slightly, but it has the effect of swamping capacitance caused by the ground. It is also very important that a small gap is left in the Faraday shield, otherwise it will act as a short circuit turn. Once the foil is wound around the coil, it must be connected to the circuit ground. The coil then needs to be made very rigid and preferably potted in epoxy compound to stop vibration, otherwise the slightest movement of the coils will cause the detector to go out of tune. Additional ways to counteract the ground capacitance effect is to use a search coil with a smaller number of turns. A coil with fewer turns will have a lower inductance and will require a larger associated capacitance for a tuned circuit of the same resonant frequency. This larger value of capacitance will also assist in swamping the ground effect. Lowering the frequency of the search oscillator can also be beneficial.

You might be wondering whether the Faraday shield affects the capability of the detector in any way, considering the proximity of such a large amount of foil to the search coil. Although the presence of a metallic shield will change the search coil's inductance, the detector's ability to locate metal objects is unaffected, since it is based on relative (not absolute) inductance changes.

Unwanted drift caused by temperature changes can be more difficult to resolve. Although regulating the power supply voltage can help considerably, as can the use of low temperature coefficient capacitors.

The layout of the printed circuit board (PCB) can also have an important effect on both the sensitivity and stability of the design. For example, if the search and reference oscillators are designed to operate at almost the same frequency (giving a nulling effect when no metal is present) then poor layout may induce coupling which can cause the oscillators to "lock" together. Sensitivity is reduced when this occurs because it requires a larger metal target than normal to force the oscillators out of this locked condition. Bad layout can also cause instability if PCB tracks delivering high voltage signals are routed near to other more sensitive input circuitry.

By now it should be clear that the major challenge when designing a BFO is the stability of the two oscillators. In practice this can be very difficult to achieve. In the real world there are many random frequency shifts, caused by non-targets, that require the user to make continual adjustments to keep the detector in tune.

Before we end this chapter on BFO principles, it is worth examining the discrimination capabilities in more detail. Firstly it is important to note that there are only two ways that a conductive target can alter the inductance of the transmit coil:

1. Permeability Effect
   This effect occurs when an iron target has a cross-sectional area that is small relative to its mass. These are compact iron targets (as opposed to an flat sheet, which has a large surface area relative to its mass). The magnetic field from the transmit coil is able to permeate the mass of the iron target, which in turn increases the inductance of the coil, thus lowering the frequency of the oscillator. In the case where the transmit frequency is tuned higher than the local oscillator, this will produce a lowering of the audio tone.
2. Eddy Current Effect
   All non-ferrous targets, such as gold, silver, copper, brass, etc., are detected using the eddy current effect. The electromagnetic field from the transmit coil induces circulating currents to flow in the conductive material. These currents act like shorted turns on the transmit coil, which has the effect of lowering the inductance, and consequently increasing the frequency. The result is an increase in the audio tone. Unfortunately there are a couple of gotchas here, which ruin what might otherwise have been an ideal method of discrimination.

Gotcha #1: For Iron targets which have a cross-sectional area that is large compared to their mass (a tin lid, for example) the eddy current effect dominates over the permeability. So these targets will give a rise in the audio tone.

Gotcha #2: Small iron targets (such as a nail, or a small piece of wire) exhibit both eddy current and permeability effects, but this is dependent of both position and orientation. These targets produce erratic signals that make target identification impossible.

Although there are many drawbacks with the BFO when compared to more modern technologies, it has the advantage of simplicity and low cost. For the electronics amateur interested in metal detectors, this is a good place to start.

CHAPTER 3

Designing a BFO Detector

"You see things as they are and ask, 'Why?'
I dream things as they never were and ask, 'Why not?"
— George Bernard Shaw

As with all design projects we need to begin with a specification. Let's start our design for a simple BFO detector by deciding to use a search frequency of 100kHz.

So why choose 100kHz?

Actually this is not an arbitrary decision, and involves something that was not mentioned in Chapter 2: "skin effect". When current with a high frequency is flowing in a conductor it has a tendency to move along the outside surface. The higher the frequency the more pronounced this effect becomes. This "skin effect" starts to become noticeable above the magic figure of 100kHz. If the magnetic field generated by the search coil only penetrates the metal target to a shallow depth, then the true nature of the target may be misinterpreted and ferrous objects can be mistaken for non-ferrous, and vice-versa. This means that the BFOs ability to distinguish between good and bad buried objects is compromised. Therefore, the highest recommended frequency we can specify is 100kHz.

Why not use a lower frequency than 100kHz?

This is a good question, but it requires some mathematics to explain. For an oscillator that uses an inductor and a capacitor to form the resonant circuit (often called a tank circuit) the frequency of oscillation can be calculated using a simple mathematical expression:

$$f = \frac{1}{2\pi\sqrt{LC}} \qquad \text{Eq 3-1}$$

If, for example, the inductance (L) is 1mH and the capacitance (C) is 2.533nF, then the frequency (f) will be 100kHz. Assuming that the presence of a metal object causes the inductance to increase by 1%, then the frequency will be reduced to:

$$f = \frac{1}{2\pi\sqrt{1.01\text{mH} \cdot 2.533\text{nF}}} = 99.504\text{kHz} \qquad \text{Eq 3-2}$$

If the frequency of the reference oscillator is also set to 100kHz there will be a difference (beat) frequency of (100kHz – 99.504kHz) which equals 496Hz.

Examining the above formula shows that the frequency is inversely proportional to the square root of the capacitance. This simply means that increasing the capacitance will decrease the frequency, and decreasing the capacitance will cause an increase in frequency. i.e. the capacitance value has an inverse relationship to the frequency of oscillation, and exactly the same argument applies to the inductance.

Since the capacitance and inductance values are enclosed by a square root, this also means that to decrease the frequency by half requires an increase in capacitance of four times. Likewise, if we wished to decrease the frequency to a third, the capacitance would have to increased by 9 times.

Let's try decreasing the frequency to 50kHz (i.e. divide the original 100kHz by 2). In this case the capacitance needs to be increased (because of the inverse relationship) by 4 times (because of the square root) which equals 10.132nF, assuming the coil remains unchanged. If a metal object is again brought close to the coil causing an increase in inductance of 1%, then the frequency will decreased to:

$$f = \frac{1}{2\pi\sqrt{1.01\text{mH} \cdot 10.132\text{nF}}} = 49.752\text{kHz} \qquad \text{Eq 3-3}$$

This yields a difference frequency of 248Hz, compared to the previous value of 496Hz.

It should be clear from these simple calculations, that a BFO detector will therefore have a greater sensitivity if it operates at a higher frequency. But what is the best frequency to use?

Low frequencies are better able to penetrate the soil than high frequencies, and are also able to handle greater levels of mineralization. However, higher frequencies provide a better eddy current response up to a level where skin effect starts to become an important factor. This simply means that the real reason BFOs use a high oscillator frequency (normally not higher than 100kHz) is to provide a better frequency sensitivity at the mixer output.

In the calculations above, we assumed that both the search and reference frequencies were identical when no metal was present below the coil. This would give a zero frequency difference at the mixer output and no audio tone would be heard until a metal target was detected. Unfortunately this has the disadvantage that discrimination between ferrous (iron containing) and nonferrous targets is no longer possible, as both target types will produce the same response. This leads us to another question.

What is the optimum audio frequency to use when no metal target is present?

If we choose an audio frequency (when a metal target is not present) of 500Hz, a frequency shift of 10Hz will move the audio frequency to 510Hz, a change of +2%. However, if we choose an audio frequency of 50Hz, the same frequency shift of 10Hz will produce a change of +20%. Which change do you think will be the easiest to detect?

As with all design decisions there will be some element of compromise. In this case the obvious choice is to provide as low an audio frequency as possible to increase sensitivity, but unfortunately there are some limitations to consider. Opting for no audio response in the absence of metal causes two problems – potential oscillator locking (as mentioned in Chapter 2) and no ability to discriminate between ferrous and nonferrous targets. Likewise, selecting too high an audio frequency can drastically reduce sensitivity to small objects. The main limitation is the response of the human ear, which drops off rapidly below 50Hz. A situation can also occur where a large ferrous object enters the influence of the coil. This object may cause a decrease in beat frequency, but the audio tone may be set so low that the search oscillator frequency is moved from a position above the reference oscillator frequency to one that is below. This will result in an increasing audio tone rather than a decreasing one, and the BFO's discrimination capabilities will again have been compromised. Therefore, as the sensitivity of the BFO is so reliant on the operator's hearing, it is important that some adjustment is provided to allow the most suitable audio tone to be selected. Users should then be able to adjust this tone to be as low as possible (for maximum sensitivity) within the limits of their own hearing, but without affecting the ability to discriminate between target types.

Reference Oscillator Design

The next step in the design process is to develop the reference oscillator, mainly because we can choose to use a standard value inductor. Whereas the search oscillator requires a custom made coil over which we have to make other design decisions.

As far as the type of oscillator circuit to use, there are many configurations to choose from, such as Armstrong, Pierce, Hartley and Colpitts. By far the most common circuit used in metal detectors is the Colpitts oscillator (see Figure 3-1). In comparison to the other types, the Colpitts possesses fairly good frequency stability, and can be designed to operate over a wide range of frequencies. It also has the advantage that only a single coil is needed, and does not require a centre tapped coil or conventional transformer to feed back the output signal to the input.

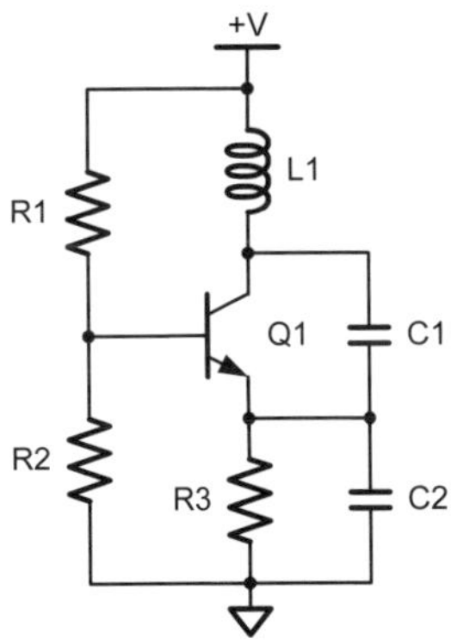

Fig. 3-1: **Colpitts Oscillator**

In this oscillator the tuned circuit consists of L1, C1 and C2. A portion of the collector voltage is returned back to the emitter via the capacitor divider formed by C1 and C2, in order to create feedback.

The frequency of oscillation may be calculated using:

$$f = \frac{1}{2\pi\sqrt{LC_T}} \qquad \text{Eq 3-4}$$

$$\text{where } C_T = \frac{C_1 C_2}{C_1 + C_2}$$

In other words, the inductance L1 in parallel with the series combination of C1 and C2 forms the resonant circuit.

If we decide to use a standard RF inductor with a value of 4.7mH, then the capacitance C_T can be calculated from the following equation:

$$C_T = \frac{1}{\omega^2 L} \qquad \text{Eq 3-5}$$

where $\omega = 2\pi f$

Whenever the frequency (f) is mentioned in technical books, it is measured in Hertz (cycles per second). But when the term $2\pi f$ is used, it is known as the angular frequency ω (pronounced omega) and is measured in radians. This expression is derived from the trigonometrical analysis of the motion of a sine wave. For example, 360 degrees of a circle (one cycle) equals 2π radians. Hence 10Hz is equivalent to $10*2\pi$ which equals 62.83 radians. It's just a different way of expressing the same thing. Now, back to our calculation for C_T:

$$C_T = \frac{1}{(2\pi \cdot 100\text{kHz})^2 \cdot 4.7\text{mH}} = 539\text{pF} \qquad \text{Eq 3-6}$$

Since $C_T = 539$pF, C1 and C2 can both be set equal to twice C_T. The nearest standard value is therefore 1nF.

If we recalculate using L = 4.7mH and C1 = C2 = 1nF, we have an expected frequency of 103.8kHz. As a first attempt at a reference oscillator for our experimental BFO detector, this is close enough.

Unfortunately the assumption that C1 is equal in value to C2 does not lead to an optimal design for the reference oscillator. This is because the amount of feedback from the capacitive divider is too large, which causes the circuit to produce an output waveform that is not a clean sine wave. The result is a transmitted signal that contains many harmonics along with the fundamental of 100kHz. Harmonics are additional (and generally unwanted) signals that exist at multiples of the fundamental frequency. For the reference oscillator, these harmonics exist at 200kHz, 300kHz, 400kHz, … up to infinity. In an ideal oscillator only the fundamental frequency would be present in the output spectrum, and the harmonics would not exist. However, in the real world, the oscillator circuit contains many nonlinearities that result in the introduction of these harmonics, and it is the job of the designer to minimise these as much as possible. By designing the oscillator to produce a clean sine wave results in maximum power being transmitted at the desired frequency, rather than being wasted in providing unnecessary harmonic content.

Actually, there is a special case where harmonics are deliberately introduced to provide increased sensitivity, but this will be discussed later on in this chapter.

If you examine other published metal detector designs that use a Colpitts oscillator, you will notice there is a rule-of-thumb which dictates that C2 is approximately equal to 3 times C1. Since we know that the total capacitance (C_T) of the parallel combination of C1 and C2 is:

$$C_T = \frac{C_1 C_2}{C_1 + C_2} \qquad \text{Eq 3-7}$$

then it is a simple process to substitute C2 with 3*C1 and (with some algebraic manipulation) arrive at the following result:

$$C_1 = \frac{4C_T}{3} \quad \text{Eq 3-8}$$

Note – if you want to work this out for yourself, then it is easier to use the following formula for C_T in order to avoid solving a quadratic equation.

$$\frac{1}{C_T} = \frac{1}{C_1} + \frac{1}{C_2} \quad \text{Eq 3-9}$$

If the required value for C_T is 500pF, then C1 = 680pF, and C2 = 2nF (using standard capacitor values). This gives an operating frequency of 101.9kHz.

Search Oscillator Design

Unlike the reference oscillator, it is necessary to construct the inductor ourselves. This inductor is known as the search coil and is located at the end of the detector shaft.

There are two possible approaches to creating a suitable search coil:

1. Simply start winding a coil – e.g. 50 turns on a 6 inch (~150mm) former. Then measure the inductance and calculate suitable values to use for C1 and C2.
2. Specify an inductance for L1 to suit standard values for C1 and C2. Then wind L1 to match the specified inductance.

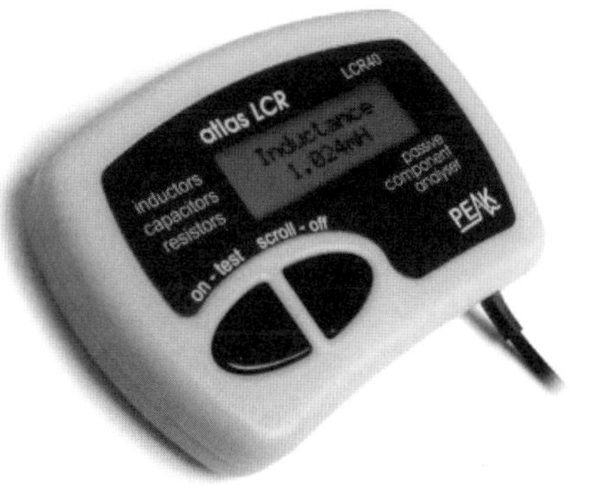

Fig. 3-2: **Passive Component Analyser**

The main problem in both cases is the measurement of the inductance of L1. If you have access to an inductance meter, then either method can be used. Otherwise it is easier to construct the search coil first, build a Colpitts oscillator using this coil, and measure the frequency using an oscilloscope. Then simply calculate the inductance, as you already know the frequency of operation and the values of C1 and C2. From this point you can then calculate new values for C1 and C2 to provide the required frequency. This is such a long-winded and error-prone method that it is strongly advisable to invest in a simple inductance meter. Figure 3-2 shows a photograph of an excellent tool that is almost indispensable for this type of design work. It is called a Passive Component Analyser, and is capable of measuring the values of inductors, capacitors and resistors. This device is manufactured by Peak Electronics Ltd., and made in England. The make and model is an Atlas LCR40, costing around £70 at the time of writing this book.

As we discussed earlier, the search oscillator frequency needs to be slightly higher than the reference frequency. This is to allow a decrease in frequency for ferrous objects and an increase for non-ferrous. The simplest way to achieve this situation is to add a trimmer capacitor across the C1 / C2 network in the reference oscillator. Examination of the standard values available for miniature trimmer capacitors indicates that a range from 2.2pF to 22pF would be suitable for this purpose. This would allow the reference oscillator to be adjusted from 99.8kHz to 101.7kHz. If the search oscillator is designed to work at 101.7kHz, then the beat frequency would be adjustable from zero up to 1.9kHz.

For a practical example, let's say we construct a search coil with 143 turns on a 5" (127mm) diameter former using 0.56mm enamelled wire. The measured inductance yields a value of 5.5mH. Since the required frequency of oscillation has been determined as 101.7kHz, this means we have the following information to use for calculating the values of C1 and C2:

$L = 5.5\text{mH}$
$f = 101.7\text{kHz}$

Using $C_T = \dfrac{1}{\omega^2 L}$ (where $\omega = 2\pi f$) then $C_T = 445\text{pF}$.

Next, using: $C2 = 3C1$ and $C1 = \dfrac{4C_T}{3}$ gives $C1 = 560\text{pF}$ and $C2 = 2.2\text{nF}$. This results in the final circuit for the search oscillator as shown in Figure 3-3.

One point worth mentioning here is the circuit topology used to represent a Colpitts oscillator. Both Figure 3-1 and Figure 3-3 have used the standard representation for such a design, and consequently it is difficult to understand how C1 and C2 form a resonant circuit with L1. If the Colpitts circuit is redrawn, as shown in Figure 3-4, the L1 / C1 / C2 tank circuit becomes more obvious, and also shows how part of the output is fed back to the emitter of Q1 via the capacitive divider network of C1 and C2. The only change is that the end of C2 is connected to the positive supply rather than 0V, which makes no difference as far as the a.c. signals are concerned.

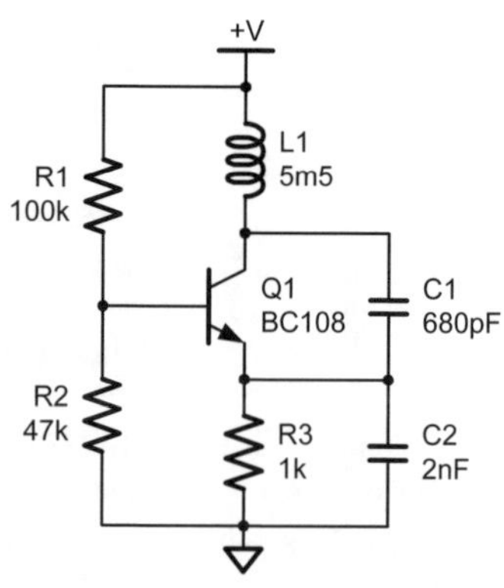

Fig. 3-3: **Search Oscillator**

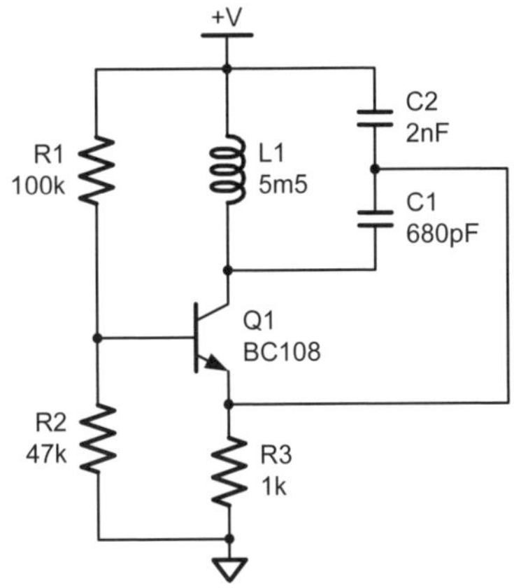

Fig. 3-4: **Search Oscillator Redrawn**

But, for some unknown reason, the circuit is usually drawn with C2 connected to ground. C'est la vie!

Mixer and Detector Design

The job of the mixer circuit is to "mix together" the outputs from the reference and search oscillators in order to generate the sum and difference frequencies. Of course, we are only interested in the difference (or *beat*) frequency, and any higher frequencies at the output of the mixer will need to be removed.

The two oscillator outputs can be mixed together quite simply using two capacitors and one resistor, as shown in Figure 3-5. A signal exists at the connection labelled "mix" that comprises both the search and reference oscillator and the sum and difference frequencies. Unfortunately, this simple mixer circuit provides very little isolation between the two inputs, and use of this circuit can easily result in oscillator "locking".

The transistor-based design shown in Figure 3-6 gives superior isolation, and provides sufficient gain to allow the use of a simple headphone amplifier. The value of R9 can be modified to adjust the modulation depth of the output signal.

So what does the output of the transistor-based mixer look like? Figure 3-7 will give you the answer.

The mixer output consists of a high frequency signal with an amplitude modu-

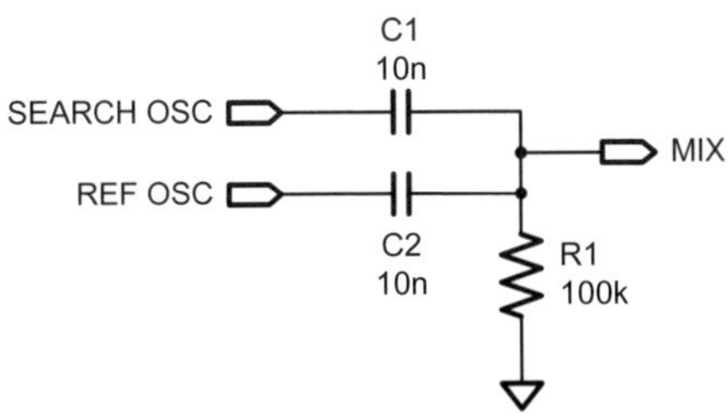

Fig. 3-5: **Simple Mixer**

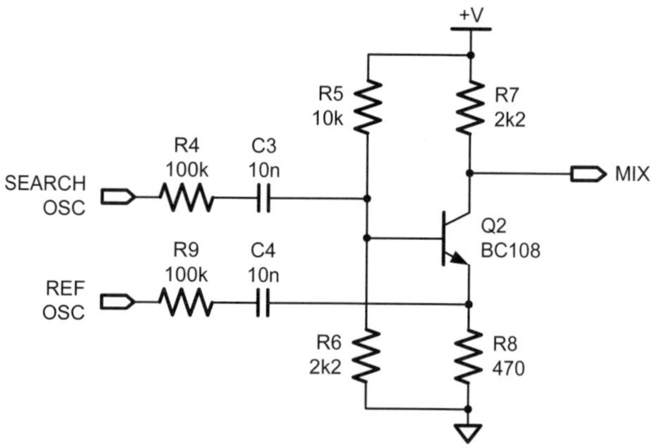

Fig. 3-6: **Transistor-based Mixer**

lated envelope. The frequency of the amplitude fluctuations in the envelope represents the *beat* (or difference) frequency between the search and reference oscillators. It can be readily seen from the plot of the mixer output, that the reference oscillator has been adjusted to give a *beat* frequency equal to 100Hz.

Headphone Amplifier

At this point in the design you may be thinking that we need to consider how to demodulate the complex waveform shown in Figure 3-7 in order to extract the slower moving *beat* frequency. This is where the limited range of human hearing comes to the rescue, allowing the use of comparatively simple circuitry to provide an audio output suitable for detection purposes. In fact, the requirement for a demodulator can be simply ignored, and the signal in Figure 3-7 can be fed directly into a simple head-

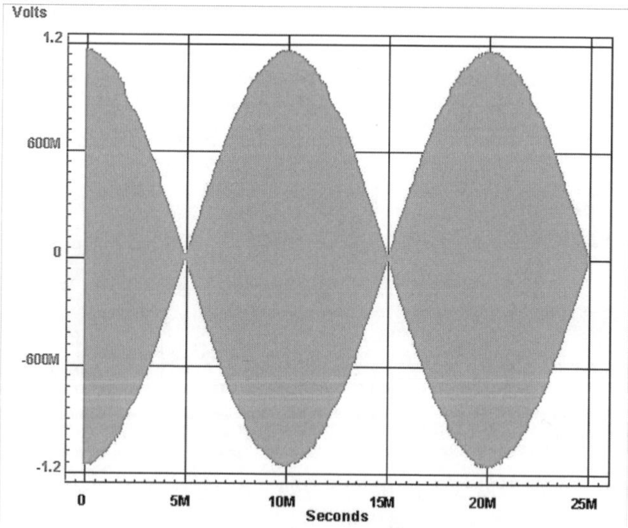

Fig. 3-7: **Mixer Output**

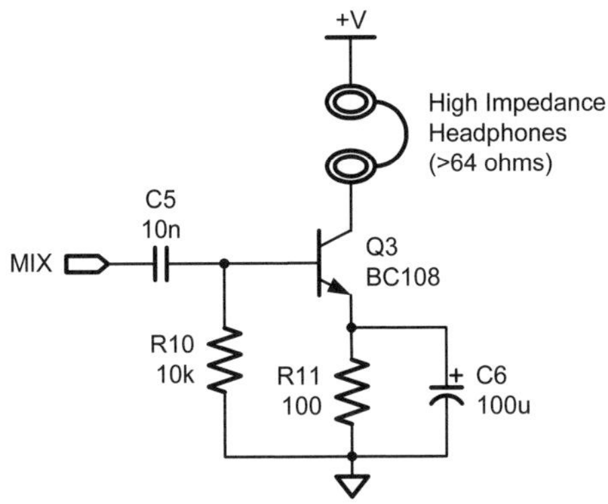

Fig. 3-8: **Headphone Amplifier**

phone amplifier. Any signal content that exists in the 100kHz region of the spectrum will be filtered out by the hearing of the user, since the frequency range of a person with good hearing extends from 20Hz to 20kHz.

The circuit for a suitable headphone amplifier is shown in Figure 3-8, and is capable of driving either high impedance headphones, or a high impedance speaker (e.g. 64 ohms). The completed BFO detector was tested using a set of headphones with a DC resistance of 268 ohms per earpiece, which was ideal for this application.

The Complete BFO Detector

At this stage of the design we now have all the ingredients required to create a working BFO detector. The only thing we have not considered, is how to adjust one of the oscillators to provide the required *beat* frequency at the output. Although it is theoretically possible to calculate the capacitor values required (including any trimmer capacitors) in reality, the loading caused by other circuitry will produce different results than expected.

As an example, here are the results obtained when testing this design:

> For the reference oscillator (with C1=680pF, C2=2.2nF and L=4.7mH) we can calculate an operating frequency of 101.9kHz.
>
> By adding a trimmer capacitor with a range of 2.2pF to 22pF, the frequency can then be adjusted from 99.8kHz to 101.7kHz.

So that's the theory — but in practice it was found that without the tuning capacitor the actual measured frequency was 90.8kHz, which was lower than expected.

Here are a number of reasons why the measured value could differ from the calculations:

1. All components have a tolerance associated with them, which must be taken into account during the design process.
2. Even if you measure each component individually to obtain a more accurate result, the actual layout of the design can introduce stray (or parasitic) capacitance that cause the measured and calculated results to be different.

3. When an oscillator circuit (for example) is driving a mixer, the input impedance of the circuit being driven will affect the frequency of oscillation.

The result is that you do not always get what you expect.

It is the job of the designer to be aware of these potential pitfalls, and produce a design that can continue to operate within a given range of values. For projects where only a small number of units will be produced, then tolerance becomes less of an issue. However, if you were designing something that could potentially have a production run reaching into the 10's of thousands, you would not be very popular if half the manufactured units failed to function, because of poor tolerancing in the design.

Now let's turn our attention to the search oscillator. The search coil was wound using 0.56mm enamelled wire, with 143 turns on a 5" former — and before anyone makes a comment — I'll apologise now for mixing metric and imperial measurements[1]. The result was a 5.5mH coil with a DC resistance of 4.4 ohms.

As we calculated previously, the nearest standard values for these components are: C1=560pF and C2=2.2nF.

With all the separate sections of the BFO design (search osc, reference osc, mixer and headphone amplifier) assembled and connected for testing, the measured values were as follows:

Search Osc: 75.8 kHz
Reference Osc: 87.7 kHz (max) and 85.5 kHz (min)

So we have a problem... (and - can you spot it?)

The search oscillator needs to be higher in frequency than the reference oscillator, otherwise ferrous/non-ferrous discrimination will not work as expected. Therefore we need to set the maximum reference oscillator frequency to 75.8 kHz, allowing the trimmer capacitor(s) to lower the beat frequency to an appropriate level.

With L=4.7mH, then:

$$C = \frac{1}{\omega^2 L} = \frac{1}{(2\pi \cdot 87.7\text{kHz})^2 \cdot 4.7\text{mH}} = 700.7\text{pF} \qquad \text{Eq 3-10}$$

The value of capacitance calculated in the above equation represents the total capacitance in the real physical circuit. Compared to the original calculated value (C1 in series with C2) of 519.4pF, there has been an increase in capacitance of nearly 35%. This is too large a discrepancy to be attributed to component tolerance alone, and capacitance introduced by other components and the circuit layout must be the main cause of the problem.

To compensate we must calculate new capacitance values for the reference oscillator, as follows. The reference oscillator frequency needs to be adjusted to 75.8kHz:

$$C = \frac{1}{\omega^2 L} = \frac{1}{(2\pi \cdot 75.8\text{kHz})^2 \cdot 4.7\text{mH}} = 938\text{pF} \qquad \text{Eq 3-11}$$

1. See Chapter 15 for a discussion of metric and imperial issues in electronics, as well as the minor nuances involved in having a Brit and a Yank collaborate on a book involving circuits.

This is the value of C required for operation at 75.8kHz, and represents an increase in value of 237.3pF. The recommended approach here would be to add a standard value capacitor (e.g. 220pF) in parallel with a trimmer capacitor (e.g. with a range of 5.5pF to 65pF). This would also provide the reference oscillator with a coarse and fine frequency adjustment that could cope with variations in both component tolerance and circuit layout.

The complete circuit of the BFO detector can be seen in Figure 3-9 and the actual circuit constructed using stripboard is displayed in Figure 3-10.

If you wish to construct this design you will notice that the use of a transistor mixer provides good isolation between the search and reference oscillators, which allows a low difference frequency to be used without the two oscillators locking together.

The completed design was tested against various metal targets with the following results:

Metal Target	**Detection Distance**
Victorian Penny	3" (76.2mm)
Iron Ring (2" diameter")	5" (127mm)
British £1 coin	2.5" (65.5mm)
PP3 Battery	3" (76.2mm)
U.S. Cent	1.5" (38.1mm)
British Penny	1.5" (38.1mm)
Metal Waste Paper Basket	10" (254mm)
Copper 1p (Cartwheel) 1797	4" (101.6mm)
Small Roman Coin (17mm diam)	1.25" (31.75mm)
Roman Coin (22mm diam)	2" (50.8mm)
Roman Coin (33mm diam)	3" (76.2mm)
Hammered Silver Coin	1" (25.4mm)
Horse Shoe	7" (177.8mm)

As you can readily see from the test results, the BFO detector is not particularly sensitive, although the circuit presented here is at least as good as many commercial designs. The inductance of the search coil has little effect on the sensitivity, but the diameter has a major impact. A larger diameter coil will achieve greater depth, but will reduce the BFO's ability to detect small objects. Over the years, several people have attempted to improve on the basic BFO concept with limited success, and we will now briefly explore one of these techniques.

Using Oscillator Harmonics

Near the beginning of this chapter, it was mentioned that harmonics in the oscillator output were undesirable. However, there is one instance where this is not the case.

As we have seen, one of the main problems with the BFO is that the inductance change in the search coil (due to the presence of a metal target) is very small. We also noted that one solution was to adjust the reference and search oscillators such that their frequencies were very close together, but this has the drawback of eliminating any possibility of ferrous/non-ferrous discrimination.

If an oscillator does not produce a pure sine wave, then there will be harmonics of the fundamental frequency (f) present in the output spectrum. These harmonics will occur at 2f, 3f, 4f, 5f and so on, up to infinity. In practice, the easiest way to introduce intentional harmonics is to use a square wave instead of a sine wave oscillator. There is however one important difference here — the square wave output only consists of the fundamental frequency plus the odd harmonics. This means that a 100kHz square wave oscillator will have a spectrum consisting of 100kHz + 300kHz + 500kHz, etc. The trick is then to run the reference oscillator close to one of the harmonic frequencies. For example, if we choose to set the reference frequency to 300.1kHz, then this will beat with the 3rd harmonic of the search oscillator producing a 100Hz beat signal at the headphones. A shift of 10Hz in the fundamental will be detected as a 30Hz shift in the 3rd harmonic, or (if you want to make the design even more sensitive) a 50Hz shift in the 5th harmonic.

As you might expect, you cannot get something for nothing — and this is definitely the case here, as instability becomes a serious design problem. It is not only the frequency shift that is multiplied by the harmonic number, but also the jitter in the transmitted signal. The disadvantages with such a design can easily outweigh the advantages, such that no improvement is gained in practice. There are also major issues with using harmonics above 5, as the amplitude of the harmonic is equal to the amplitude of the fundamental divided by the harmonic number. This means that the signal of interest becomes vanishingly small as you move up to higher harmonic numbers.

The bottom line is simply that the BFO technique is relegated to the lower end of the metal detector market, and only used in the construction of beginner's machines aimed mainly at children.

Note (if you do construct this BFO design) — do not forget to add a Faraday shield to the search coil. This is easily constructed using self-adhesive aluminum tape, and must not form a complete circuit, otherwise it will act like a shorted loop. The shield must also be connected to the circuit ground.

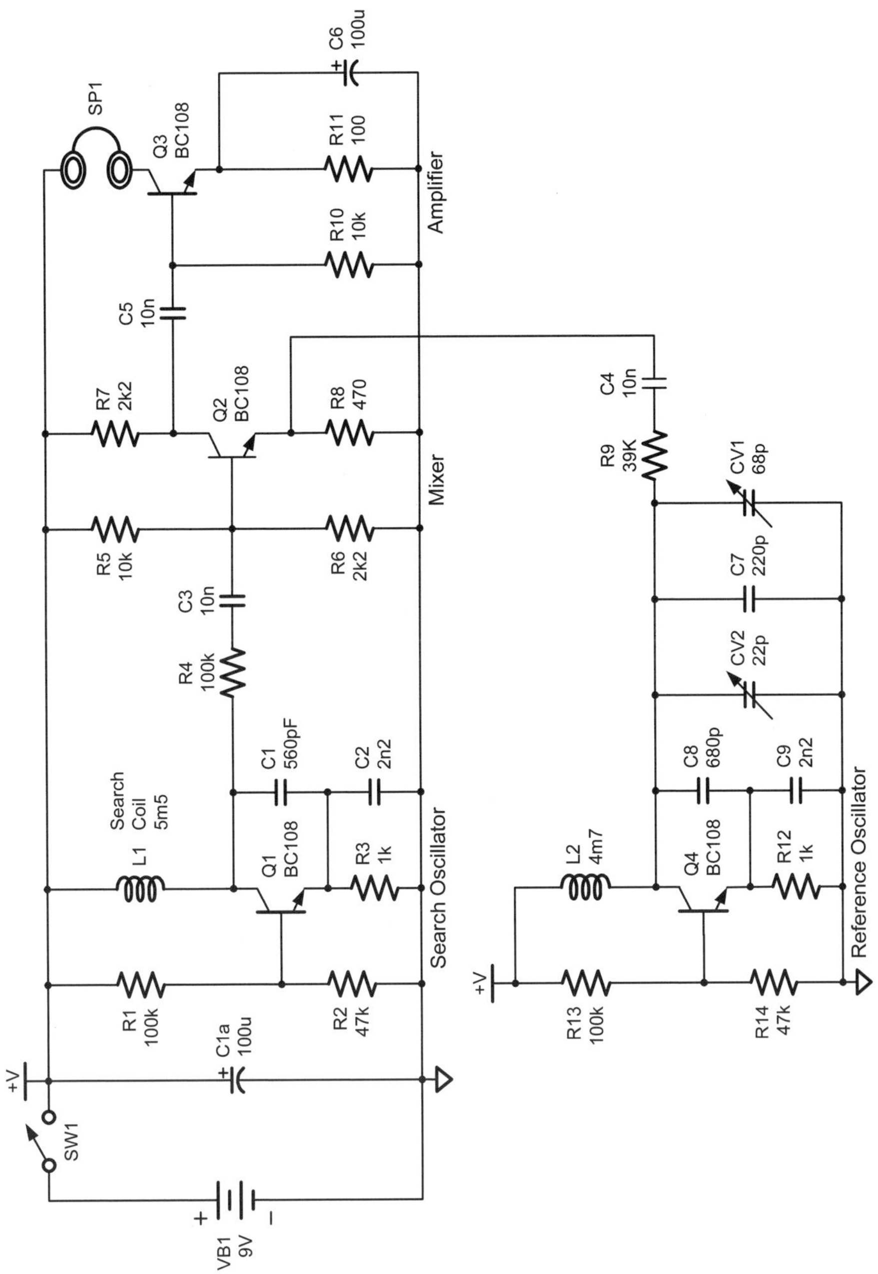

Fig. 3-9: **Complete BFO Circuit**

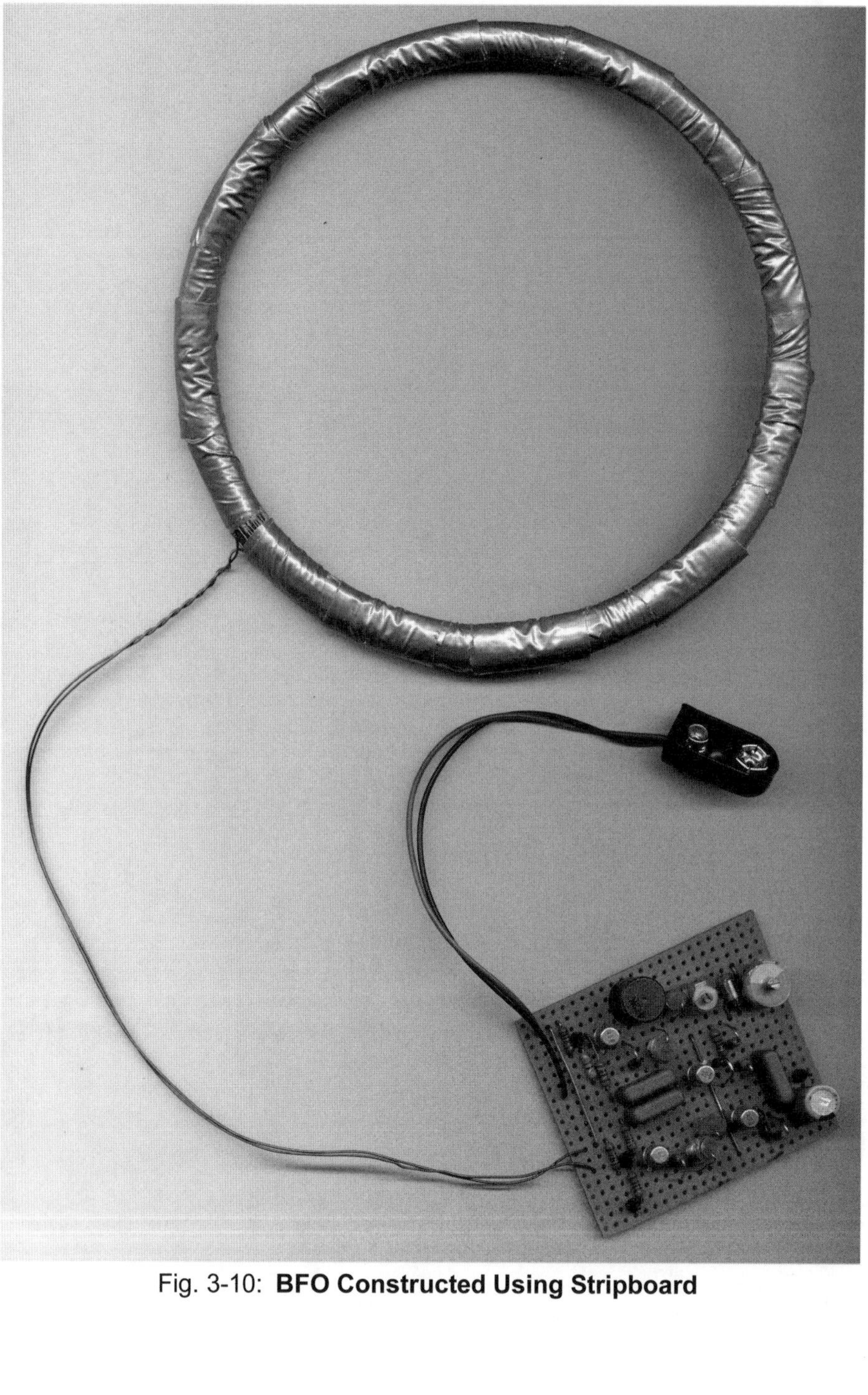

Fig. 3-10: **BFO Constructed Using Stripboard**

Construction Details

The component placement and PCB layout are for illustrative purposes only. You may need to adjust the layout to accommodate components available in your area. In particular, please note that transistor pinouts can vary depending on the country of manufacture, even for what appears to be an identical part number. See Chapter 15 for more details.

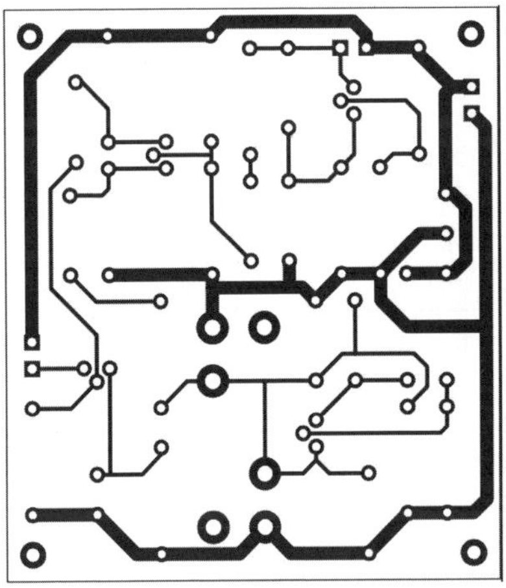

Fig. 3-11: **PCB Layout**

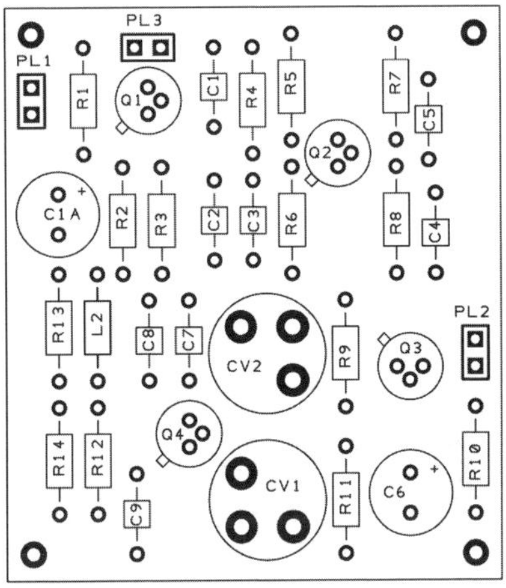

Fig. 3-12: **PCB Parts Placement**

The PCB layout view (Figure 3-11) is from the underside of the board. The actual size of the PCB is 1.9" x 2.15" (48.3mm x 54.6mm). There are no jumpers required in this layout. The parts placement (Figure 3-12) is shown from the top-side. The 3-dimensional view (Figure 3-13) provides an idea of what the final product will look like in real life. Note the connectors are designated as follows:

PL1 = battery connector
PL2 = speaker
PL3 = search coil

Parts List

Resistors: (5% 1/4W)

R1, R4, R13	100k
R2, R14	47k
R3, R12	1k
R5, R10	10k
R6, R7	2k2
R8	470
R9	39k
R11	100

Capacitors:	
C1A, C6	100u elect., 10v
C1	560p
C2, C9	2n2
C3, C4, C5	10n
C7	220p
C8	680p
CV1	68p variable
CV2	22p variable
Inductors:	
L1	Search coil: see text for details
L2	4m7
Transistors:	
Q1, Q2, Q3	BC108 (or any general purpose NPN)
Misc:	
Speaker	64 ohms

Fig. 3-13: **3D View**

CHAPTER 4

A Simple Probe That Works!

"It was all too complicated and, where it is too complicated, it meant that someone was trying to fool you."

— Terry Pratchett (The Fifth Elephant)

In our exploration of BFO technology, it was discovered that the presence of a metal target was capable of altering the inductance of the search coil. In order to detect this small change in inductance, a reference oscillator was used to produce a beat tone that could be heard in a set of headphones. Perhaps you have been wondering whether there are any alternative methods that could be used to detect this change in inductance, and thereby eliminate some of the instability problems of the basic BFO design. If so, then you would be correct.

There is a type of metal detector known as "off-resonance", which uses a different technique to sense the change in inductance. An oscillator is used to drive the search coil (similar to the BFO) but via a high resistance, and the output voltage is rectified before being applied to the input of a comparator, where it is compared to a reference voltage.

In some "off-resonance" designs, the rectifier output is applied directly to the input of a voltage-controlled oscillator (VCO) (Figure 4-1) so that the operator can effectively "hear" the change in inductance caused by the presence of a metal target. This method has the advantage of providing some elementary ferrous/non-ferrous discrimination. Unfortunately there is little (if any) improvement in performance/depth over the standard BFO. Perhaps the only real advantage is the simplicity of implementing this type of design, given that it only uses one tuned oscillator.

However, there is one very good use for the "off-resonance" detector — as a simple handheld probe, where performance/depth is not an issue. Also, the ability to compare the search oscillator output voltage to a reference source, and give a yes/no indication of a metal target, is a distinct advantage. The reason why this method works is very simple. The search oscillator is designed to produce a sine wave output

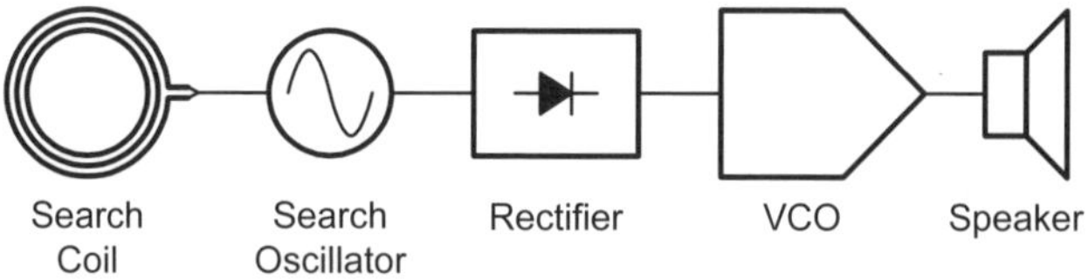

Fig. 4-1: **"Off-resonance" Metal Detector Block Diagram**

that has an amplitude which varies with frequency. Therefore, we only need to monitor the amplitude of the oscillator, rather than directly comparing its frequency to a reference source.

Search Oscillator

The BFO in Chapter 3 was designed to oscillate close to 100kHz. If you have constructed this circuit, it will have become apparent that the discrimination capabilities are less than perfect. The main culprit is our old friend the "skin effect", which tends to make all metal targets look alike regardless of their thickness. In order to combat this problem, the "off-resonance" detector uses a much lower frequency, which allows the signal to penetrate the target below the surface layer. Search frequencies from 30kHz down to only a few kilohertz can be used in practice.

For this design we are going to choose an operating frequency close to 25kHz, and (just to be different) we will use a Hartley oscillator, instead of the Colpitts we implemented last time.

Since we are designing a probe circuit, there is no need for a large search coil. In fact, a large search coil would be a disadvantage, as the probe search head must be capable of being inserted into holes where a normal sized detector coil would not fit. If we also choose to use two standard RF chokes (similar to the one used in the reference oscillator of the BFO) then it just so happens that these will fit quite nicely into the end of the oval pvc tubing that is readily available from most DIY stores. Hence we have a simple and sturdy construction for the probe head.

The RF chokes used in this design were both 1mH. Note that the 4.7mH choke used in the BFO reference oscillator has too large a diameter to fit within the tubing. Next, (referring to Figure 4-2) we need to calculate a value for C1 that will provide an operating frequency of approximately 25kHz.

The Hartley oscillator equation is:

$$f = \frac{1}{2\pi\sqrt{(L_1 + L_2)C}} \qquad \text{Eq 4-1}$$

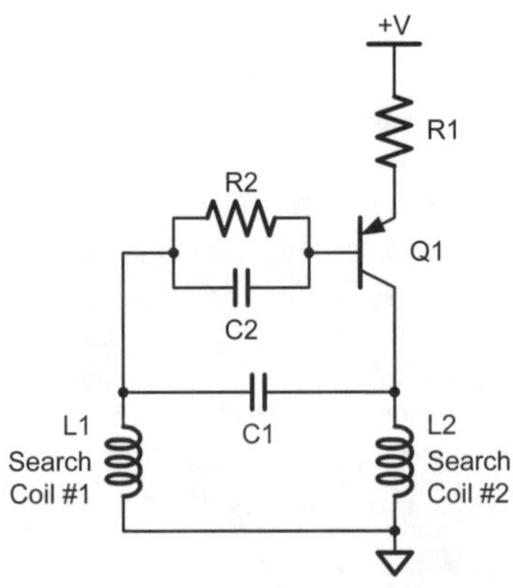

Fig. 4-2: **Hartley Oscillator**

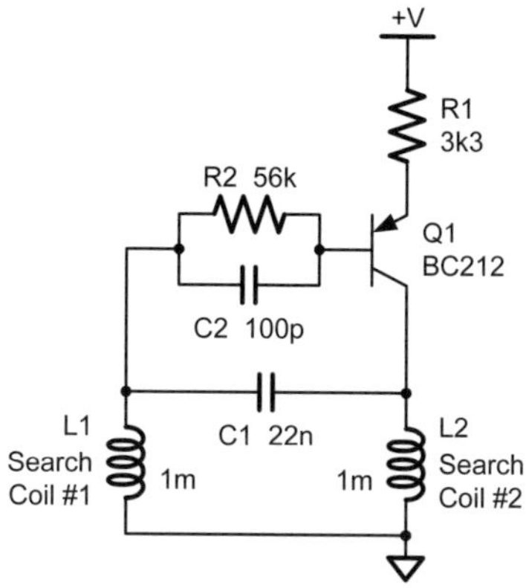

Fig. 4-3: **Search Oscillator for Probe**

Which can be rearranged as follows:

$$C = \frac{1}{(2\pi f)^2(L_1 + L_2)} \qquad \text{Eq 4-2}$$

For a frequency of 25kHz, and L1 = L2 = 1mH, then C = 20.3nF.

If we decide to use the nearest standard capacitor value of 22nF, we can recalculate using the first equation to arrive at an operating frequency of 24kHz. The final circuit for the search oscillator is shown on Figure 4-3.

You may have noticed that the biasing arrangement for our Hartley-based oscillator is rather unusual. The passive components R2 and C2 have been carefully chosen to give a frequency response that peaks at a higher frequency than the tank circuit formed by L1, L2 and C1. Consequently the bias voltage on Q1 is sensitive to any changes in the inductance of L1 or L2, with the result that the collector voltage of Q1 will vary in amplitude. This means that a decrease in inductance gives an increase in frequency and also an increase in amplitude at the collector of Q1. Likewise, an increase in inductance will result in a decrease in frequency and also a decrease in amplitude.

For the circuit values shown, the peak is set to 33.9kHz, whereas the search oscillator is designed to resonate at 24kHz. Hence the name "off-resonance".

At this point we could simply rectify the oscillator output, and feed this DC signal into a comparator (for comparison against a reference voltage) before driving an audible or visual warning device. Although this would be one possible solution, there are a number of potential issues to consider. Firstly, the oscillator is liable to drift due to changes in component tolerances, or simply due to temperature. Secondly, we would also have the problem of supplying a stable reference voltage. The reality is that these "drift" problems means that the user is constantly having make manual adjustments to compensate.

Automatic Tracking Circuit

What if there was some way to automatically track these slow moving changes, such that the user never has to make an adjustment?

Let us first consider the circuit shown in Figure 4-4:

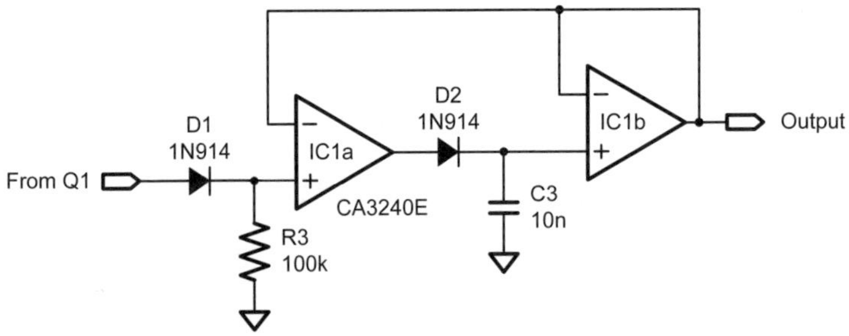

Fig. 4-4: **Envelope Follower**

The diode D1 is required to block the negative-going transitions from the search oscillator, so that a half-wave rectified signal appears at the non-inverting input of IC1A. The CA3240E is a dual version of the popular CA3140 operational amplifier. This opamp has a very high input impedance, and very low input current, and can operate from a supply voltage from 4V to 36V (either single or dual supply). Therefore this device can be connected directly to the oscillator output with virtually no loading, and the amplitude of the positive voltage swing will be tracked at the output (pin 1) of IC1B.

Now comes the clever bit ->

If we construct a second identical circuit of the envelope follower, but increase the value of C3 (in Figure 4-4) to 10uF. This will provide two outputs that track the oscillator amplitude, but they will do so at different speeds. In other words, one of these circuits will tend to track the amplitude in real time, whereas the other slower circuit will lag behind the first and track at a more leisurely pace. If the outputs of these two circuits are fed into a comparator, then the slow tracker will act a reference for the fast circuit. Whenever a metal target comes within the influence of the search coil, the fast circuit will respond first; and the voltage at the inverting input of the comparator will rise above the voltage at the non-inverting input (i.e. the output from the slow tracker). This will cause the comparator output to go low, which can then be used to activate a warning device.

If we now combine all these parts of the design together, the result will be as shown in Figure 4-5.

The sensitivity of the final design can be adjusted using VR1. After setting this 18turn potentiometer there will be no need for further adjustment, since any variations due to circuit tolerance, temperature or supply voltage will be automatically handled by the combination of slow and fast tracking circuits. Note that although the slow tracking circuit is much slower than the fast one, it will eventually catch up. Which means that the LED will turn off if you hold the search head next to a metal target for a sufficiently long time. This arrangement means that the probe really is a true “switch-on and go” detector.

IC4 is an LM7805 3-terminal regulator that provides a stable 5V supply, giving a very long life from a single PP3 9V battery.

The on-off switch is of the DPDT type that not only connects the positive terminal of the battery to the regulator input, but also resets the slow tracking circuit when turned off.

Figure 4-6 shows the completed unit, together with a close-up of the probe head assembly containing the two 1mH RF chokes. The complete circuit will fit comfortably within a handheld ABS case that features a battery compartment suitable for one PP3 battery. The case used for the prototype shown also included a set of clip- in battery terminals that were supplied as standard.

There is little point to be gained in compiling a chart of metal targets against distance, given that the probe head is a miniscule 6 x 12mm. Typical detection distance for all targets is between 1” and 1.5”, which is perfect for this application. If you would prefer to have an audible rather than visual indication of a target, then it is possible to replace the LED with a piezoelectric buzzer.

Of course you may wish to connect a larger search coil to this circuit, and produce a simple full-size “off-resonance” detector, that in many ways is much simpler to construct than the standard BFO design. The search oscillator could also be redesigned using a Colpitts, so that the search head requires only a single coil without a centre tap. This design has a number of advantages for the beginner who wishes to construct their first detector, such as:

1. Simple search head construction
2. Simple to build and setup
3. Absolute minimum of adjustment
4. Switch-on and go
5. Very stable design
6. Low power consumption

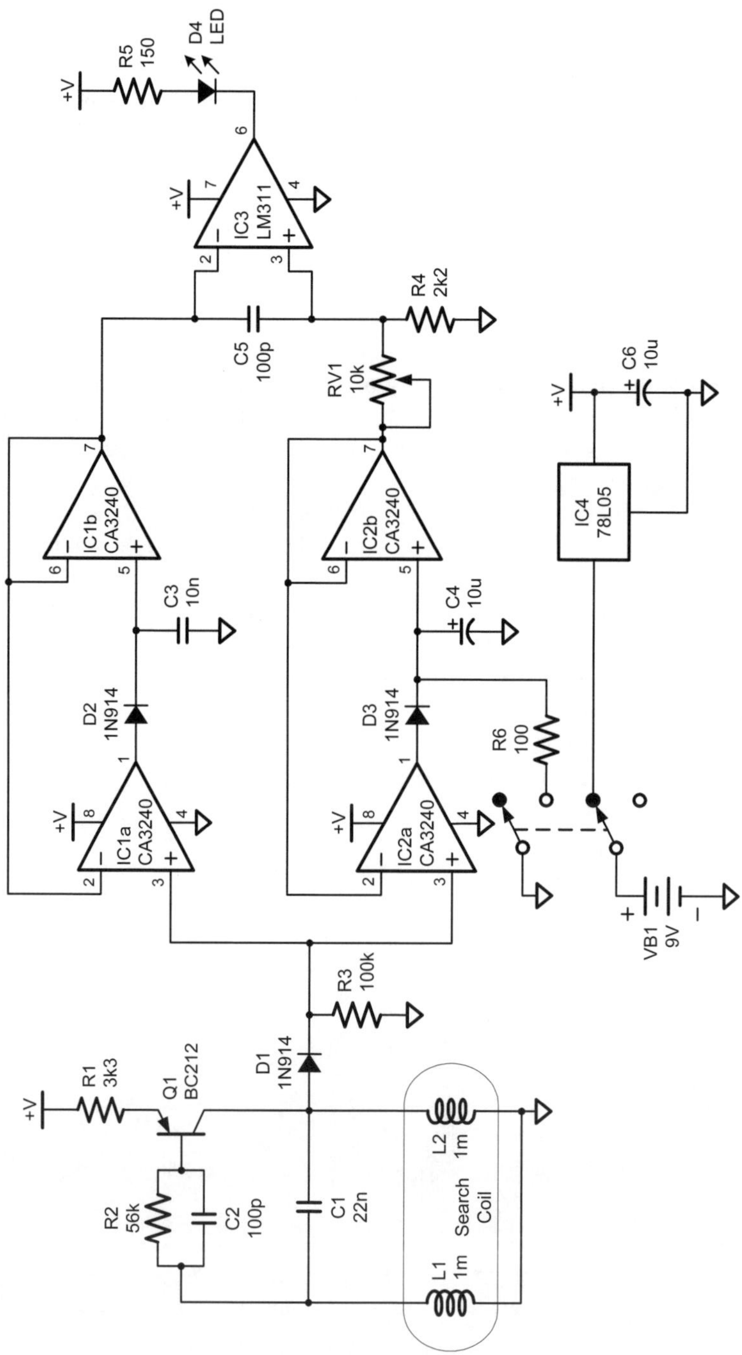

Fig. 4-5: **"Off-Resonance" Probe Circuit**

Construction Details

The component placement and PCB layout are for illustrative purposes only. You may need to adjust the layout to accommodate components available in your area. In particular, please note that transistor pinouts can vary depending on the country of manufacture, even for what appears to be an identical part number. See Chapter 15 for more details.

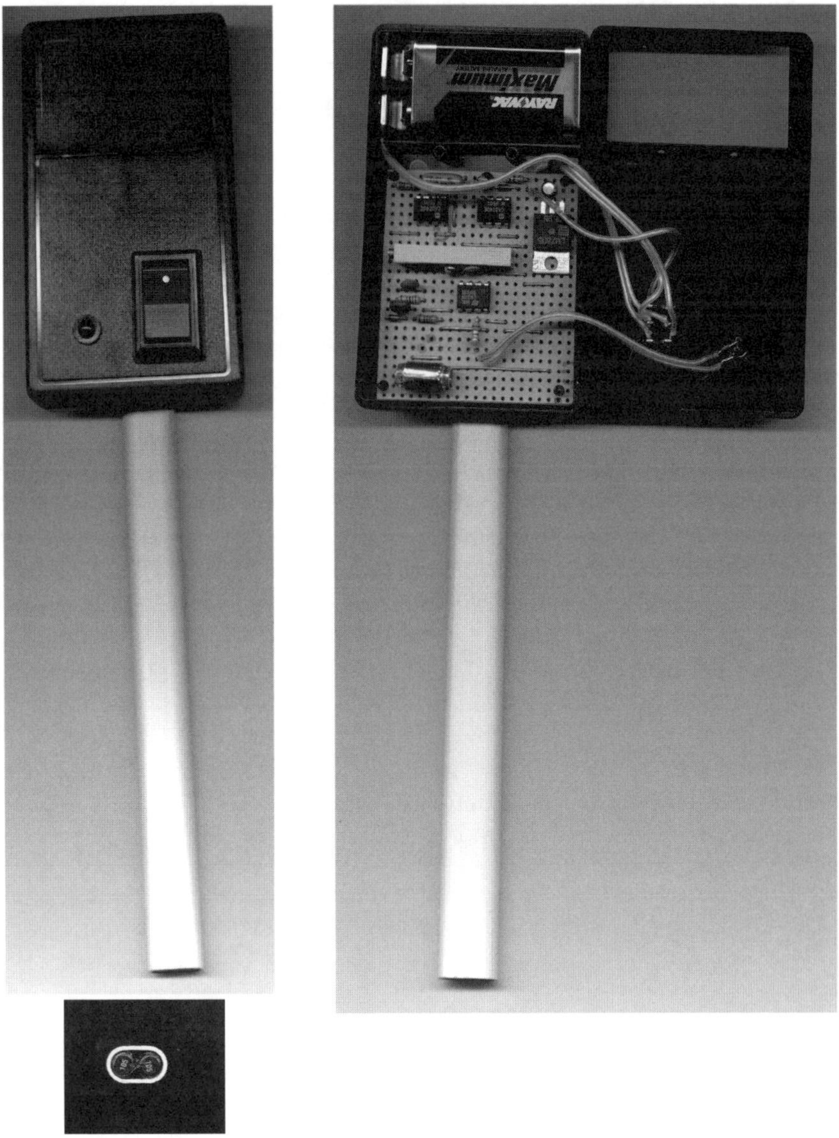

Fig. 4-6: **Probe Constructed Using Stripboard**

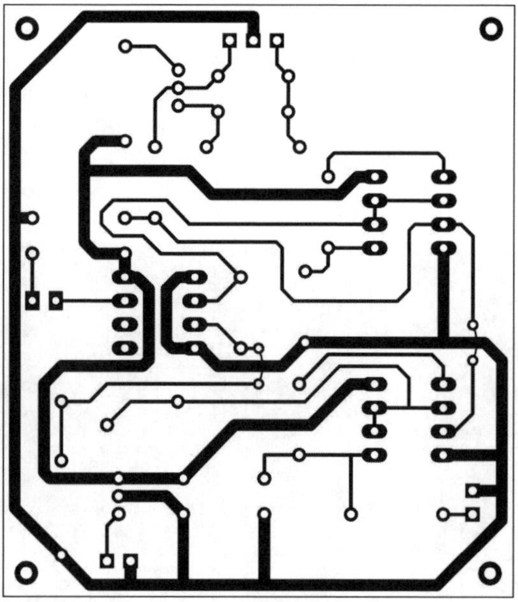

Fig. 4-7: **PCB Layout (single-sided)**

The PCB layout view (Figure 4-7) is from the underside of the board. The actual size of the PCB is 2.2" x 2.5" (55.9mm x 63.5mm). Note that there are 2 jumpers required in this layout. The parts placement (Figure 4-8) is shown from the top-side. The 3-dimensional view (Figure 4-9) provides an idea of what the final product will look like in real life. Note the connectors are designated as follows:

PL1 = search coils
PL2 = SW1a
PL3 = SW1b

Parts List

Resistors: (5% 1/4W)

R1	3k3
R2	56k
R3	100k
R4	2k2
R5	150
R6	100
RV1	10k multi-turn pot

Capacitors:

C1	22n
C2, C5	100p

C3	10n
C4, C6	10u elect., 10v
Inductors:	
L1, L2	1m RF choke
Diodes:	
D1, D2, D3	1N914
D4	LED
Transistors:	
Q1	BC212 (or any general purpose PNP)
ICs:	
U1, U2	CA3240
U3	LM311
U4	LM7805
Misc:	
SW1	DPDT switch

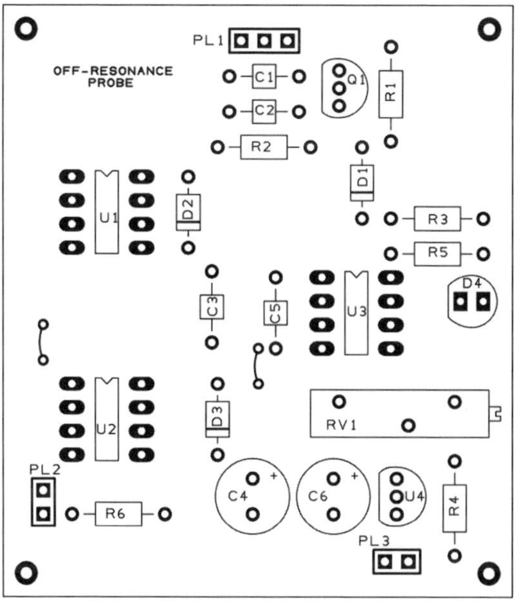

Fig. 4-8: **PCB Parts Placement**

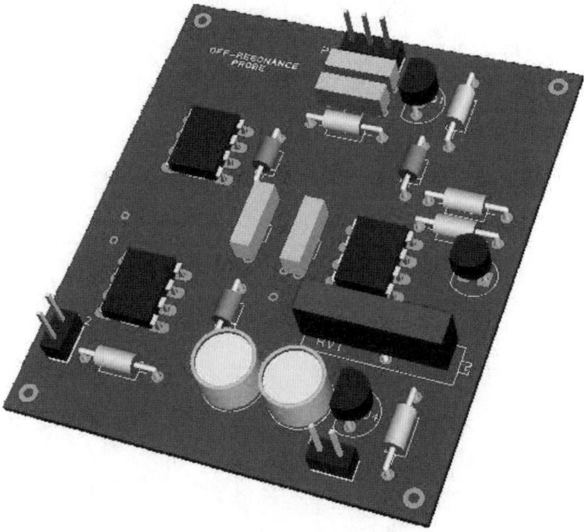

Fig. 4-9: **PCB 3D View**

CHAPTER 5

Induction & Coils

"I am busy just now again on electro-magnetism, and think I have got hold of a good thing, but can't say."
— *Michael Faraday*

The information in this chapter is really a prerequisite for completely understanding any kind of metal detector design, but we've held off presenting it until just now. This was to give the reader a chance to dive into some real metal detector circuits before embarking on a journey into the dry, technical presentation of theory and other uninteresting details.

It is true, that you can pretty much skip this chapter if all you want to do is build metal detector circuits without really understanding how everything works. But you might find some of the information in this chapter fun and interesting, and it will certainly help lay a better foundation for those who want to move beyond the projects in this book.

Magnetic fields

Metal detectors operate via the principle of *induction*. Induction is the method of coupling two circuits though an alternating magnetic field. To clarify, we'll step back in time and look at some of the very same experiments mentioned in the *History* chapter. Let's start with basic magnetic fields first.

When you pass a direct current (DC) through a wire a *static* magnetic field develops around the wire[1]. This is a very simple test that is often demonstrated in grade school science class, and one that anyone can do at home.

Experiment 5-1: Demonstrate that a current through a wire produces a magnetic field.

Required: Battery (C-cell is fine); a length of small-gauge wire (18 in. or 1/2 m will do), and a needle-style magnetic compass.

Figure 5-1 illustrates the setup. Place the compass on a wooden table, and form the wire into a large circular loop. While holding the battery, *briefly*[2] short the wire across its terminal. If you move the loop close to the compass, you will see

1. As discovered by Øersted.
2. *Briefly* is the operative word here. The wire will get quite warm if you hold it on too long. "Too long" is about a second.

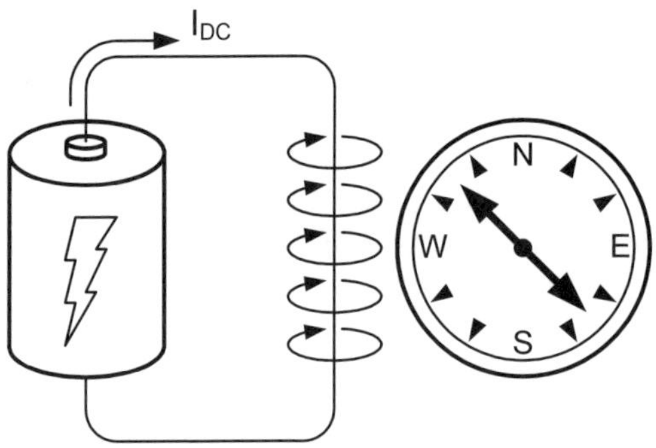

Fig. 5-1: **Magnetic Field Experiment**

the needle jump. Try to determine which orientation of the wire, as compared to the normal direction of the needle, causes the greatest deflection.

What happens when the wire is held above the compass? What happens when the wire is held below the compass? What happens when the battery polarity is reversed?

If you wrap the wire in a long tight coil (Figure 5-2) you can enhance the magnetic field, because the coil "focuses" the field though its center. Each turn of wire adds more "flux" to the field. It ends up that the strength of the magnetic field depends on the current through the coil (I), the dimensions of the coil (length l and radius r), the number of turns of wire (N), and the material used as the core of the coil. The maximum field strength of such a coil is

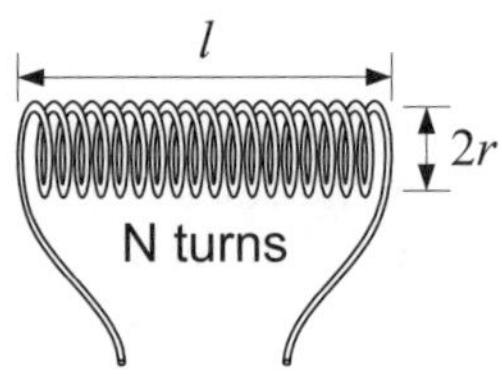

Fig. 5-2: **Basic Coil**

$$B = \frac{\mu NI}{\sqrt{4r^2 + l^2}} \qquad \text{Eq 5-1}$$

where μ is the "permeability" of the core material the coil is wrapped around. This is the field strength exactly at the center point *inside* the coil. The lines of flux that make up the magnetic field exit each end of the coil and wrap around the outside to form a complete path. The field strength reduces slightly as you move towards the ends of the coil, and is weaker around the outside of the coil because it is not confined and can spread out. Figure 5-3 illustrates. Also, the magnetic flux lines have direction, and the field external to the coil has a direction opposite to the flux inside the coil. This will be important in the design of metal detector coils.

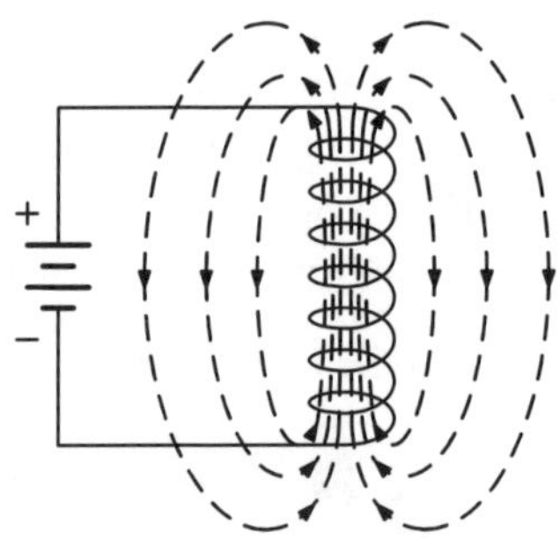

Fig. 5-3: **Coil Field**

So what is "permeability" in the equation above? It is the magnetic equivalence to conductivity[3] for current. A material that has more permeability is more "conducive" to the presence of magnetic flux. The coil in Figure 5-2 appears to have no core

3. Conductivity is, of course, the reciprocal of resistance.

material other than air, and the permeability of air is $\mu = 4\pi x 10^{-7}$ Henries/meter (H/m). Air is often considered to be the norm by which other permeabilities are compared[4], and is given the special symbol μ_0.

If we were to wrap our coil around a steel core, such as a nail, then the magnetic field would be enhanced because steel has a much higher μ than air, about 100 times higher. A material's permeability as compared to air is called *relative permeability*, μ_r, so for steel $\mu_r = 100$. It is the iron content in steel that gives it a high permeability, and pure iron is even higher; 99.8% pure iron has a $\mu_r = 5000$. Certain alloys are made to have a $\mu_r = 1,000,000$ or more. A coil wrapped around a nail forms a simple electromagnet, also often demonstrated in elementary school science class.

Experiment 5-2: Demonstrate that winding the wire into a coil increases the magnetic field strength.

Required: Battery (C-cell is fine); a length of small-gauge wire (36 in. or 1 m will do), a needle-style magnetic compass, a wooden pencil, and a large nail.

Wrap the middle 18 inches (1/2m) of wire around the pencil to form a coil with 9-inch "pigtails" on each end. See Figure 5-4. For a normal-sized #2 pencil, you should end up with about 15 turns. You may remove the pencil; the coil will probably have a tendency to unspring slightly, that is nothing to worry about.

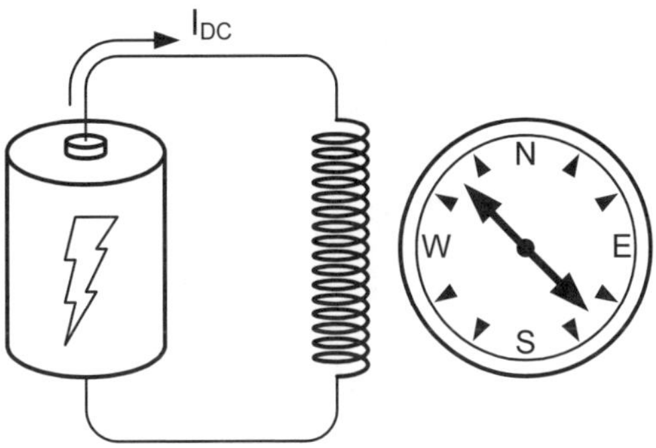

Fig. 5-4: **Coil Experiment**

Repeat the actions in Experiment 5-1, holding the coil close to the compass. Try different orientations of the coil relative to the at-rest needle, such as on top of, and along the edge of, the compass. Slide the nail inside the coil and repeat the experiment.

In performing these last two experiments you should have noticed a difference in how the orientation of the wire (or coil) affects the deflection of the compass needle. In the case of the straight wire, the magnetic field "wraps around" the wire, as shown in Figure 5-1. It is always true that the magnetic field attempts to "wrap around" the flow of current, so when we wind the wire into a coil, the magnetic field tries to wrap around each individual coil winding, and ends up focused through the center on the coil and "wrapping" back around the outside of the coil, like we saw in Figure 5-3.

Let's go back to Equation 5-1 which gave us the strength of the magnetic field at

4. Actually a vacuum is the norm, but air is almost identical.

the center of a coil. If the coil is much longer than its radius (long & skinny), then $l >> r$ and this simplifies to

$$B = \frac{\mu NI}{l} \qquad \text{Eq 5-2}$$

But metal detector coils are not long & skinny, they are short and large, where $r >> l$:

$$B = \frac{\mu NI}{2r} \qquad \text{Eq 5-3}$$

Again, this is the field strength exactly in the center. Assuming the strength of the magnetic field is an important issue in designing a metal detector[5], what we see from this is that for a given coil size (radius r) we can either increase the number of turns of wire, increase the current through the coil, or use a core material with higher permeability. If you've looked at a lot of detector coils, you will notice that the core material is always air, or some material (plastic, fiberglass, resin, etc.) that has a $\mu_r = 1$ (that is, equal to air). So we can assume that choosing a different μ is not a viable option or companies would be doing that. This leaves us with only windings and current. It also appears that large radius coils should have a weaker field, but we'll see later why this generally is not true.

So far, we have only considered the effect of a DC current[6] through a wire. As long as the current is constant, the magnetic field will be constant. If the current is varied, then the magnetic field will vary as well. An alternating current (AC) in a wire produces an *alternating* magnetic field around the wire. That is, as the current continuously reverses direction, the polarity of the magnetic field reverses, too. Clamp-on ammeters use this principle to determine the current in household wiring, without having to actually cut the wire.

Experiment 5-3: Demonstrate that an AC current[7] produces a changing magnetic field.

Required: The coil, compass, and nail from Experiment 5-2, plus a variable signal generator and a 100Ω resistor.

Instead of driving the coil with a battery, we will now drive it with an oscillator. Use a signal generator that can go down to about 1 Hz; a sine wave is preferable but a square wave will work. Since the generator probably has a more limited current drive than the battery, so may need to use the nail as a core for the coil.

With the generator set to oscillate at about 1 Hz or so, if you place the coil near

5. And we haven't really established this yet, have we? We'll get to that later...
6. "DC" means "direct current", so "DC current" means "direct current current", which clearly is redundantly repetitive. But in electronics, we also use the term "DC voltage", so DC has become synonymous with "constant".
7. Likewise, "AC current" is redundant, so just pretend AC means "alternating". Usually, AC refers to a periodic (repeating) signal, most often a simple sinusoid.

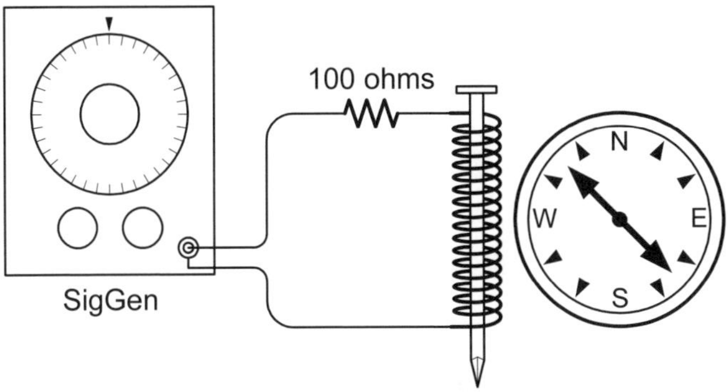

Fig. 5-5: **AC Test**

the side of the compass (try the north or south position, with the coil aligned east-west), you should see the needle move back and forth slightly. As you slowly increase the generator's frequency, the needle will move faster, but at some point mechanical inertia will begin to limit its movement. It's possible, if you can develop a strong enough magnetic field, to get just the right frequency so that the needle starts spinning in one direction; this is a crude motor.

Induction

If a current through a wire can create a magnetic field, an obvious question to ask is, can a magnetic field create a current in a wire? The answer is yes, but only in the case of a *changing* magnetic field. In other words, you can place a wire (generally as part of some closed-loop circuit) in an alternating magnetic field and generate an AC current in the wire. It might seem that placing a wire in a static magnetic field would generate a DC current[8], but not so. To generate a current with a static field, you have to move the wire *through* the field, for which you get a *transient*[9] current. If you move the wire through the field in a periodic manner, then you will get a periodic (AC) current. This is how a generator (alternator) works. But a wire sitting motionless in a static magnetic field will get you zilch.

Curiously, an alternating magnetic field gives rise to an alternating electric field (and vice-versa). Although it's really not important to this discussion, the basic idea is that a changing magnetic field produces an electric field which can move electric charge and, conversely, electric charges in motion (whether they are inside a wire or not) create a magnetic field. So they mutually support one another, and the combination of the two is known as an electromagnetic (EM) field. The EM field is how radio signals travel through space.

8. That would be nice, because you could use it to generate free electricity.
9. A transient signal (voltage or current) is not constant, nor is it periodic, so it generally is not considered DC or AC.

The reason for bringing this up is because we will often refer to the alternating magnetic field as an "EM field", and you will see others call it this as well. But metal detectors only make use of the magnetic field portion, and the electric field part does nothing for us. So when you see the term "EM field" applied to metal detectors, it really means the alternating magnetic field[10].

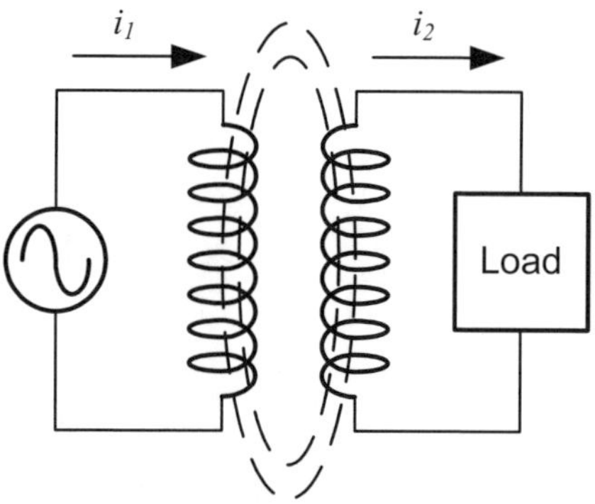

Fig. 5-6: **Inductive Coupling**

As with the DC experiments, if we wind the wire up into a coil we can intensify either effect. Pass an AC current through the coil and we get a stronger EM field; put the coil in an EM field and we get a stronger AC current. If we take two coils and place them close to each other, then driving one coil with AC will create the EM field, which will *induce* an AC current in the other coil. And here you have a *transformer* (Figure 5-6). This is the stuff that was figured out by Michael Faraday. It's called *induction*, because something (current) is getting *induced.*

What magic is doing the inducing? With wires it's electron motion, where the electron motion (current i_1) in one coil produces a magnetic field, and where that same magnetic field cutting across conductors in a second coil causes electrons to move, and produces a secondary current (i_2). In transformers, the driven coil is called the *primary* coil, and the induced coil is called the *secondary* coil. In metal detectors, the driven coil is called the transmit (TX) coil, and the induced coil is called the receive (RX) coil.

Experiment 5-4: Demonstrate the principle of induction.

Required: Small-gauge magnet wire, 100Ω resistor, a signal generator, an oscillo-

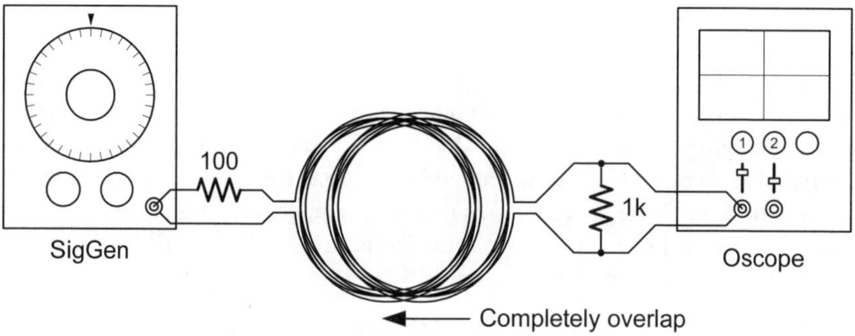

Fig. 5-7: **Induction Experiment**

10. A static magnetic field is not considered an electromagnetic field, even when it is produced by an "electromagnet".

scope.

Wind two coils of wire from small-gauge magnet wire (such as 26-30AWG), 50-turns each, about 4 inches (10cm) in diameter. For each coil leave ~10 inch (25cm) pigtails. Connect one coil via the 100Ω resistor to the signal generator, connect the other coil to the oscilloscope with a 1k load resistor. Arrange everything as shown in Figure 5-7, with one coil laying directly on top of the other. For the signal generator, a sine wave at 100kHz works well; set the output amplitude to fairly high level. Observe the signal amplitude on the oscilloscope.

Now place a sheet of paper or card stock between the two coils. Does this have any effect? Replace the paper with aluminum foil. What happens now? Finally try a piece of steel sheet metal.

Induction Balance

The above experiment shows that the EM field produced by one coil will induce electron motion in another coil, thereby generating a current. Again, this is called a transformer. Normally, transformers are designed for maximum coupling between the primary and secondary coils, in order to get the highest efficiency possible. Usually they include an iron-based core material to help maximize efficiency. With metal detectors, we actually want the opposite: we want to minimize the *direct* coupling between the primary and secondary coils. When a target enters the field an indirect coupling occurs that is discernible.

In Experiment 5-4 we laid the secondary coil directly on top of the primary, so that they were coaxially aligned. In this manner the magnetic field from the primary coil coupled with the secondary coil uniformly. If we slide the secondary coil sideways so that they overlap only partially, the secondary coil will begin to couple with a portion of the magnetic field on the interior of the primary coil and, simultaneously, with portion of the magnetic field on the exterior of the primary coil. Figure 5-8 illustrates. Since the magnetic field on the outside of the primary coil is opposite in direction to the interior field, there will be some cancellation in field coupling to the secondary coil. As we continue to slide the secondary coil over, at some point there will be a position where the interior and exterior fields exactly cancel, and there will be no coupling between the primary and secondary coils. This state is called *induction*

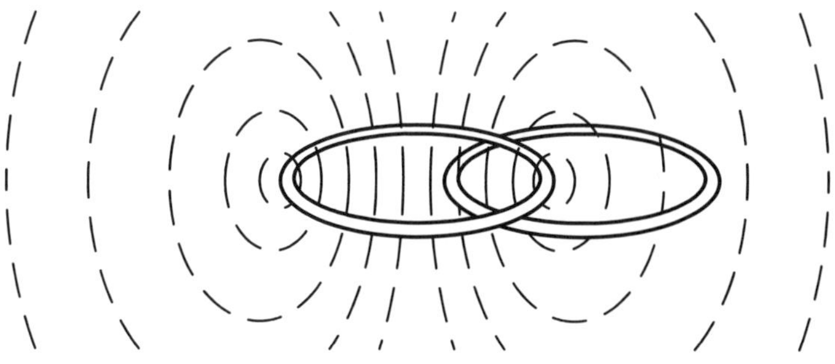

Fig. 5-8: **Induction Balanced Coils**

balance.

In metal detectors, we call the primary coil the *transmit* (TX) coil and the secondary coil the *receive* (RX) coil.

Experiment 5-5: Demonstrate induction balance.

Required: The setup from Experiment 5-4.

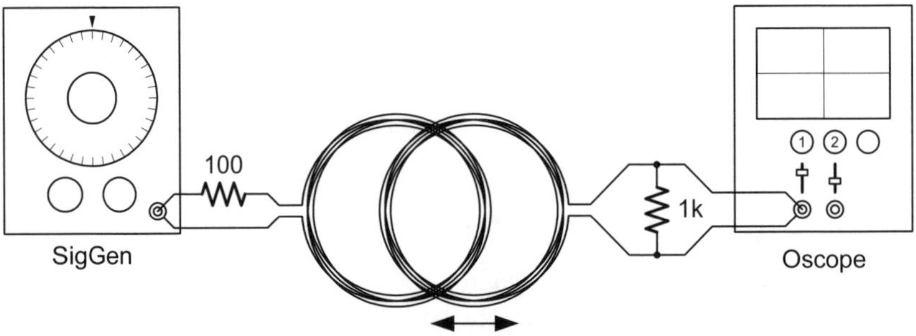

Fig. 5-9: **Induction Balance Test**

Using the same setup as in Experiment 5-4, place the receive coil directly on top of the transmit coil. Adjust the oscilloscope until the signal amplitude reasonably fills the screen. Begin sliding the second coil off the first coil. What happens to the signal? Try to achieve zero coupling; is it possible? (Increase the sensitivity of the oscilloscope as the signal decreases.) Continue sliding the second coil beyond this point; what happens to the signal?

Return the position of the second coil to the point where there is minimal coupling. Now introduce a metal target close to the coils. What happens to the signal?

Eddy Currents

Experiment 5-5 demonstrates the basics of an induction-balance metal detector. Alexander Graham Bell built this arrangement in 1881, using a mechanical clapper for the signal generator and a telephone earpiece in place of the oscilloscope. The final step in Experiment 5-5 was to move a metal target close to the balanced coils, which should have resulted in an increase in the received signal at the oscilloscope. The question is, what causes the signal to increase?

We saw from Experiment 5-4 that a changing magnetic field induces current in a secondary coil, as long as the coil is connected to some sort of useful circuit; in the experiment our *useful circuit* was just an oscilloscope. But what if it's not connected? What if, instead of a secondary coil connected to a circuit, we introduce a disconnected piece of metal to a changing magnetic field? Will the EM field still try to push the electrons around? Yes, but when you have a disconnected piece of metal the electrons have no where to go. So they do something really odd: they just go around in circles, something called *eddy currents* (Figure 5-10). If you have a *really* strong EM field, you can get really high eddy currents, enough to make the metal heat up. Maybe you've seen an induction cooktop where eddy currents are generated in the metal pot to heat the food. Foundries also use induction crucibles to melt steel.

So with a metal detector, we drive an AC current through a coil, which produces (transmits[11]) an EM field around the coil. When a metal target is near the coil, the EM field induces eddy currents in the target. How is that useful? It turns out — per Faraday's Law of Induction and Lenz's Law — that the induced eddy currents have a 90° phase shift[12] from the incident EM field, and that the eddy currents themselves create their own EM field, again with another 90° phase shift. The end result is that the eddy currents produce a *reverse* EM field; that is, 180° out-of-phase with coil field, because there are *two* 90° shifts between the coil's transmitted EM field and the target-induced EM field. See Figure 5-10.

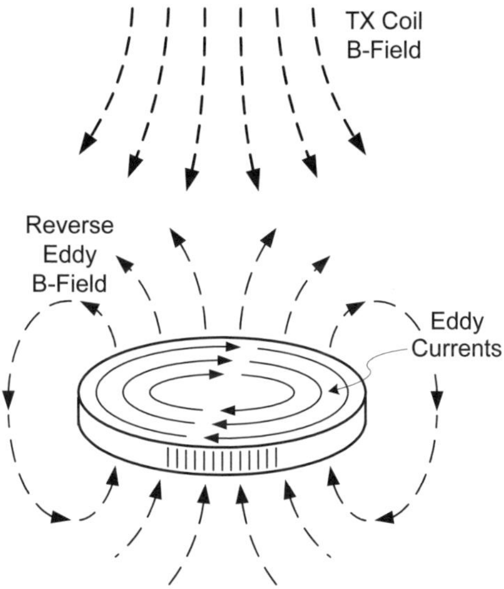

Fig. 5-10: **Eddy Currents & Field**

This reverse EM field gives us something to look for in order to detect the presence of a metal target. The problem is, it's an extremely weak EM field compared to the one that was produced by the primary (TX) coil. If we introduce our secondary (RX) coil right on top of the TX coil (as in Experiment 5-4) the resulting induced RX signal will be swamped by the TX signal, and the target signal will be impossible to distinguish[13]. What we need to do is eliminate the presence of the TX field *at the RX coil*, so that the RX coil only "sees" the target field. This was the purpose of Experiment 5-5.

11. Again, the "EM field" is just a local magnetic field around the coil, so it doesn't really get "transmitted". But this is popular jargon with metal detectors.
12. Not really; only a perfect superconductor will exhibit a 90° phase shift. More later.
13. It's like trying to hear someone whisper at a rock concert.

With the RX coil positioned so there is minimal coupling to the TX field, we can now detect the presence of a metal target by its eddy-induced reverse field. This reverse target field inductively couples with the receive coil and generates the signal that says you've found something. You can either think of this reverse target field as separate from the transmit field, or as a *distortion* of the transmit field — it has been described both ways by various authors, and either way of thinking is fine.

We mentioned before that the induced eddy current in a metal target is 90° out-of-phase with the incident magnetic field, and the counter-magnetic field of the target is 180° out-of-phase. The phase shift actually depends on some parameters of the metal, and only a perfect conductor (a *superconductor*) gives a 180° (total) phase shift. Everything else produces a lesser phase shift, and it turns out that variations in the total phase shift is how we can discriminate between different targets. Thicker, higher conductive targets, like silver dollars, are closer to 180°. Thinner, lower conductive targets, like aluminum foil, are closer to 90°.

Experiment 5-6: Demonstrate target phase shifts.

Required: Everything from Experiment 5-5, various targets.

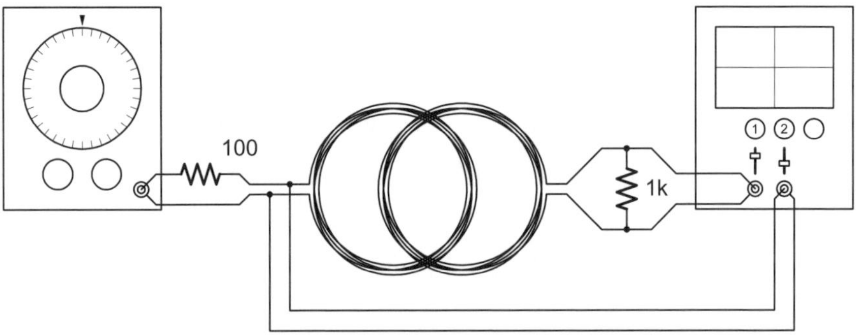

Fig. 5-11: **Induction Balance Test**

Connect the transmit signal to the second channel of the oscope and put the scope in "XY" mode. With the coils inductively balance you should get a minimal line or ellipse on the screen. Now introduce silver, copper, nickel and gold coins. What happens to the line? How does the line vary with target distance?

Let's briefly return to eddy currents. Eddy currents tend to be circular in motion, which is why they're so named, just like the eddies you find in streams. The circular nature of eddy currents means that circular targets will more efficiently support them. In fact, given a round target and a square target of identical surface area, thickness, and metal type, the round target will give a slightly stronger response.

Would there be a difference between a solid round target like a coin and an open round target like a ring? It turns out there is. A ring will give a *stronger* response, even though it might have far less surface area. That's because it appears more like a shorted loop of wire, in which the induced current flows around the loop (Figure 5-12). The coin can be thought of as having multiple concentric rings, in which each ring supports an independent eddy current. Each eddy will produce a counter-magnetic field, and each counter-magnetic field cuts across other rings, effectively creating an eddy drag. So the extra metal in the middle of the coin actually hurts its response.

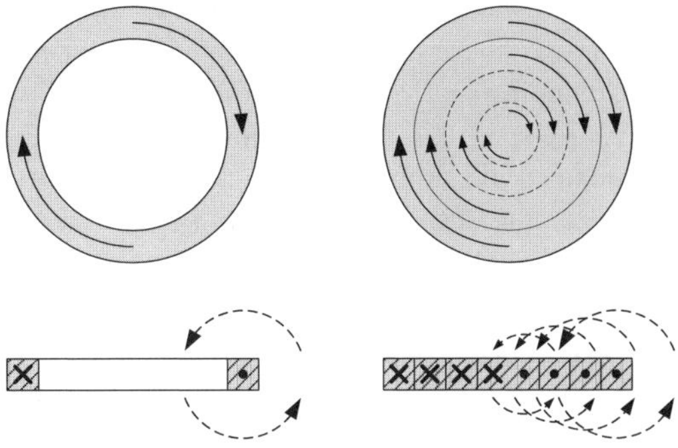

Fig. 5-12: **Ring vs. Coin**

Experiment 5-7: Determine whether a solid coin or a holed coin gives a stronger response.

Required: Two coins, metal detector or the setup from Experiment 5-6.

Drill out the center of one coin; try to remove at least half its diameter. Use a metal detector or the setup from Experiment 5-6 to determine which coin can be detected at the greatest distance[14].

In a version of this test, two pre-zinc U.S. Lincoln cents were used, with a 5/8" hole drilled through one. Both a VLF (all-metal mode) and PI detector indicated the drilled-out cent 3/4"-1" deeper than the solid cent. The experiment was repeated with silver Washington quarters[15] with the same result. So, with the conductivity variable removed we see that the ring shape will slightly edge out the solid coin, even when the metal and diameter are identical. Also, while a ring response is a tad stronger than a coin, a broken ring (Figure 5-13) is *really lousy.* Now the induced currents can't travel all the way around the ring so they move in tiny surface circles. By superposition, the resulting overall eddy

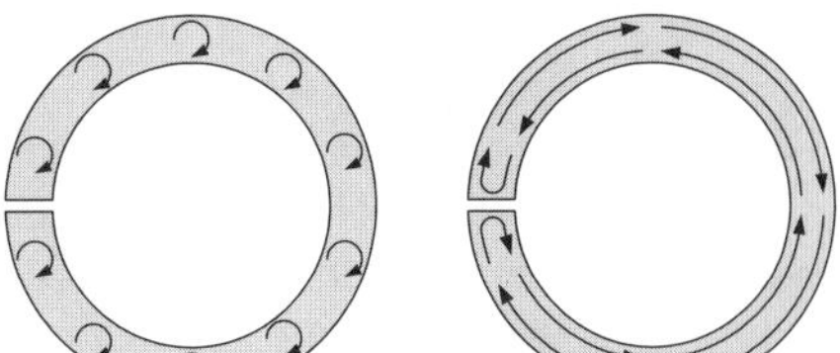

Fig. 5-13: **Broken Ring**

14. Interestingly a drilled-out silver coin, while producing a stronger signal, will have a (very) slightly higher resistance than a solid coin and this shows up in the target phase.
15. Based on the current price of silver, drilling holes in cupro-nickel coins is probably preferred.

movement is in one direction at the outside edge, but in the opposite direction at the inside edge. These produce opposing counter-magnetic fields that nearly cancel each other, resulting in a very weak overall field.

We've spent a little more time discussing the principle of induction because it is so fundamental to modern metal detector design. Nearly all standard general-purpose metal detectors are designed around the induction balance (IB) principle. We've seen that the principle behind IB is actually quite simple. There are two coils in the search head; one (TX) driven by an oscillator, and the second (RX) used to receive signals. The coils are set up to achieve a null in their coupling, with the simplest solution being a partially overlapped arrangement with just the right amount of overlap. When a metal target is in close proximity to the search head, the magnetic field pattern of the transmit coil is disturbed, and the coupling between the transmit and receive coils is increased. This technique is far more sensitive than the BFO type of detector we saw in previous chapters, but it has the disadvantage that careful alignment of the coils is required.

Search Coils

We saw in the design of BFO detectors that all we need for a search coil is a single coil[16] of wire. We will see in the later chapters that many pulse induction designs also use a single coil of wire. Induction balance designs require two or more coils, and there are numerous ways to set up an induction balanced coil system, many of which have been used by metal detector manufacturers over the years. So let's take a look at several.

The method of partially overlapping two round coils to achieve induction balance was used by Alexander Graham Bell in his second attempt to locate President Garfield's bullet (1881), and possibly predates Bell by a few years. Figure 5-14 shows illustrates this, with the transmit (TX) coil in grey and the receive (RX) coil in black. The RX coil is slightly overlapped with the TX coil, so that part of the inner field of the TX coil goes through the RX coil, and part of the outer field of the TX coil also goes through the RX coil. The inner and outer fields of the TX coil are of opposite "polarity," so if the RX coil is precisely positioned, it is possible to get the effects of the opposing fields to cancel. This type of coil is often referred to as "coplanar" because the TX and RX coils lie in the same plane. Another common name for this configuration is the double-O (or "OO") coil.

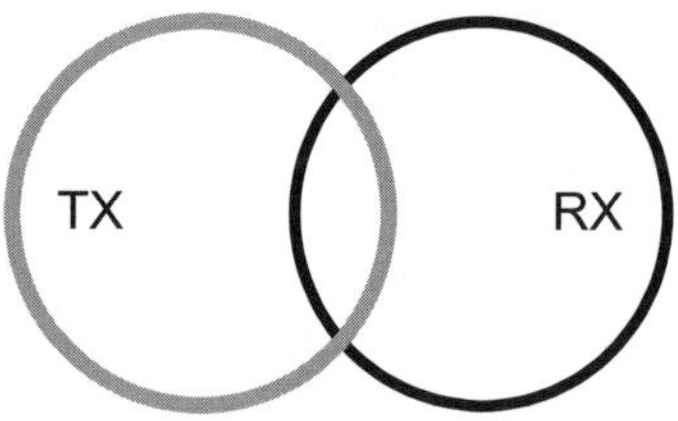

Fig. 5-14: **Double-O Coil**

16. There is some confusion when talking about a "coil", as to whether we mean the total "search coil" or one of the "coils" contained in the search coil. Some people call the search coil the "loop" or the "search head". When talking about the complete search coil, we will use the term "search coil," or use a specific type such as "mono coil" or "concentric coil." We reserve the word "coil" for individual coils of wire, and will furthermore use the terms "TX coil" and "RX coil" to denote specific coils.

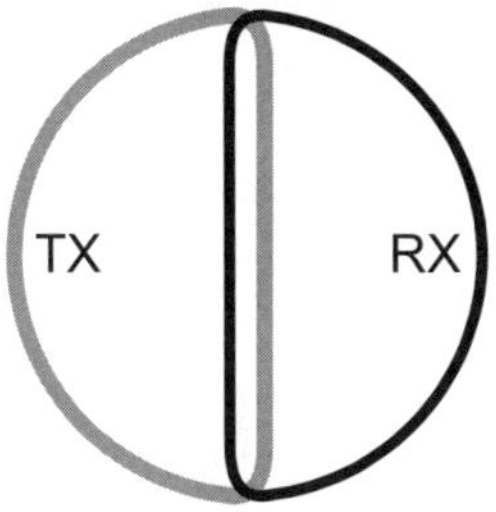

Fig. 5-15: **Double-D Coil**

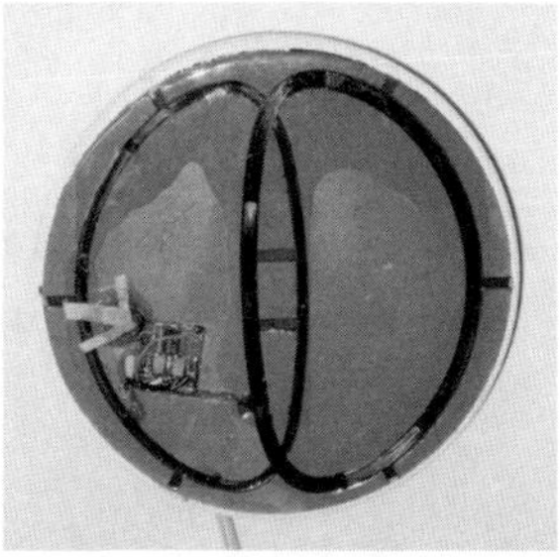

Fig. 5-16: **Minelab DD**

Figure 5-15 shows a variation of the overlap design which has become very popular in modern detectors. The overlap portions of the coils are somewhat flattened, which results in an overall circular search head that many people find easier to maneuver. In this variation, each coil looks like a letter "D", so the overall type is commonly referred to as the Double-D (or "DD"). The DD was first seen in a 1938 U.S. patent awarded to Charles Heddon. It was not seen in commercial detectors until Compass began using them in the 1970's. DD coils are widely regarded as the best configuration for highly mineralized soil, so they are very popular in areas like Australia where other coils have problems. Figure 5-16 shows a photograph of a Minelab DD loop. Note the heavy gauge wire used in the TX coil versus the much smaller gauge RX coil wire.

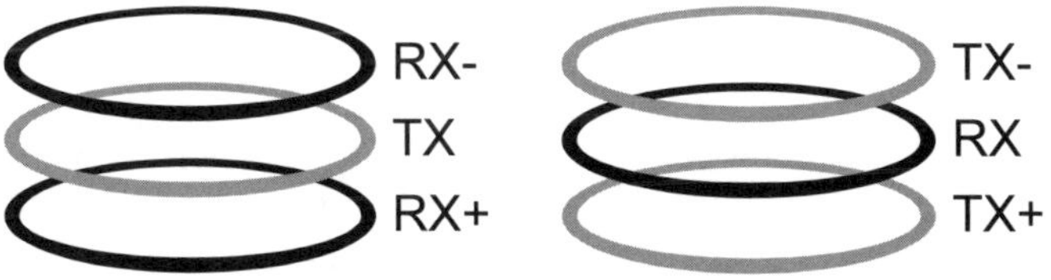

Fig. 5-17: **Coaxial Coils**

Figure 5-17 shows another method of attaining balance in a pancake coil. This is commonly called a *coaxial* coil because all the coils lie along the same center axis. The left version shows the TX coil sandwiched between two RX coils wired in opposition, such that the induced RX signals cancel each other. An alternative method, shown to the right, has the RX coil placed exactly between two equal TX coils that are wired in opposition so that their fields precisely cancel at the RX coil. In practice, the left version is more widely used.

Variations of coaxial coils date back to the work of David Hughes and Alexander Bell in the late 1800's. In hobby detectors, it was used in early Garrett VLF machines and in detectors from C&G Technology (maybe others as well) in the late 1970's. The C&G units used a modified version of the left-hand stack, where the receive coils were smaller than the transmit coil. This still achieves induction balance and, according to Jack Gifford, was done to reduce the magnitude of the ground signal for a better target-to-ground ratio.

The coaxial coil arrangement usually carries a slight depth penalty, because of

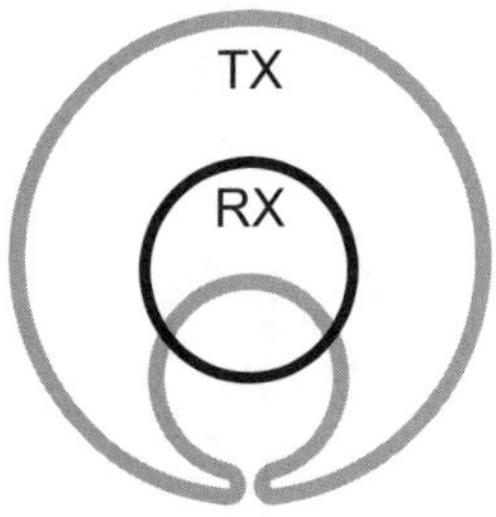

Fig. 5-18: **"4B" or "Omega" Coil**

Fig. 5-19: **Red Baron Coil**

partial signal cancellation between the "+" coil and the "–" coil. However, it does have a couple of distinct advantages. Because induction balance is achieved in the vertical direction, it makes the overall coil less sensitive to metal targets on the edge of the stack. This allows a coaxial coil to get closer to objects such as metal fence poles without detecting them. Since the left version has opposite-polarity receive coils, it is also exceptionally good at canceling EMI, and so does well under power lines. Today, the only coaxial coils produced are from aftermarket vendors.

Figure 5-18 shows a coplanar coil design that was very popular in the 1970's and early 80's. It is commonly called the "4B" loop and was widely used in TR and VLF detectors from White's and Bounty Hunter. You may also hear this type referred to as an "omega" coil because of the lower shape of the TX coil. The earliest omega coil was in a 1969 U.S. patent awarded to Robert Penland.

A small part of the transmit coil is folded inward, and the receive coil lies across this section. The folded portion produces a reverse transmit field about the receive coil that cancels the larger transmit field. As with the double-O coil, nulling is easily achieved by sliding the RX coil. Figure 5-19 shows a photograph of a Bounty Hunter Red Baron search coil. Note the small epoxied circuit board, used to trim the balance.

Many detectors today use the concentric coil, shown on the left of Figure 5-20. Although not widely used in commercial detectors until the 1980's, it dates back to at least 1948 in a U.S. patent awarded to Harold Wheeler. Normally, if you place a receive coil concentrically within the transmit coil you will get an enormous amount of inductive coupling. The trick here is to add a bucking coil (also called *nulling* or *feedback* coil), which is another transmit coil placed very close to the receive coil. The bucking coil gets a 180° out-of-phase transmit signal and, just in the vicinity of the receive coil, cancels the main transmit field without much effect on target depth. The bucking coil can be placed just outside the receive coil, or just inside, or even right above it.

Another way to configure the concentric is to use a counter-receive coil that is placed close to the transmit coil, but wired out-of-phase. See the right side of Figure 5-20. This configuration is rarely used; it is possibly the method used in Troy Shadow detectors, which exhibit a slightly wider detection cone at depth. Figure 5-21 is a concentric coil from White's.

With the other search coils we've looked at so far, any residual imbalance can be removed simply by moving the receive coil around slightly. Residual imbalance is not

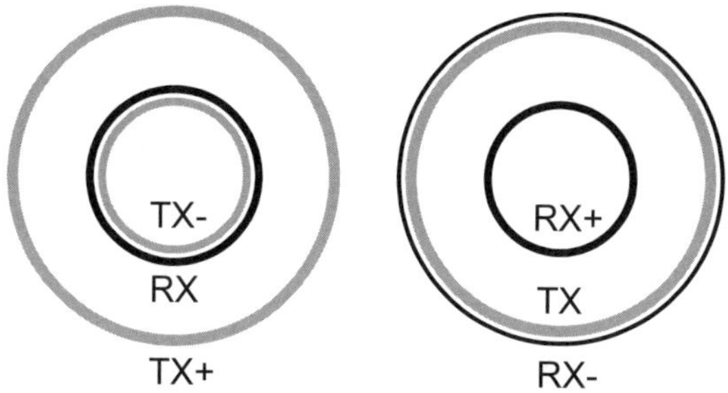

Fig. 5-20: **Concentric Coils**

Fig. 5-21: **White's Coil**

so easily removed in the concentric coil. In the White's search coil, the single strand of wire (taped down) is a small loop from the TX bucking coil that is moved around to fine-tune the balance. The black object is conductive rubber that presses against the shielding that is sprayed into the other half of the coil shell.

The concentric has become the standard search coil for practically all detectors, primarily because they have excellent depth in low-to-moderate mineralization, have about the best pinpointing capability of all coil types, and are relatively easy to manufacture. In most cases, the receive coil is half the diameter of the transmit coil, though it does not have to be. In Tesoro search coils, the RX coil is smaller than half. In older Fisher concentric search coils, the receive coil is elliptical. In Garrett imaging search coils, there are two receive coils; one about half the diameter of the TX coil, and the other at about ¾ the TX diameter.

A final interesting coplanar configuration is the "Figure-8." The earliest depiction of this loop is in Jullig's 1915 patent, where he used it as an "equivalent" diagram of an orthogonal configuration. There are many ways to make an equivalent Figure-8 coil. For the left most version in Figure 5-22, the RX coil is twisted in a figure-8 and placed inside the TX coil. It "sees" an equal amount of positive and negative transmit field and is therefore inductively balanced.

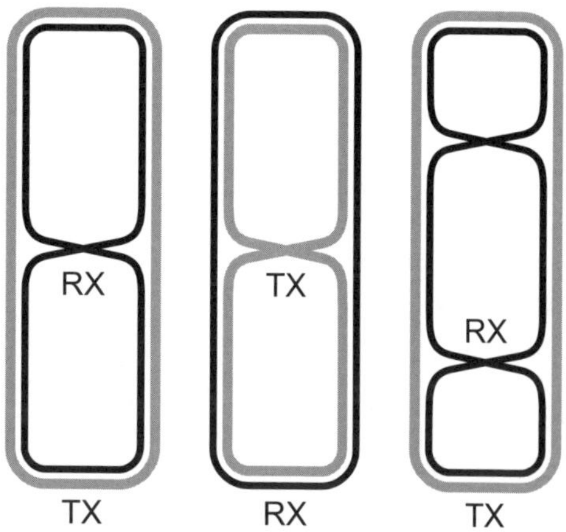

Fig. 5-22: **Figure-8 Coils**

This coil configuration has two quirks. First, the RX loop has a null at the crossover, so theoretically it will have poor sensitivity right at the center. Second, the received signal at the front half of the coil will be 180° degrees out-of-phase with that of the rear half. This means that in a phase discriminating design, proper target ID will work only in one half unless the detector is specifically designed to handle opposite phase quadrants.

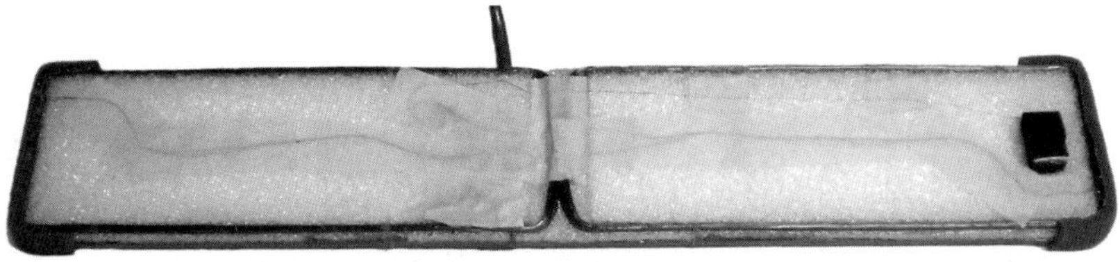

Fig. 5-23: **"Bigfoot" Coil**

But the figure-8 also has some advantages. It does an excellent job of rejecting electromagnetic interference (EMI) because the twisted RX coil sees the same amount of EMI on the in-phase half as on the out-of-phase half. It also has excellent inherent ground rejection for the same reason: ground signals are canceled by each half.

Figure 5-22 shows two more ways to make a figure-8. In the center version, the TX coil is twisted and placed it in the middle of the RX coil. This will work, but is inferior to the first method as it can no longer cancel EMI and ground signals, and it will still have the funny front-back target phase problem and the dead spot in the center. The right most version is an attempt to get the benefits of the figure-8 while improving target detection. The center portion of the search coil will detect targets correctly, and there are now two dead spots but they are moved closer to the ends of the coil. Commercially, the left-most version of the figure-8 shows up in aftermarket "Bigfoot" coils. See Figure 5-23.

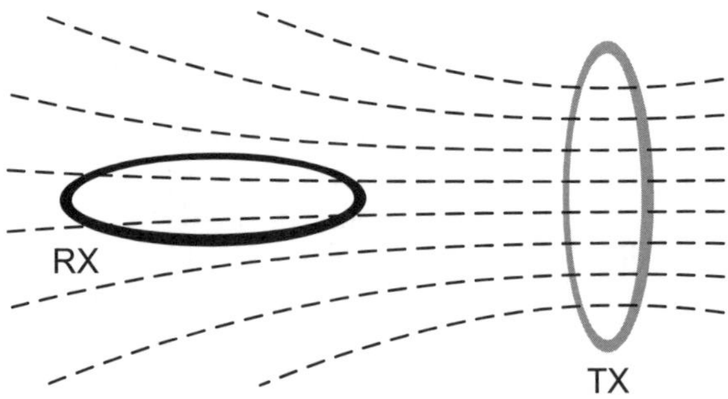

Fig. 5-24: **Orthogonal Coils**

Figure 5-24 shows an orthogonal arrangement, which seems to be one of the earliest methods of induction balance used in commercial detectors. You should recognize it as the two-box locator, or what some people call the "RF" detector. In this case the receive coil is turned 90° to the transmit coil and placed so that it lies exactly along the isomagnetic lines of the transmit field. This way, the receive coil does not have any magnetic flux cutting *across* it, and therefore no induced current. Inductive coupling is theoretically zero.

The earliest account of this coil arrangement is in a 1915 U.S. patent awarded to Max Jullig of Austria-Hungary. It began showing up on the earliest commercial detectors, including those sold by Fisher, and was popular throughout the 1920's and 30's, judging from both commercial designs and magazine construction projects. Figure 5-25 shows an early Fisher unit being used by Fisher himself.

There are at least a couple of problems with this approach. You can shrink the coils down and put them on the end of a shaft, but the orthogonal coil orientation

Fig. 5-25: **Fisher Two-Box**

results in a bulky search head. A 1969 article in *Popular Electronics* describes such a design using 3 coils (Figure 5-26).

A second problem is that coil balancing is tough in mineralized soil. As soon as the coils are moved near the ground, any mineralization will start to compress the transmit field, but on one side only, and upset the balance. Older two-box locators included an adjustment screw for tweaking the coil balance by mechanically tilting the RX coil. Newer designs electronically ignore ground signals.

There are many other interesting ways to make inductively balanced loops, but these cover the vast majority you will see on the market. With any IB coil, good bal-

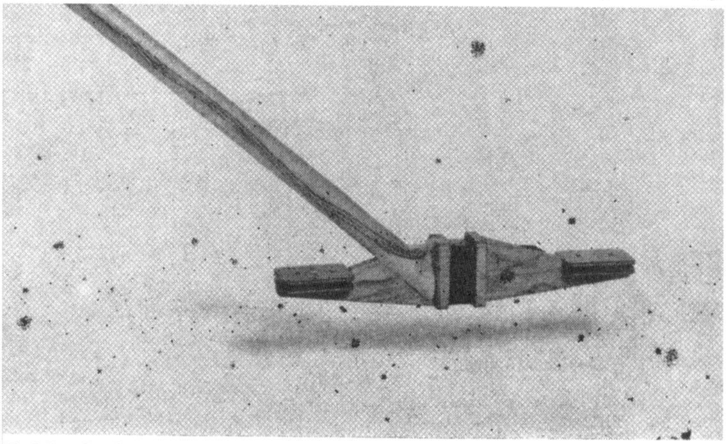

Fig. 5-26: **Orthogonal Search Head (Popular Electronics Feb 1969)**

ance is very sensitive to an exact placement of the receive coil, and moving the receive coil even slightly will upset the balance. Normally, all the coils are placed in a form or shell, adjusted for near optimal balance, then epoxy is poured over them. To fine-tune the balance after the epoxy sets, a single loop of wire can be left hanging and moved around, then glued in place.

There is, of course, more to building a good search coil than just getting the TX and RX coils well-balanced. Good shielding is important, and detector companies typically put a lot of effort into this. When a search coil is placed on wet grass, the moisture can create capacitive coupling between the TX and RX coils that causes falsing. Shielding presents a constant capacitance to the coils so that wet grass has no effect. Shielding can also help reduce external EMI-induced falsing.

We've covered quite a few ways of designing induction-balanced search coils, so the question that naturally arises is, "Which one works the best?" There is no one best design, each has advantages and disadvantages. For example, if EMI is a problem then the coaxial or the figure-8 are the best choices. The concentric has about the best pinpointing ability, and the double-O and double-D are better for use in highly mineralized ground.

Mineralization Effects

Ground mineralization is one of the biggest variables in metal detecting and is poorly understood. Why is a double-D better than a concentric in mineralization? The answer lies in how the minerals affect the transmit field. Most mineralization consists of iron-based products, either magnetite (Fe_3O_4) or maghemite (Fe_2O_3). They are both iron oxides. As we saw in our basic experiments, iron concentrates a magnetic field, so iron-based mineralization will distort the transmitted magnetic field. In particular, mineralization compresses the TX field so that it does not penetrate as deeply. Figure 5-27 illustrates.

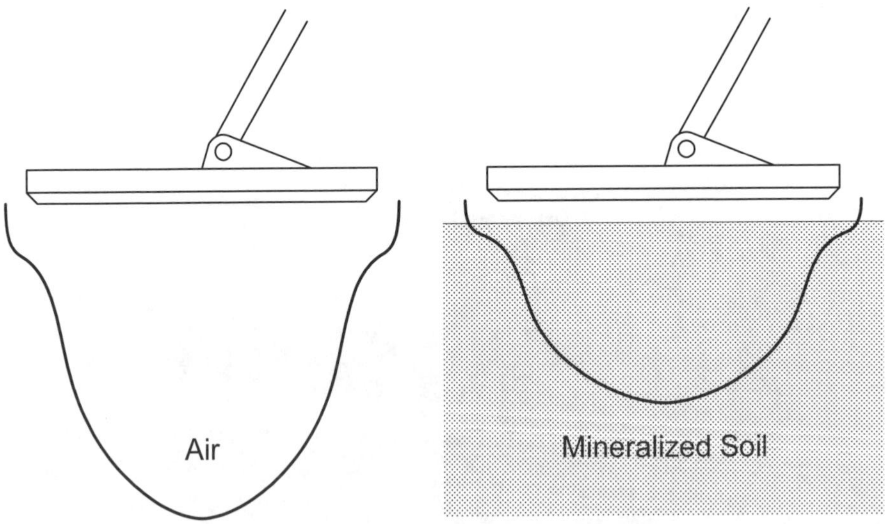

Fig. 5-27: **Ground Effect on the TX Field**

Recall from our discussions of search coils that the RX coil is carefully positioned in the TX field such that induction balance is achieved, and that it doesn't take much to throw off the balance. In the case of ground mineralization, the compression effect on the TX field can be strong enough to unbalance the search coil. The result is a large, unwanted signal on the RX coil due to a null shift.

This effect is more pronounced in some designs and less in others, depending on the behaviors of symmetry. The figure-8 coil (Figure 5-22, left most) is almost immune to ground distortion because the RX coil always sees a 50% positive TX field and a 50% negative TX field, and a homogeneous ground distortion will not alter this. The double-O search coil is also fairly immune, because the equally-sized circular TX and RX coils maintain good field symmetry with ground distortion.

The more popular double-D search coil is not as good because each coil has three different radii: the broad outer radius, the almost-flat overlap radius, and the two tight transition radii. The magnetic field produced by the different radii is very non-uniform, and therefore ground distortion will affect the field in a non-uniform manner. However, the double-D search coil still performs better than the concentric.

Concentric search coils that use a TX feedback coil (Figure 5-20, left most) create two different TX magnetic fields. The main TX coil produces a larger, stronger field while the bucking TX coil produces a smaller weaker field. As mentioned before, the bucking coil is placed in close proximity to the RX coil so its field will cancel the main TX field, achieving induction balance. However, as the search coil is lowered to the ground, mineralization will compress the larger, stronger field first (the main TX field), causing an imbalance in the null. As the search coil is further lowered, the weaker field of the bucking coil begins to get compressed as well, causing some amount of re-balancing. This effect is known as *lift-off* because, with the search coil firmly on the ground, lifting it upwards will cause an initial imbalance followed by a re-balance. In some grounds this effect can be quite severe, and in detectors with manual ground balance, lift-off makes ground balancing very difficult.

As you can readily see from our brief exploration of the various coil configurations, this represents an important discipline of metal detector design in its own right. The pros and cons of the different coil types can even become somewhat contentious, and is probably one of the areas where the most misconceptions and myths reside. We have already mentioned that we will be exploring coil construction in some depth in a later chapter, and this is where we intend to do some myth busting as well.

For the IB projects in the following chapters, we are going to take the easy route and use a Tesoro concentric coil for our initial IB design. The simple reason is that these coils are readily available, and do not contain any components in the search head apart from the TX, RX and bucking coils. Tesoro coils are also designed to be more or less interchangeable between similar detectors, which will assist in making the design more reproducible. Building a metal detector already has some potential gotchas, so the least we can do is start out with a reliable coil. Once everything is working, we can then look to build our own coils, and see if things continue working.

CHAPTER 6 *Principles of Induction Balance*

'Tis a favourite project of mine
A new value of pi to assign
I would fix it to 3
For it's simpler you see
Than 3 point 1 4 1 5 9
— Author unknown

Nearly all standard general-purpose metal detectors are designed around the induction balance (IB) principle. This is true for both the serious treasure hunter and the detecting hobbyist.

The principle behind IB is actually quite simple. There are two coils in the search head. One is driven by the oscillator, and the second is used to pick up signals from the first. These two coils are positioned in an overlapping fashion so that coil two is very loosely coupled to coil one. When a metal target is in close proximity to the search head, the magnetic field pattern of the transmit coil is disturbed, and thus the coupling between the transmit and receive coils will be increased. This technique is far more sensitive than the BFO type of detector, but it has the disadvantage that careful alignment of the coils is required.

In a similar manner to the BFO detector, the ground has an effect on the coils in the search head. If the ground is non-conducting, and has a permeability different to that of the air, the coil coupling will be affected. A ferrous metal target will distort or concentrate the field lines to a much greater degree, producing a larger detectable signal, whereas non-ferrous targets will tend to deflect the magnetic field lines. The eddy current effect is highly dependent on the electrical and physical characteristics of the target. Good conductors produce higher eddy currents than those with a high resistivity. For example, gold and silver are good conductors, whereas iron (especially if it is oxidised) is less conductive.

There are many possible coil arrangements (some with more than two coils) but most detectors available today use one of two configurations:

- Widescan – so called because its most sensitive area extends right across the two coils.
- Concentric – which has a smaller sensitive area, that greatly assists with "pinpointing". We will investigate the various coil configurations later in this chapter.

The frequencies used by IB detectors are also much lower than those used by the BFO variety. For example, commercial detectors use frequencies from as low as 1.5kHz to as high as 71.1kHz. There is also a type of detector known as a "2-box locator" that operates at 82kHz. Although this is a form of induction balance (using orthogonal loops; see Chapter 5), it is often referred to as a TR (transmit/receive) detector.

Transmit Frequency

A question that is often asked is, "What is the best frequency for metal detecting?"

From Chapter 3 you will know that maximising sensitivity, while attempting to avoid the problems caused by skin effect, were the main concerns for BFO designs. The result was that the transmit frequency was quite high.

However, with the IB technique, lower frequencies can be employed without compromising sensitivity. This is why IB detectors are also commonly referred to as VLF (very low frequency). If you examine the specifications of a number of commercial models, it becomes clear that frequencies at the higher end (around 70kHz) are used for gold detection, whereas the 10kHz to 15kHz range is claimed to be good for silver (in particular, hammered silver coins). Certain major manufacturers have opted for an even lower transmit frequency of 7kHz, which is a compromise that allows the detector to work over a large range of conditions, including coin, beach, and field hunting.

But – should we simply take the word of the marketing material from these commercial organisations? Of course not!

In order to perform a simple experiment to determine the sensitivity at different frequencies, four different transmitter (Colpitts) circuits were built, and used to drive a spare Garrett Crossfire concentric coil. The coil was first disassembled to remove the internal resistor and capacitor and allow direct access to the transmit and receive coils. At each frequency, the residual signal from the receive coil (which was tuned by using an additional capacitor(s) to match the frequency being transmitted) was measured using an oscilloscope without the presence of a metal target. Each of the targets was then placed close to the coil, and the received signal was measured once again. A relative measurement of the received signal caused by the target, as a percentage of the transmitted signal amplitude, was recorded for 3, 10, 20 and 70kHz.

As a result of this somewhat crude experimental setup, it was evident that 3kHz was the least sensitive of all the frequencies, whereas 70kHz was the most sensitive for small targets. The response to different targets, such as silver and gold, were somewhat subjective, but the range from 10kHz to 20kHz did appear to offer the best compromise.

What The Experts Say

According to the Whites Electronics Metal Detector DFX Engineering Report (freely available on the internet) 50kHz is good for finding tiny flakes of gold, but is poor for salt water hunting. Whereas 1.75kHz is good for silver and gold jewelry and has good salt water rejection. It is also claimed that 1.75kHz is a poor choice when you need to distinguish between nickels and pulltabs. [We have not examined discrimination yet, but it is worth questioning the importance of this capability anyway. Since Roman coins also fall into the same category, most UK detectorists will turn off this fine level of discrimination to increase sensitivity. Particularly since there are few pull tabs to be discovered in a ploughed field.]

There is also more power line interference at 3kHz. The Whites' report concludes that 15kHz has a better sensitivity for smaller objects than 3kHz. These two frequencies are both used in the DFX (actually 2.98kHz and 14.91kHz) and hence most of the discussion concerns the attributes of these two frequencies for metal detecting. For many years Whites have built detectors running at 6.6kHz, so the DFX is a departure from this approach. Apparently this was chosen to provide good sensitivity and discrimination for a wide range of targets and environments. However, it is claimed that for jewelry hunters and prospectors it is too low, and for beachcombers and cache hunters it may be too high.

Other detector manufacturers seem to have differing thoughts on the correct choice of transmit frequency. Some prefer to design detectors for operation at 10kHz, 12kHz, 15kHz and 20kHz. Whereas others prefer the 2.4kHz to 5.7kHz range. Some detector models try to cover all possibilities by allowing the use of multiple simultaneous frequencies[1]. All gold detectors operate around 50kHz to 70kHz, as mentioned in the Whites report.

So What's Best?

This raises the question as to which frequency is really the best one to use, assuming we wish to design a single frequency machine. A number of years ago some tests were performed on a comparatively cheap detector from the low end of the market, and a 1797 George III copper coin was found with this detector at a depth of 11 inches (28cm). This detector was transmitting at 20kHz and was surprisingly sensitive for the price, although the discrimination capability was poor. This latter problem was probably more to do with the design rather than the choice of frequency, and it was also bad at salt water rejection. In fact, practically unusable, which agrees well with the conclusions of the Whites report. Other (much more expensive) detectors have improved little if any on the depth of detection, even though the discrimination capabilities bear no comparison. In fact, this is a point not covered in the Whites report. That is, how is detection depth affected by the choice of frequency? According to theory, the lower the frequency the greater the depth. Although there must be some trade-off against sensitivity otherwise all detectors would be running at 10Hz or less! The choice of frequency probably has more to do with effective discrimination, than it has to do with either depth or sensitivity.

Firstly we need to answer the question concerning the choice of frequency (or at least a range of frequencies to use for our IB detector. Clearly the "experts" are not completely in agreement, otherwise they would all use the same frequency.

For an initial design decision (based on both real measurements and "expert" opinion) it is clear that a frequency between 10kHz and 20kHz would be a good safe choice.

Power Supply

How to power our IB detector is an issue that needs careful consideration. Commercial detectors have a typical battery life of between 10 and 30 hours, and most can use either alkaline or rechargeable batteries. In some cases a special battery pack is required, but this usually contains standard rechargeable batteries repackaged in a

1. These are known as "multifrequency" detectors, which are briefly discussed in Chapter 15.

custom unit. Although alkaline batteries tend to give a longer operating time, these need to be thrown away at the end of their life, and can be expensive to replace. Therefore many detectorists use either nickel-cadmium (NiCd) or nickel-metal hydride (NiMH) batteries. One advantage of using NiMH cells is that they do not suffer from the "memory effect" that plagues NiCds. This problem is caused by recharging the NiCds without first discharging them completely, which explains why some charging units now incorporate a discharge-before-charge facility. If you use a fast-charge unit, then it is important to check that it is suitable for charging NiMH batteries, as the charging characteristics are not the same as for NiCds.

Apart from their recharge capability, both NiCds and NiMHs have an excellent power to weight ratio. Also, NiMHs tend to have a capacity related performance that is normally 30 to 50% better than the equivalent NiCd cells, and it is the absence of cadmium that make NiMHs free of the undesirable memory effect.

Unfortunately there is one feature of both NiCd and NiMH batteries that can be very frustrating to a detectorist. Imagine you discover an unexpected opportunity to go detecting. You reach for your trusty metal detector and power it on, only to find that the batteries you recharged only days before have mysteriously lost their charge. This can be a very frustrating problem. In fact, so much so that many homemade detectors are powered by a sealed lead acid battery, which has a low self-discharge of only 3% per month. Unfortunately there are two major disadvantages with this solution: weight and lack of capacity. A reasonable solution would be to use two 6V sealed lead acid batteries with a capacity of 1.3Ah and a total weight of 0.6Kg. An equivalent solution using NiMH batteries would require 8 cells with a total weight of 0.208Kg plus battery case. However, one disadvantage of the NiMH solution is that each cell only has a voltage of 1.2V, giving a total voltage of 9.6V, as opposed to 12V. But, although specifying lead acid batteries instead of NiCd or NiMH will greatly reduce the annoyance of self-leakage, it will provide less capacity (possibly as much as 50% less) and an increase in weight (approximately 288% more).

At the end of the day, it is probably down to personal choice, but for our own IB detector (purely based on convenience, and power versus weight) we will use 8 x AA alkaline batteries. The intent is that the detector electronics will not be power hungry, and the alkaline batteries will last some considerable time before they need replacement. In this case, cost will be minimal, and the problem of unexpected flat batteries will not be an issue. Of course, there will be nothing stopping anyone wishing to build this design from using whatever power supply they desire.

Evaluation of Initial Design Decisions

In this chapter we have examined the basic principles behind the induction balance concept. As a consequence we have also explored the importance of selecting a suitable transmit frequency, the advantages and disadvantages of various coil configurations, and the merits or otherwise of using rechargeable batteries.

Here, once again, are the design decisions we have decided to use:

1. Initial design to work with standard Tesoro coil.
2. Transmit frequency of 10kHz (determined by coil).
3. 8 x AA alkaline batteries to provide a 12V supply.

The decision to use a standard Tesoro coil will remove a tremendous amount of uncertainty from the design process. An in-depth investigation into the coil construction process will be performed later, after successful completion of the first design.

So onto the next step in our exploration of IB technology – target discrimination.

CHAPTER 7

Target Discrimination

How Does It Work?

"..... the only 100% reliable discriminator is a shovel."
— White's Electronics Metal Detector XLT Engineering Report

In the previous chapter when the IB principle was explained, it was stated that current flowing in a nearby metal target will cause an associated current to flow in the receive coil. What was not mentioned was the fact that the received signal is delayed slightly when compared to the transmitted signal. This delay is caused by two factors:

1. The target object impedes the flow of the induced current. This is due to the "resistive" nature of the material that makes up the target.
2. The target object also impedes any changes in the flow of current (either rising or falling) due to the "reactance" of the material.

These two properties of the target result in a delay, commonly called a "phase shift". Luckily for detectorists, nature has arranged things such that metal objects are mainly reactive; and in particular, such objects as gold, silver and copper. Other objects that are mainly resistive exhibit smaller phase shifts. Such objects are usually smaller and thinner, and are less conductive.

Since different targets exhibit different phase shifts, it is easy to surmise that it may be possible to classify the different types of target, and therefore discriminate between them.

There is a U.S. patent 4,486,713 entitled *"Metal Detector Apparatus Utilizing Controlled Phase Response To Reject Ground Effects And To Discriminate Between Different Types Of Metals"* dated Dec 4, 1984. In this patent, the abstract reads:

> *"Metal detector apparatus includes a controlled phase responsiveness to allow operation with reduced ground effects while discriminating between different types of metals. A phase relationship is predetermined that will reject unwanted signals from mineralized ground and trash metals and will be responsive to the signals from desired metals. The algebraic relationship between two signals is determined so that only output signals falling within the desired algebraic relationship are used to provide an output signal."*

The most interesting part of this patent is the excellent phase diagrams, namely Figures 1, 2 and 3. The information contained in Figure 1 of the patent can be summarised as follows:

Target	Phase Angle (degrees)
Mineral	-90
Iron Nail	-50
Steel bottle cap	-30
Nickel	+30
Pulltab	+60
Coin	+80

Phase Demodulation

Modern metal detectors use a sampling technique whereby the amplitude of the received signal is measured at the zero and ninety degree phase positions relative to the transmitted signal. These two measurements are usually termed the X and R channels respectively. Strictly speaking the X and R channels represent the in-phase (I) and quadrature (Q) components of the received signal. The term "quadrature" means one quarter of a circle – that is 90 degrees. In other words the R channel is a measurement of the amplitude that can be attributed to the resistive aspect of the target, and the X channel is associated with the reactive part. Phase demodulation is a technique favoured by detector manufacturers, as it simplifies the process of determining the phase angle when the received signal strength can fluctuate over a wide range.

Mathematically the demodulated signals can be represented by:

$$X = A\cos\theta \qquad \text{Eq 7-1}$$

and

$$R = A\sin\theta \qquad \text{Eq 7-2}$$

The process of dividing R by X yields tan θ, where θ is the phase angle of the received signal. This results from the trigonometrical analysis:

$$\tan\theta = \frac{\sin\theta}{\cos\theta} \qquad \text{Eq 7-3}$$

Consequently

$$\frac{R}{X} = \frac{A\sin\theta}{A\cos\theta} = \tan\theta \qquad \text{Eq 7-4}$$

Then in order to extract θ we just need to find the arctangent of R/X.

Note. Although this is often described as the method used by metal detectors to provide discrimination, the reality can actually be somewhat different. It's really only appropriate when the detector is capable of some digital processing, whereas analog-based detectors need to use an alternative technique. All be revealed later when we start constructing our own design.

See Figure 7-1 for an example, which shows a typical response for a pulltab. As you can see the R channel returned 50mV, and the X channel was –85.73mV. From these two samples the phase angle can be determined using tan θ = R/X.

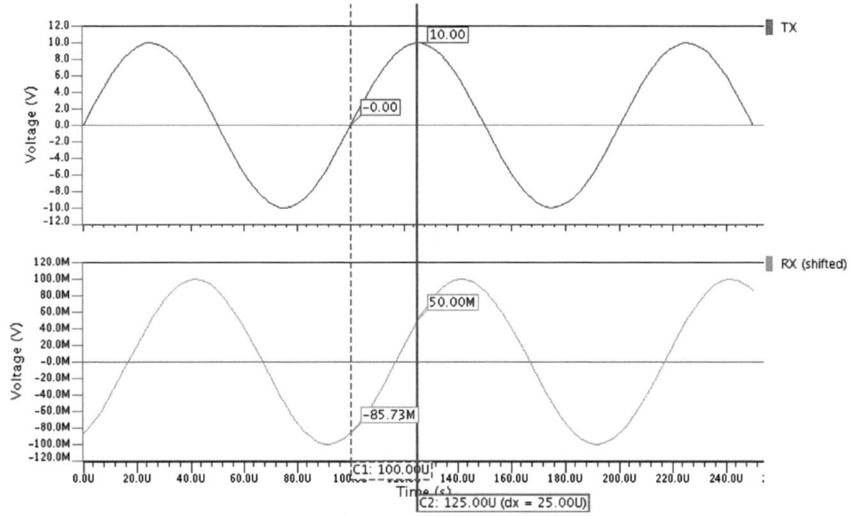

Fig. 7-1: **Sampling Technique to Extract Phase Angle**

In order to obtain the value of θ we need to rearrange the equation to give

$$\theta = \arctan\left(\frac{R}{X}\right) \qquad \text{Eq 7-5}$$

Therefore θ = arctan (50 / -85.73) = -30.25 degrees.

Then add 90 degrees to get it in the right quadrant, so that θ = 59.75 degrees … and — voila! — a pulltab.

Unfortunately we must point out that this phase angle measurement is not an exact science, and is dependent on many factors. If the discriminator is set to reject pulltabs (for example) you may also reject that Roman coin you were hoping to find.

We also need to point out here that an assumption is made in the patent that non-ferrous targets will shift the receive signal to the right (relative to the TX signal) and ferrous targets will shift the signal to the left. It also shows the TX and RX signals are initially in phase with no targets present. In reality this may not be the case, as the phase-shifts are also dependent on the coil geometry.

The simplest way to implement the calculation of the phase angle is in software using a microprocessor, and to create a lookup table for the arctan function. An alternative method would be to measure the time delay, by monitoring the zero-crossing points of both the TX and RX signals, then (since the transmit frequency is known) directly calculate the phase angle. In detectors that are not microprocessor controlled, the phase angle has to be determined by analogue circuitry. Whichever method is used, the accuracy of the results tends to be less than perfect. This lack of uncertainty in the phase angle causes "chatter" from the audio output, and (if the detector has a visual display indicator) some "jitter" in the display as the circuitry attempts to track the signal received from the target.

There are a number of reasons why identification can be difficult, including the presence of junk, ground mineralisation, and depth of the target. Some of the identification problems may also be attributed to the concept of motion detection. This prevents the discriminator circuitry from obtaining a stable reading and can lead to incorrectly identified targets. Certain objects exhibit different responses as the search coil moves across them, and can often give different results if swept in different directions. The discrimination circuitry will include some method (either implemented in hardware or software, depending on the detector) to average the X and R values over a set period of time. This averaging is usually achieved by a process called "integration", and is a necessary part of discrimination because of the imprecise nature of the sampling process. If the signal is varying (both in amplitude and phase) as the search coil passes over the target object, the signal may change sufficiently between sample times to generate X and R values that are not consistent. Integrating these results over time can remove some of the uncertainty, but due to the finite time available as the target is scanned, a significant error in the calculated phase angle can still occur, especially near the edges of the target. This is one of the drawbacks of using a motion mode of detection.

Motion Versus Non-Motion

The idea behind so-called "motion" detectors is that the ground signal tends to remain fairly constant when compared to the target signal. When the search coil passes over a metal target the signal amplitude rises quickly and then falls. Hence, if a detector can look at the rate-of-change of the received signal (rather than the signal itself) this is a simple way of distinguishing a real target from a change in ground mineralisation, which tends to change more slowly.

When the concept of motion detection was first introduced, the user was expected to swing the coil at quite a high speed, but in recent years the idea of "slow-motion" detectors has become popular, meaning that the search coil does not require such vigorous sweeping to be effective. The ease with which ground effect can be eliminated has made motion detection the most popular method in use today.

The non-motion detector, on the other hand, does not require the coil to be in motion, which can be a major advantage when pinpointing, and may help improve target discrimination due to the stability of the received signal. This is one of the reasons for the introduction of the slow-motion detector, which attempts to combine the benefits of both motion and non-motion, and improve discrimination by allowing the search coil to remain over the target for a longer period during the scan.

When a motion detector needs to be used for pinpointing, it must be switched into non-motion mode, which disables the discrimination circuitry. Since the target has already been identified (at least to the extent that digging seems necessary) then only the actual position and an estimated depth reading is required.

Discrimination With an Analog Detector

Up until this point we have not yet described how an analog detector is able to provide discrimination, when it appears from the previous description that some digital processing is necessary to perform the mathematical calculations. In fact, all of the so-called metal detector books that have some technical content only skim over this part of the inner workings of detector design. Some may say that this is an intentional ploy by the detector manufacturers, who are the main authors and publishers of these books. Whereas, in reality, the method used to discriminate against different metals is extremely simple. Although this chapter deals specifically with the subject of target discrimination, we would like to delay describing how this works in an analog detector until Chapter 9. But first, in Chapter 8, we will develop a basic VLF/TR design as a crucial step beyond the BFO, which will assist in demonstrating how this logically developed into its modern day incarnation with ground canceling and target discrimination.

CHAPTER 8 Transmit-Receive Detector

"After all, the struggle between 'It's good enough' and 'I can do better', is part of the engineer's DNA."
— Bill Schweber (Editor of Planet Analog)

Before we start the design of our IB detector, we will first develop a basic Transmit/Receive (T/R) detector, in order to better understand how these designs evolved into the detectors we are familiar with today. The T/R detector existed for many years alongside the BFO, and the original T/R detectors (like the BFO) were also susceptible to ground effect. In the next chapter we will investigate ground effect in some detail, and describe how this can be eliminated. However, the original T/R detectors did not have this ability, and the design we will develop here follows the same design concept. Its simplicity has made this type of detector circuit a popular project in many hobby electronics magazines. Although there were some problems with this new technique, it was clear that the days of the BFO were numbered.

Transmit Oscillator

The transmitter used here (Figure 8-1) is somewhat different to the previous designs. It produces a series of pulses, which results in a set of decaying sine waves. This type of oscillator was popular with Heathkit and some Radio Shack designs. The main advantage of this type of transmit oscillator is that it eliminates the requirement for a separate audio oscillator.

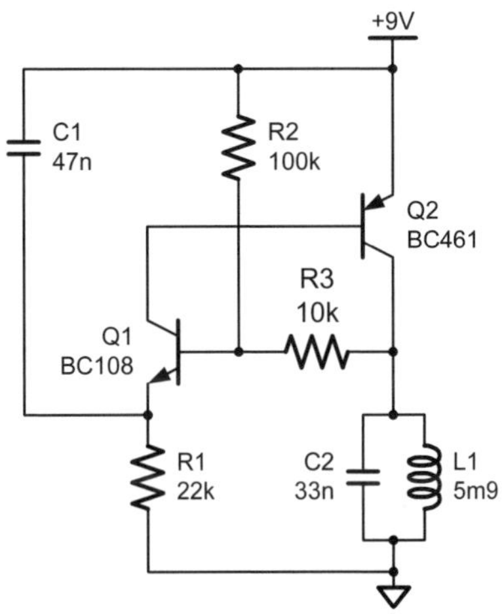

Fig. 8-1: **Transmit Oscillator**

The decaying sine wave signal is generated at a frequency 11.4KHz, and a pulse rate is 476 pps (pulses per second) with a maximum peak-to-peak amplitude of approximately 25V. If you remove the transmit coil and associated capacitor from the circuit, and replace it with a resistor, the oscillator will still produce a series of pulses. Therefore it is clear that the transmit frequency depends only on the values of L and C. In the Heathkit and Radio Shack designs the pulse repetition rate is heard in the headphones or speaker as an audio tone. As the amplitude of the received signal varies in the presence of a metal target, this in turn produces a change in volume of the audio tone.

If you examine the commercial designs that use this approach, you will notice that the oscillator we are using here is significantly simpler, but produces the same result. The Heathkit designs either make use of feedback windings in the search coil, or have multiple windings, and the same is true for Radio Shack. Hence the design was modified to allow the use of a Tesoro coil. In fact, most of the testing was performed with a 4" diameter coil.

The other significant difference is that the commercial designs have the coil connected to the supply rail, but here it is referenced to 0V. None of the components are critical, and the transistors used were just ones that were available at the time.

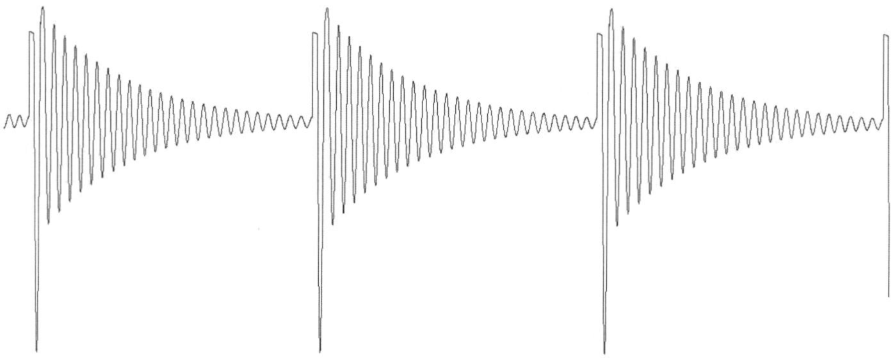

Fig. 8-2: **Transmit Oscillator Waveform**

RX Amplifier

The RX signal is amplified by an opamp (U1) operated from a single supply of 5V. A virtual ground of 2.5V is provided by a potential divider (R4 and R5) connected to the non-inverting terminal. This acts as the reference for the input and output voltages.

This amplifier has a gain of 100, and the output is decoupled from the following audio circuit by capacitor C6.

Due to the construction and balancing of the Tesoro coil, the RX signal increases in amplitude in the presence of non-ferrous targets, and decreases for ferrous. Therefore this detector is capable of ferrous/non-ferrous discrimination. Unfortunately, because the ground itself produces a decrease in amplitude, it is important that the search head is maintained at a constant distance above the ground. This is one of the drawbacks of the T/R design, but its increased sensitivity and excellent discrimination, when compared to the BFO, resulted in these detectors becoming very popular. The problem of keeping the search head at a constant height is not such a problem as you might imagine, as this detector is of the non-motion type, enabling the user to hover over the target while adjusting the threshold control.

Note that, in Figure 8-3, the lower connection of L2 is actually connected to 0V inside the search head.

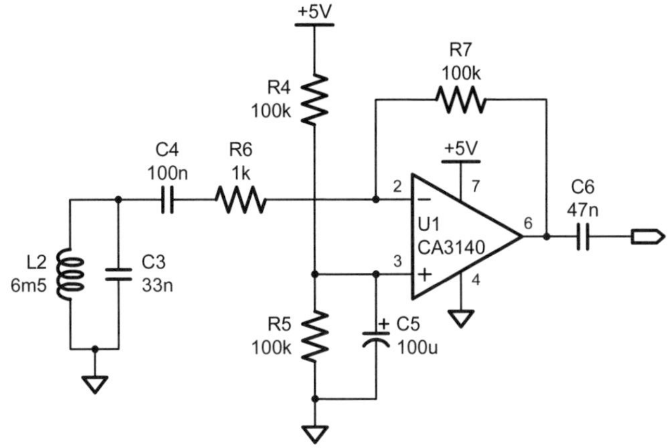

Fig. 8-3: **RX Amplifier**

Audio Output

The audio circuit is very simple, and consists of a 555 timer (U2) configured as a monostable (Figure 8-4). The output from the RX amplifier is connected to pin 2 of U2, which causes the monostable to trigger on a negative- going input signal when the level reaches 1/3 of the supply voltage.

A piezo buzzer is directly driven from the output (pin 3) of U2, and the multi-turn pot VR1 can be used to adjust the detection threshold. This pot should be adjusted until a slight clicking can be heard, similar to the sound of a geiger counter. When a non-ferrous target approaches the coil the clicking increases in intensity, but when a ferrous target approaches the clicking decreases or goes silent.

Power Supply

Finally we need a stabilized power supply that can provide both 9V and 5V. The 9V supply is used to power the TX oscillator and audio circuit, and the 5V supply powers the RX amplifier. The requirement for a stabilized supply is very important in

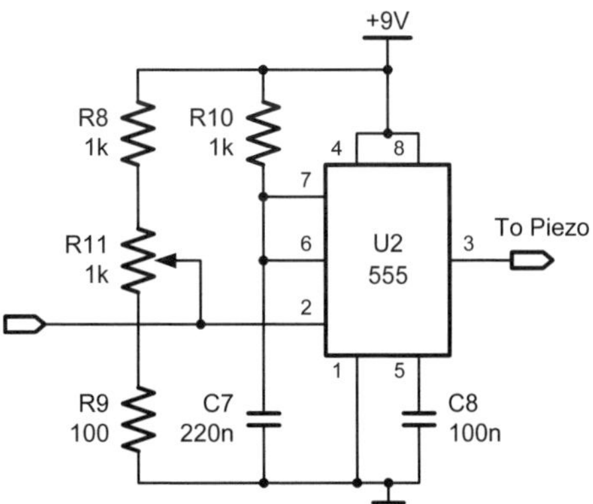

Fig. 8-4: **Audio Stage**

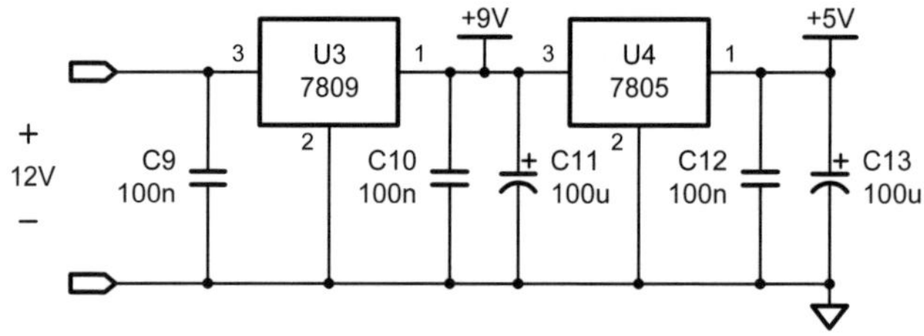

Fig. 8-5: **Power Supply**

this design, as the TX oscillator output is affected by changes in supply voltage. Also, using a separate supply for the RX amplifier provides some isolation from the noise present on the 9V power rail. See Figure 8-5.

Testing

The gain of 100 in the RX amplifier gives the optimal result. You might think that increasing the gain would increase the sensitivity, but that is not the case, It becomes increasingly difficult to adjust the threshold, with the result that sensitivity is actually reduced.

This simple TR detector (when correctly adjusted) can detect a Victorian penny at 15cm (5.9"), and a soft drinks can at 26cm (10.2") with a 4" Tesoro concentric coil. Ferrous rejection is very good. A large rusty nail is not detected until it is within 35mm (1.4") from the coil.

As mentioned previously, detectors of this type have no method for ground rejection. Both ferrous targets and the ground matrix will cause the input signal to be reduced in amplitude, and there is no way to distinguish between the two. Phase-shifts caused by the target are completely ignored.

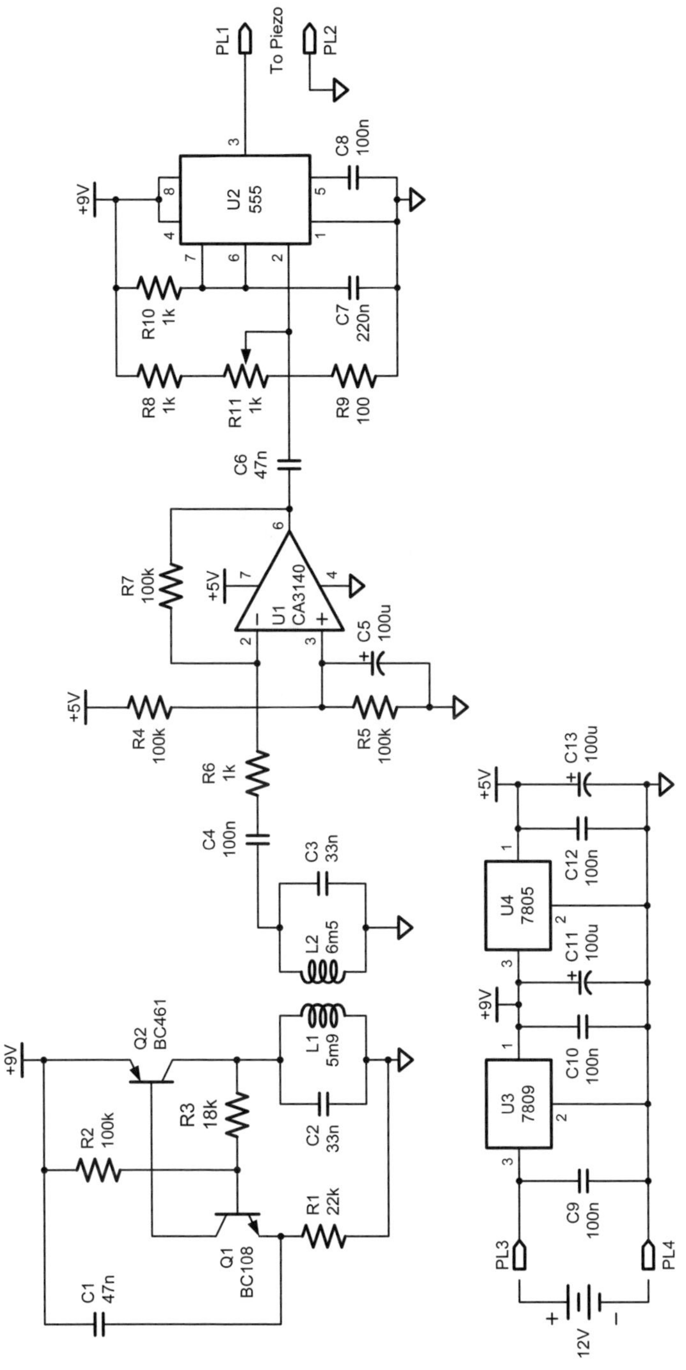

Fig. 8-6: **Complete Schematic of TR Detector**

Construction Details

The component placement and PCB layout are for illustrative purposes only. You may need to adjust the layout to accommodate components available in your area. In particular, please note that transistor pinouts can vary depending on the country of manufacture, even for what appears to be an identical part number. See Chapter 14 for more details.

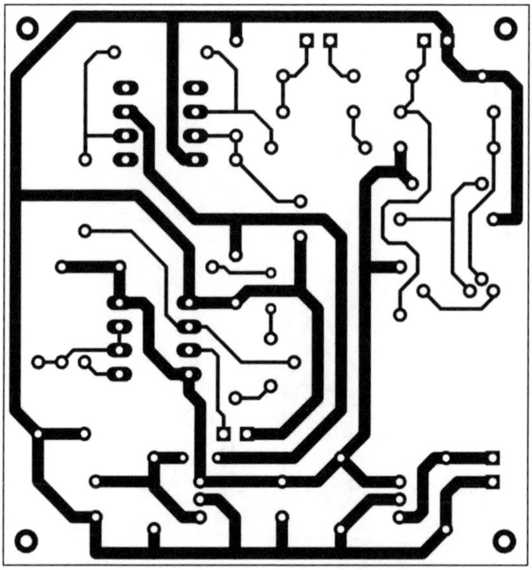

Fig. 8-7: **PCB Layout (single-sided)**

The PCB layout view (Figure 8-7) is from the underside of the board. The actual size of the PCB is 2.25" x 2.35" (57.2mm x 59.7mm). Note that there are 2 jumpers required in this layout. The parts placement (Figure 8-8) is shown from the top-side. The 3-dimensional view (Figure 8-9) provides an idea of what the final product will look like in real life. Note the connectors are designated as follows:

PL1 = TX coil
PL2 = RX coil
PL3 = Battery
PL4 = Piezo buzzer

Parts List

Resistors: (5% 1/4W)

R1	22k
R2, R4, R5, R7	100k
R3	18k
R6, R8, R10	1k
R9	100
R11	1k multi-turn pot

Capacitors:

C1, C6	47n
C2, C3	33n
C4, C8, C9, C10, C12	100n
C5, C11, C13	100u elect., 10v
C7	220n

Inductors:

L1, L2	Search coil (see chapter 8 for details)

Transistors:

Q1	BC108 (or any general purpose NPN)
Q2	BC461 (or any general purpose PNP)

ICs:

U1	CA3140
U2	LM555
U3	LM7809
U4	LM7805

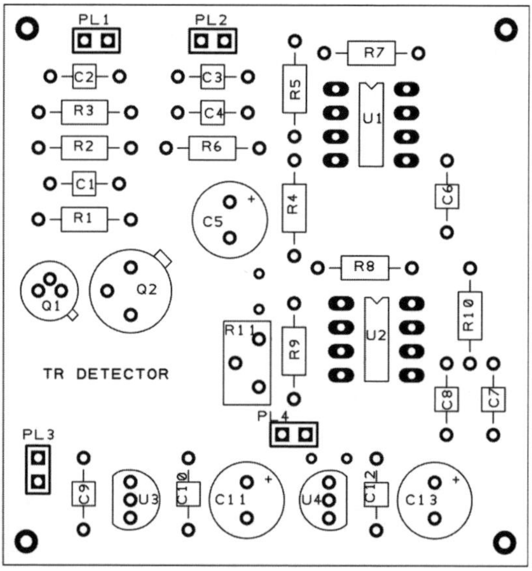

Fig. 8-8: **PCB Parts Placement**

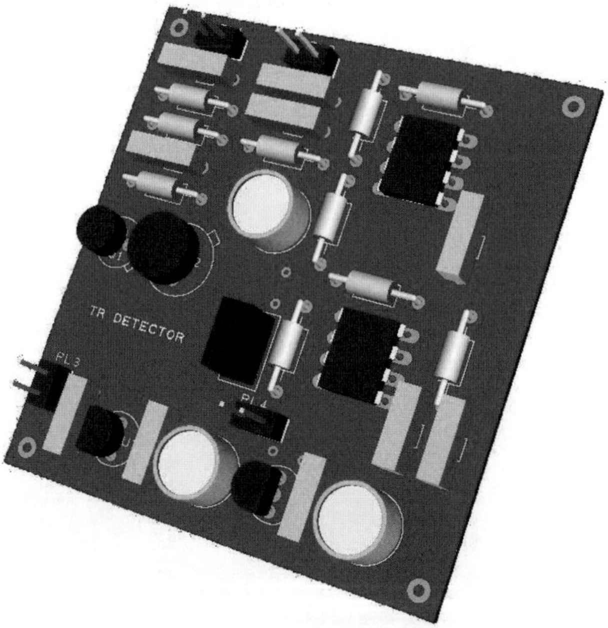

Fig. 8-9: **PCB 3D View**

CHAPTER 9

Designing an IB Detector

The Raptor Project

"In the little grey cells of the brain lies the solution of every mystery."
— Hercule Poirot

Acronyms

In the previous chapters we have stepped gradually through a number of designs, in order to demonstrate the logical progression that led ultimately to the GEB/VLF detectors that we know and love today. In the process we have managed to use a number of acronyms, such as IB, VLF, GEB and T/R, with little or no mention of their precise meaning or derivation. These terms are also freely used throughout the metal detecting literature, which can be very confusing for the uninitiated, and are often intermixed and poorly understood. Therefore an explanation of these terms will be necessary before we go any further.

GEB

Ever since the invention of the handheld metal detector certain minerals in the soil have plagued the detectorist. The two most troublesome are iron and salt. Salt is normally not a problem, except when it is wet (on the beach, for example) when it becomes electrically conductive and registers on the metal detector as a good (non-ferrous) target. It is difficult (if not impossible) to distinguish between wet salt and metal, although modern metal detectors, with a beach mode of operation, are able to eliminate the salt effect somewhat by adjusting the discrimination setting. Fortunately the effect of iron minerals can be cancelled by a detector without compromising its ability to detect metal. The term GEB stands for *ground exclusion balance* which refers to the capability of the detector to ignore (or cancel) the effect of the ground. We experienced this problem firsthand in the previous chapter when we designed the non ground cancelling TR design. GEB detectors have transformed the treasure hunting industry, and it is this type of detector that we will looking at in this chapter.

IB

The term IB (induction balance) refers to the configuration of the coils inside the search head. Any coil arrangement where there is a transmit coil and at least one receive coil, and that is set up to minimize the pickup between the TX and RX coils, can be called induction balance. The Tesoro search head we plan to use for this design is a commercial example of an induction balance coil arrangement.

VLF

VLF (Very Low Frequency) refers to detectors that operate between 3KHz and 30KHz. This is probably one of the simplest acronyms to define.

T/R

T/R means Transmit Receive, and is usually taken as referring to a detector with separate transmit and receive coils (to distinguish it from a BFO or off-resonance type detector) and operating at a frequency higher than 30KHz. However, this is not a hard-and-fast rule, and T/R detectors can also be found that operate at much lower frequencies. For example, our previous design (in the last chapter) was operating at 11.4kHz. Of course, the fact that the T/R has separate transmit and receive coils, implies that it might also be an IB, but not necessarily. Because these acronyms refer to specific aspects of detector technology, there is often an overlap between the different terms, and it is easy to understand how confusion can occur.

As we said before, some detectors are labelled as being of the T/R type, even though they operate in the VLF range. Therefore a more appropriate description would be that a T/R detector is one that does not include GEB, or (in other words) is non ground cancelling. Even more confusing is the situation where a detector is a VLF IB detector with GEB, but can be switched into T/R mode. In fact it is also possible to create a T/R detector with GEB, but the higher frequencies generally associated with the T/R type makes this less effective than with the VLF.

Stabilized Power Supply

The most important criteria for this design are stability, ease-of-use, simplicity of construction and calibration. For maximum stability it is important to provide a good power supply. In accordance with our previous design decisions, we specified the use of a 12V battery supply. By using cheap readily available IC regulators, it is easy to provide 9V to power the transmitter and the audio amplifier, and +/-5V for the rest of the circuitry. It is common practice in many commercial designs to use a simple diode-based charge pump to generate the negative supply, but unfortunately this produces a negative supply rail at a reduced voltage compared to its associated positive rail, thus causing unwanted DC offsets and possible problems with drift.

To prevent these issues occurring in our own IB design, the power supply circuit shown in Figure 9-1 will be used. Typical capacitor values are 10uF - 100uF. The popular ICL7660 is a charge pump with built-in oscillator, needing only one external capacitor. It can be used either as a voltage doubler or a voltage inverter; we are using it as the latter. Even though the output of this chip doesn't reach the full -9V under load, we are not using the -9V for anything except to regulate down to -5V, so the accuracy of the voltage at this point is not critical.

Transmit Oscillator

Eliminating the problems associated with constructing our own coil will remove a number of unknowns from the design process. Although we will be examining coil construction in the next chapter, at this time it makes sense to have a known good starting point. since we have opted to use a Tesoro coil, this will dictate the transmit

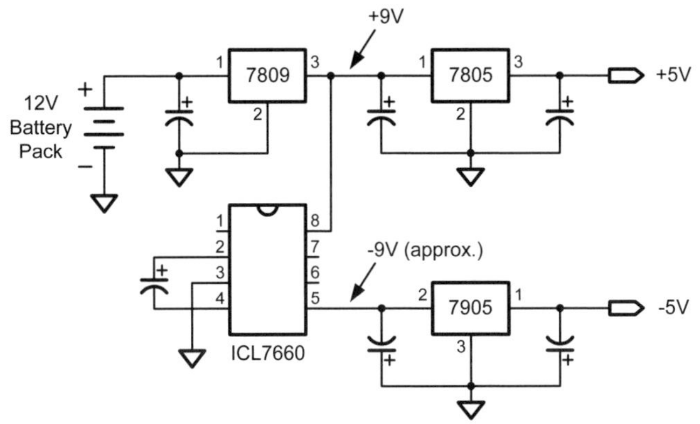

Fig. 9-1: **Stabilized Power Supply**

frequency. The Tesoro coils are particularly useful for home-brew detector projects, because they do not contain any additional components, such as tuning capacitors, in the search head, and are available secondhand at a reasonable cost. The coil wiring details are shown in Figure 9-2.

This design is initially being targeted to use a transmit frequency of 10kHz, meaning that you will need to use a coil from either a Bandido, Silver uMax, Golden uMax, Cortes (Hawkeye), Laser, or DeLeon. The prototype was tested with a 9”x8” Web coil from a Hawkeye, an 8” concentric from a Laser, a 4” concentric, and a 10” concentric.

Note that pins 3 and 5 (shown by a dotted line in Figure 9-2) are connected together in the search head. This means that the TX and RX coils share a common connection which needs to be connected to electrical ground. Tesoro transmit oscillators often make use of the Colpitts configuration, based on an PNP transistor. This enables the coil to be connected to the 0V rail, as shown in Figure 9-3.

With C1 = 470nF, C2 = 47nF, and using the equations provided in Chapter 3, you will be able to determine that the transmitter is designed to operate at 10kHz. The BFO in Chapter 3 also used a Colpitts oscillator, whereas for the off-resonance probe design in Chapter 4 we used a Hartley configuration. The Colpitts has the advantage that it only needs a single coil, but with the added complication of two tuning capaci-

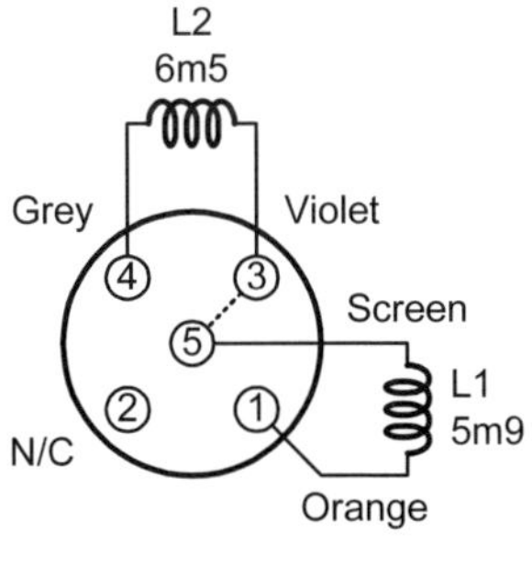

Fig. 9-2: **Tesoro Coil Details (from front of plug)**

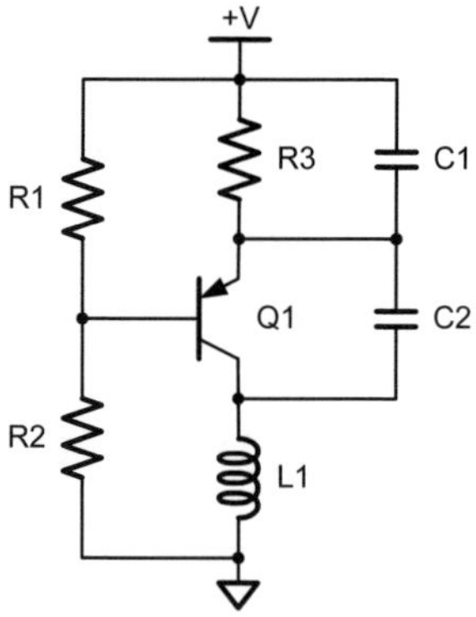

Fig. 9-3: **Colpitts Oscillator Using PNP Transistor**

tors, which can make the tuning calculations a little tricky. On the other hand the Hartley is simple to tune, with no need to calculate a serial combination of capacitors. However, this requires a center-tapped inductor, and is therefore incompatible with the Tesoro coil. So, is there an alternative? The answer is yes.

Figure 9-4 shows a 2-transistor oscillator that combines the advantages of both the Colpitts and the Hartley. That is, it only requires a single coil and a single capacitor. Thinking ahead, this will make our job easier when we start to construct our own coils. For this oscillator, the frequency calculation is simply:

$$f = \frac{1}{2\pi\sqrt{LC}} \qquad \text{Eq 9-1}$$

So that's it for the transmitter. Everything is quite straightforward so far. Now let's look at the receiver.

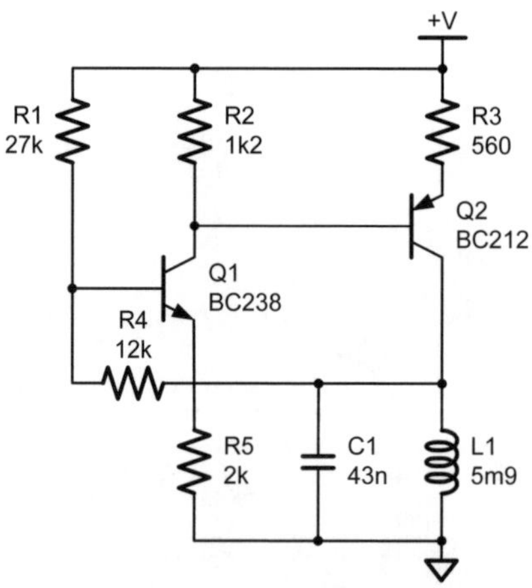

Fig. 9-4: **2-Transistor Transmit Oscillator**

Receiver Preamp

The receiver preamp is shown in Figure 9-5. By connecting an oscilloscope across the RX coil (between the violet and grey wires) with a parallel 33nF capacitor, you should be able to measure a few mV residual signal from the transmit circuit. By bringing various targets close to the coil there will be a detectable change in amplitude of the received signal. i.e. there will be an increase in amplitude for good (non-ferrous) targets, and a decrease for undesirable (ferrous) targets.

For a simple non-motion metal detector it would only be necessary to amplify the input signal and convert this to a DC level. Then the DC signal could either be used to drive a VCO (voltage controlled oscillator), or a comparator that triggers an audio oscillator when the signal goes above a preset threshold. This is the basis for many simple metal detector designs that have been published in hobbyist electronics magazines over the years. This is also the method used by the older T/R-type detectors which were developed during the same time period as the BFO. The main problem with this simplistic approach is its susceptibility to ground effects., which means that they can only be used in non-mineralized (or lightly-mineralized) soils.

The T/R design presented in the previous chapter demonstrates quite well why so much effort has been expended by metal detector manufacturers in developing designs that can exclude the troublesome ground effect. These have become known as GEB (Ground Exclusion Balance) detectors, sometimes called "ground cancelling."

Designing a VLF IB Detector with GEB

GEB is not a new concept. In 1942 the Hazeltine Service Corporation developed the ground cancelling idea as part of a national Defence Research Committee contract. As a result the first GEB detectors were used by the Army, and even today's sophisticated microprocessor controlled detectors are using the same electronic theory.

The basic idea of ground cancelling is to eliminate the false signals received from ground iron minerals, and it relies on detecting the phase shift introduced in the received signal by metallic objects close to the search coil. Without ground cancelling it is difficult to operate over iron ground mineralization. Even raising the search coil a fraction of an inch can produce false signals.

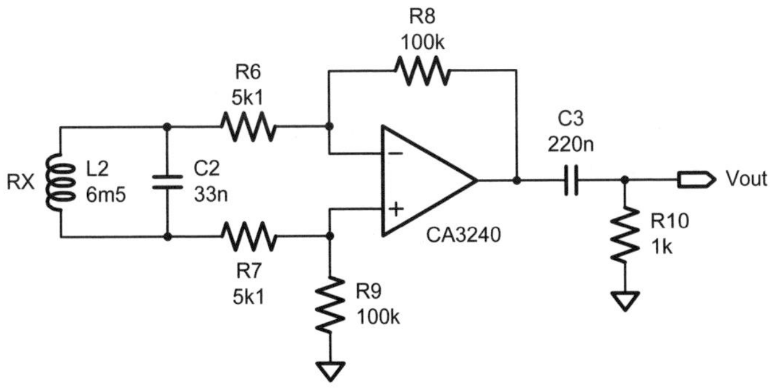

Fig. 9-5: **Receiver Preamp**

As mentioned previously in Chapter 7, the secret of discrimination is the ability to extract phase-related information from the received signal. This information is also used in the process of ground cancelling. When a target enters the electromagnetic field of the search coil, there are two effects at work - redistribution and absorption. If the target is conductive, then eddy currents are generated in the surface of the object and the magnetic field is excluded from the volume of the target. If the target is non-conductive and has a permeability greater than air, the electromagnetic field can occupy the volume of the object. These two effects cause a distortion in the electromagnetic field around the search coil, and is attributable to the reactive (or X) component of the received signal.

Highly conductive objects, such as copper, are not attracted to a magnet. The only way to produce a magnetic field from copper is to pass current through it, or (in this case) to induce eddy currents in the surface from the alternating electromagnetic field of the transmitter. These eddy currents in turn generate their own electromagnetic field, which causes the imbalance. This field is phase-shifted when compared to the transmit signal, and the amount of phase shift depends on the conductivity of the target. Non-conductive objects with a permeability greater than air produce a similar imbalance by concentrating the magnetic field within their internal volume. Salt (NaCl - sodium chloride) is a problem when it becomes wet, because ions are produced and sodium (Na) is a metal. The sodium ions cause the salt to become conductive.

Since no target is a perfect conductor then there is also a resistive (R) component to the received signal. This causes energy from the transmitted signal to be absorbed, which also produces a voltage change at the receiver. Some targets, such as conductive iron, can exhibit both redistribution and absorption. The high permeability causes the energy in the volume to be increased, whereas the conductive surface causes the energy to be decreased. There are many factors at work here, such a conductivity, permeability and transmit frequency.

Regarding discrimination, Nature has once again smiled upon the metal detectorist and given most desirable targets a low electrical resistance (i.e. high conductivity) and the opposite for undesirable targets. You might think that an object such as aluminium foil would register as a desirable target, as it possesses a moderate conductivity, However, this is not the case. Because of the foil's lack of thickness the volume of energy displaced is small when compared to the energy loss. Unfortunately some modern coinage is iron-based and these also have a high loss factor. Consequently they will register as junk. The result is that discrimination is less than 100% accurate.

After all that, the big question still remains. How can we use this information to provide ground cancelling and discrimination? Firstly, lets look at ground canceling. In general the ground is not conductive. A typical dry soil measurement gives a value of 10k ohms per metre, whereas the resistance of copper is about 0.2 ohms per metre. Therefore metallic targets should provide a significantly different response to the ground they are buried in.

Since the majority of metallic targets will produce both resistive (R) and reactive (X) components, a synchronous demodulator can be used to extract this information. This means that we need two separate channels to extract the R and X measurements.

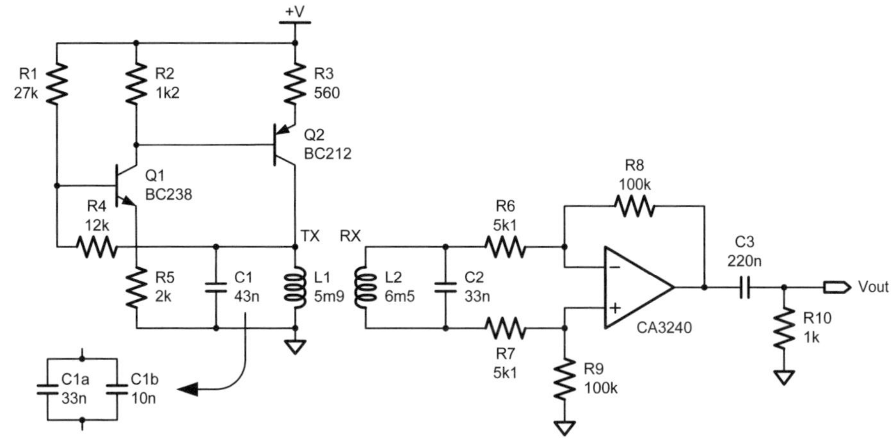

Fig. 9-6: **Test Circuit For GEB and DISC Experiment**

The ground balance control is used to adjust the phase in the X channel, and the discrimination control is used to adjust the phase in the R channel.

To better understand this statement, it's time for another experiment. On this occasion we will be using the circuit shown in Figure 9-6. Using a dual-beam oscilloscope, connect channel 1 across L1, and channel 2 to the output of the pre-amp. Set the external trigger to channel 1. Then try placing the following items in the front of the search coil one at a time:

1. Victorian penny (or similar highly conductive object).
2. Iron bolt (or similar ferrous object).
3. Pulltab.

Note that the received signal is phase-shifted to the left for all targets, but that the Victorian penny produces the most phase-shift, whereas the iron bolt produces the least. The pulltab is somewhere in between. Also note that the amplitude increases for the Victorian penny and pulltab, but decreases for the iron bolt. This phase-shifting to the left is an important factor to note, and is a function of the way the Tesoro coils are constructed. We will be examining this more closely in the coil construction chapter.

Next take a suitable container (e.g. a large ice-cream tub) and fill it full of soil. Make sure the soil does not contain any metal, and place this in front of the search coil. Note that the amplitude decreases but the phase remains unchanged. This simple experiment reveals the secret of how to implement ground exclusion balance. By taking a sample around the zero-crossing point of the receive signal, the amplitude changes are effectively ignored. However, any changes in phase will shift the received signal to the left and consequently the zero-crossing point will also be shifted. This will create a change in amplitude during the sample period. Since only metallic objects (both ferrous and non-ferrous) can produce this phase shift, the detector will indicate the presence of a target. This is the all-metal mode of operation, but with the addition of ground exclusion. So, although it is the resistive part of the signal that is affected by the ground, it is not the R channel that we need to use for GEB. In fact, we actually want to ignore the resistive part which causes the amplitude changes, and use only the reactive part which causes the phase shifts.

To provide discrimination against ferrous objects, and perhaps against pulltabs or other undesirable junk, we need to take account of both the phase of the received signal and the amplitude. To do this, a sample must also be taken at the peak of the received waveform. Again, in this case, both ferrous and non-ferrous targets will cause a phase shift to the left, and the non-ferrous targets will increase the amplitude, and ferrous targets will cause an decrease. Also, the ground will produce an decrease. So how does this help us when the dreaded ground problem still affects the discrimination channel?

Simple. If the GEB channel is used on its own to trigger the audio signal we would have an all-metal (relic) detector with the ability to ground balance. By cross-coupling the output of the discrimination channel with the output of the GEB channel, it is possible to cancel the ferrous signals. So the main purpose of the discrimination channel is basically to disable the audio signal whenever an undesirable target is detected. Since the ground signal no longer produces a response in the GEB channel, there is nothing for the DISC channel to block, and the fact that the signal amplitude in the discrimination channel is affected by the ground is irrelevant. So, this is the secret of how an analog detector can provide discrimination without the need to use trigonometry.

The phase-shift of the received signal, relative to the transmitter, is used to indicate the presence of a metal target. Whereas the direction the received amplitude changes (up or down) is used for discrimination. The ability to ignore certain higher conductivity targets, such as pulltabs, is achieved by adjusting the discrimination sample point away from the peak of the received waveform, and further to the right. In this case the phase-shift introduced by the target also has an effect. For example, although the signal from a pulltab will causes an increase in amplitude, the phase-shift to the left will cause a decrease. When the DISC control is correctly adjusted, the decrease due to the phase-shift has a more pronounced effect than the amplitude increase, and the target is rejected.

GEB and DISC Sampling

In most commercial analog detector designs, the circuitry that generates the sampling pulses is relatively simplistic. Since the main objective of this design is to act as a teaching aid and experimental platform for IB, GEB and VLF technology. then flexibility and ease of calibration are of paramount importance, and consequently we will be using a slightly more complex circuit. This will give us the ability later on to fully experiment with different coil configurations and frequencies without major circuit modifications, and to allow for correct alignment of the front panel controls during the setup procedure. Whereas the characteristics of commercial coils tend to be very consistent from coil to coil, homemade versions can vary dramatically, and a more flexible sampling circuit will provide extra leeway for the amateur constructor. It will also give us the potential to use other commercial coils, rather than just those from Tesoro. Figure 9-7 shows the proposed sample pulse generator circuit.

The DISC pot should be located on the front panel of the control box, whereas the GEB OFFSET, GEB CONTROL and DISC OFFSET pots are multiturn presets located on the PCB. We will be covering the GEB and DISC setup procedure later on in this chapter.

The circuit is driven from the TX oscillator (collector of Q1), and the output of the first comparator (LM339) provides a square wave running at the same frequency as the transmitter, but which can be offset by as much as 56 degrees by the DISC OFFSET preset. Note that the comparator associated with the DISC sample pulse is used in the inverting configuration, whereas the other three comparators are non-inverting. By adjusting the pots and presets accordingly, it is possible to position the GEB and DISC sample pulses at the correct locations on the received waveform. The LM339 comparators have open-collector outputs and therefore each one requires a pullup resistor.

CMOS Analog Switches

The next step in the design process is to provide some means of sampling the received signal and storing the result until the next sample pulse. A convenient and efficient method is to use CMOS analog switches that can be triggered by the sample pulses from the previous circuit. See Figure 9-8.

With the GEB SAMPLE pulse set to sample at the zero-crossing point of the receive signal, examination of the signal at the output of the GEB sample gate will reveal that the amplitude is zero volts. Adjustment of the GEB control to either side of this position will cause the amplitude to either rise or fall. When either a ferrous or non-ferrous target is passed in front of the coil, the amplitude will increase. However, if a bucket of soil is passed in front of the coil, there will be little or (ideally) no response. The same applies to a small piece of ferrite, such as a tuning slug.

Likewise, the DISC SAMPLE pulse must be located at the peak of the receive

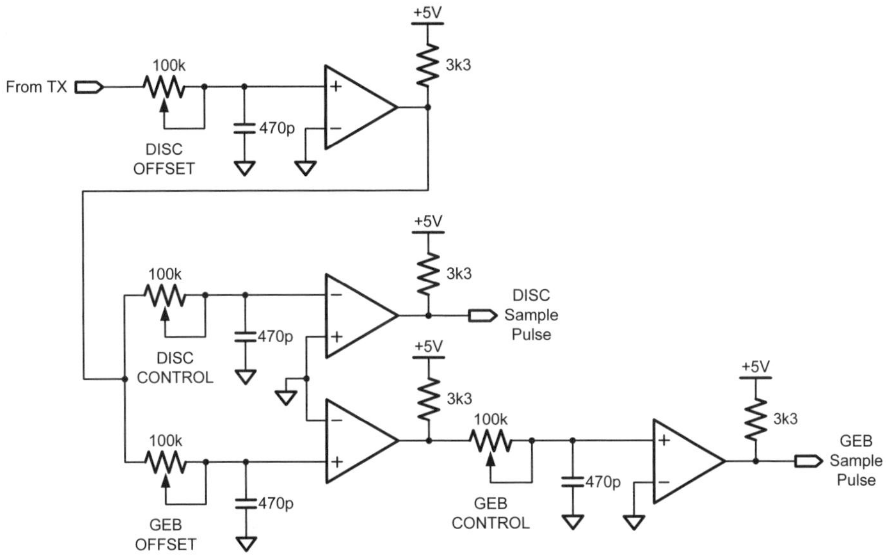

Fig. 9-7: **GEB and DISC Sample Pulse Generator**

signal, and examination of the signal at the output of the DISC sample gate will reveal a positive voltage. In this case, however, a non-ferrous target will cause this amplitude to increase, and a ferrous target will cause the amplitude to decrease.

Motion Mode of Detection

This detector is intended to be a motion detector. That is, it will respond to the rate-of-change of the received signal and not the DC level itself. The main idea behind motion detection is that the ground signal changes relatively slowly, whereas a metal target causes a fast change as it enters the electromagnetic field generated by the coil. The filter circuits in a motion detector are specifically designed to react to fast changes, while ignoring slow changes in the received signal level that may be caused by component or temperature drift and ground mineralization.

In order to react to the rate-of-change of the input signal we need to use a differentiating circuit, which is capable of producing an output that is proportional to the derivative of the input. In practice there is a problem with the differentiator, because it amplifies the input in direct proportion to its frequency. Which (in simple terms) means that it increases the level of high frequency noise at the output. What happens in reality is that the high frequency noise does not continue to infinity, since the amplifier has a limited bandwidth, but the bandwidth is large enough that noise presents a real problem. The solution in a practical differentiator is to purposely introduce a break frequency that is below the upper cutoff frequency of the amplifier.

There is no easy method of estimating the required frequency response for the GEB and DISC differentiating stages. For instance, how fast are the rising and falling edges of the target signal? Where does detection of a target object begin and end as it crosses the search coil? These are very difficult to measure in practice, but we can be certain that we need to reject the following:

1. The 10kHz transmit frequency, as this could possibly still exist in the sampled waveform.
2. The mains frequency, which could be either 50Hz or 60Hz.

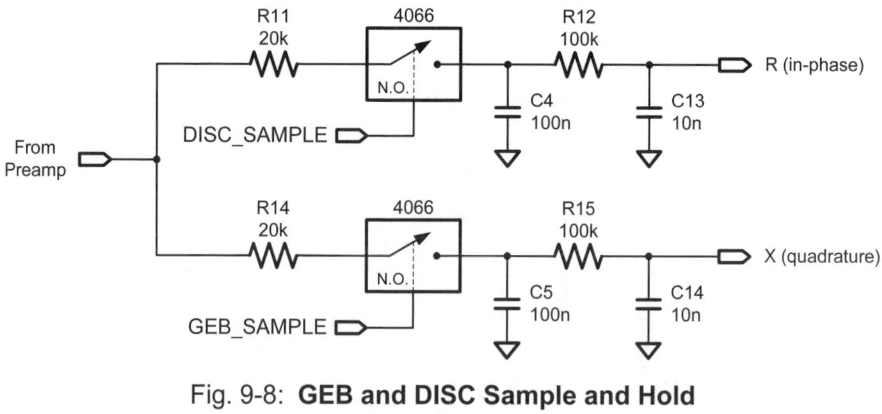

Fig. 9-8: **GEB and DISC Sample and Hold**

3. The DC level of the sampled signal. This is the whole basis of the motion detector approach, and completely removes problems caused by long-time voltage drift.

Of course, it's always possible to take a peek at some commercial schematics to get a general idea of the filter responses in these designs. Using this approach, it is relatively easy to spot that the break frequency is commonly set close to 7Hz, although some designs go as low as 3Hz.

In the low frequency region where differentiation occurs, the gain ramps up from DC (0Hz) at a rate of 20dB per decade. It is important that we also define a cutoff frequency (f_c) so that both the mains and transmit frequencies are rejected. A cutoff frequency of 15Hz should be sufficient for our needs. The gain then levels off beyond the break frequency (f_b) and starts to fall off at a rate of -20dB per decade.

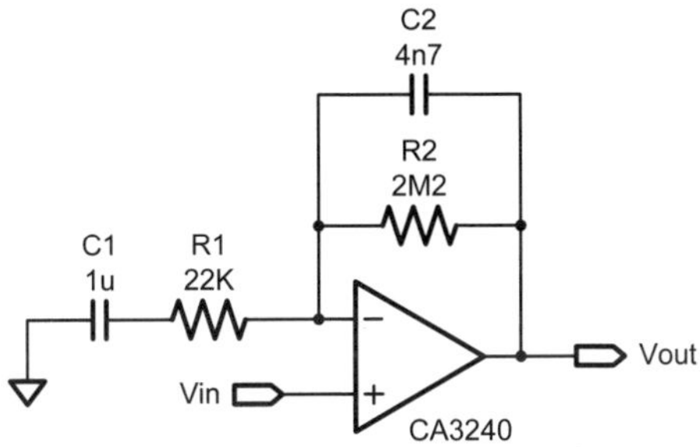

Fig. 9-9: **Practical Differentiator**

How do we calculate the required values for the filter in order to achieve our goal? In actual fact the calculations are quite simple. Referring to Figure 9-9, we can calculate the values of the passive components using the following formula:

$$f = \frac{1}{2\pi RC} \qquad \text{Eq 9-2}$$

Note that R1 and C1 define the break frequency (f_b), whereas R2 and C2 (in the feedback path) define the cutoff frequency (f_c). You should easily be able to confirm that $f_b = 7.2$Hz and $f_c = 15.4$Hz.

The frequency response of the practical differentiator (from 0.1Hz to 100kHz) can be seen in Figure 9-10. You should also note that the gain for this configuration is 36.8dB (i.e. 69x) and that the response rolls off from the center frequency (approximately 10.5Hz) at a rate of 20dB per decade.

Any unwanted sampling noise at 10KHz is effectively zero. Mains interference at 50Hz will be amplified by 29.4dB (29.5x), or alternatively at 60Hz by 28dB (25x). In practice, any interference from the mains wiring is negligible, and you have to make a real effort to pick up any signal at all, such as placing the search coil next to a power cube.

For our detector to have a reasonable sensitivity, we need to add a second differentiating stage to the design, and also create two separate channels. One for the

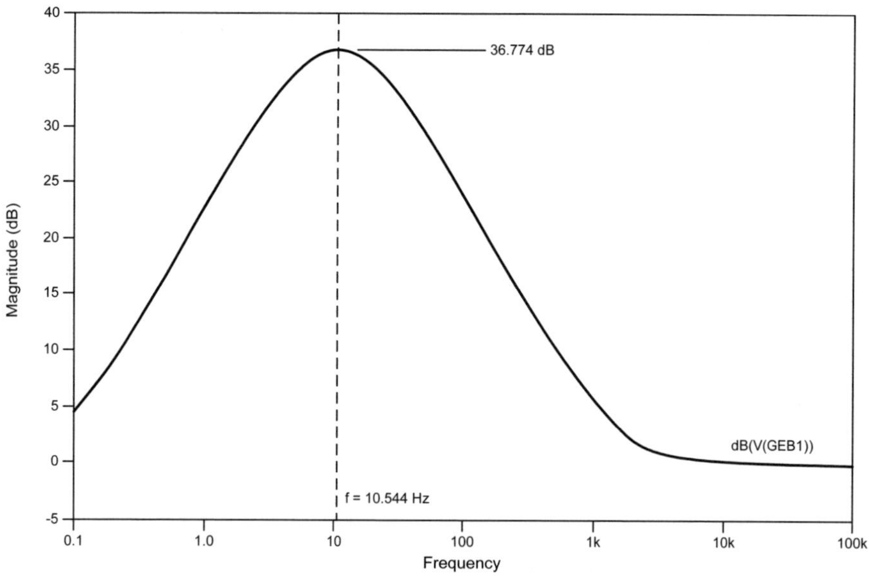

Fig. 9-10: **Frequency Response For Practical Differentiator**

ground balance (GEB) and the other for discrimination (DISC). But before we proceed any further, let's design the audio output stage, as this may affect the configuration of this second stage.

Audio Oscillator, Chopper and Amplifier

The audio oscillator is probably the most straightforward part of the design, as we can simply use a common 555 timer. See Figure 9-11.

The 555 timer is configured for astable operation. With pins 2 (trigger) and 6 (threshold) connected, it will trigger itself and free run as a multivibrator. In this mode of operation, the capacitor C1 charges and discharges between 1/3 +V and 2/3 +V. The charge and discharge times are independent of the supply voltage, as is the fre-

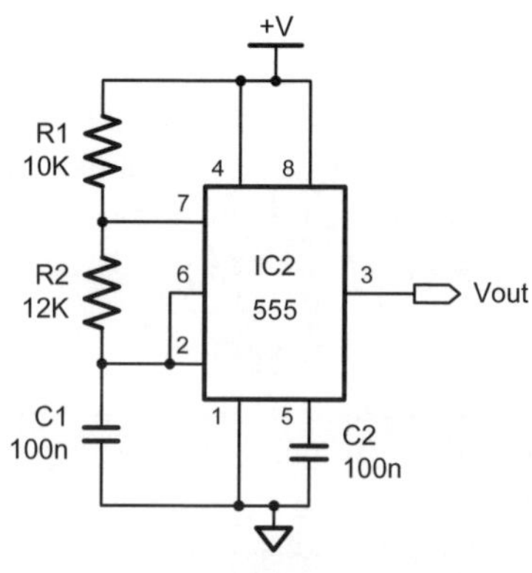

Fig. 9-11: **555 Oscillator**

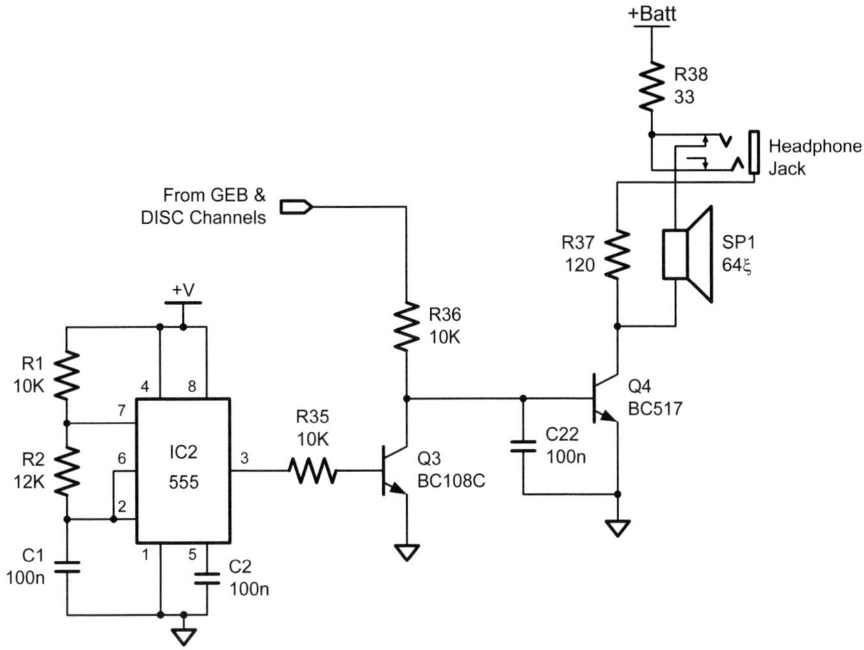

Fig. 9-12: **Complete Audio Circuit**

quency of oscillation.

The charge time (output high) is given by:

$$t1 = 0.685(R1 + R2)C1 \qquad \text{Eq 9-3}$$

and the discharge time (output low) by:

$$t2 = 0.685 \cdot R2 \cdot C1 \qquad \text{Eq 9-4}$$

Thus the frequency is given by:

$$f = \frac{1.46}{(R1 + 2 \cdot R2) \cdot C1} \qquad \text{Eq 9-5}$$

For the values provided in Figure 9-11, t1 = 1.52ms, t2 = 0.83ms, and f = 423.5Hz.

The audio frequency used by commercial detectors varies widely. Some are as low as 293Hz, whereas others are up at 950Hz. If you would prefer to use a different audio frequency, then simply recalculate the values for the passive components. Alternatively, there are some online 555 calculators on the internet to make the job easier.

The audio chopper and amplifier circuits are relatively simple, as can be seen in Figure 9-12. The 555 timer audio oscillator drives the audio output stage, which is biased on by the output from the GEB and DISC channels (more on this part coming soon). The transistor being driven by the oscillator is often referred to as a chopper, since it "chops" the bias signal. When a metal target is detected, the bias signal

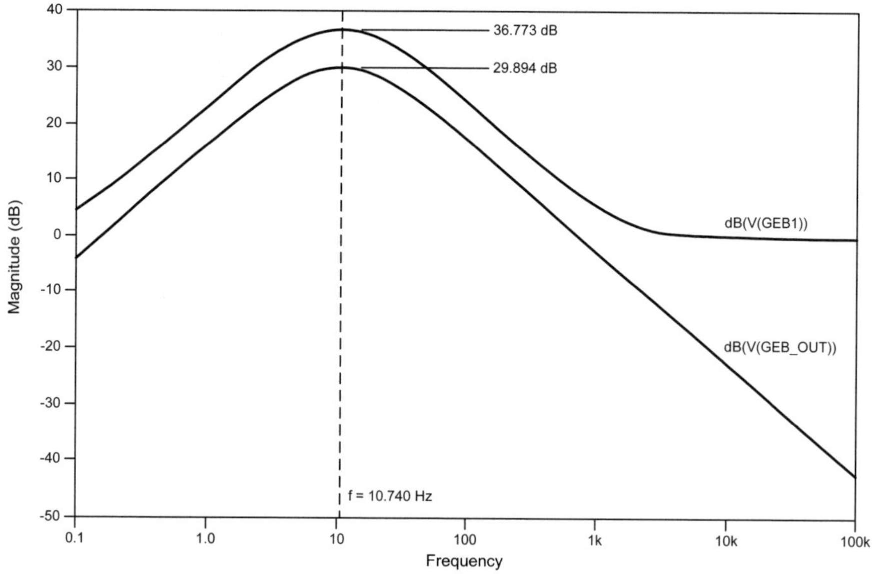

Fig. 9-14: **Comparison of Frequency Response For Differentiators**

increases and turns on the first transistor, allowing the audio signal to pass through to the final audio stage that drives the speaker or headphones. Note that the second transistor (BC517) is an NPN darlington pair.

Now – things start to get interesting!

Next we need to add a second differentiator stage to both the GEB and DISC channels to increase the sensitivity of the detector. The proposed circuit in Figure 9-13. is an inverting version of the previous differentiator, but with a gain of 29.9dB (31.2x).

You can make a comparison of the frequency response of two differentiating circuits by examining Figure 9-14. Note, in particular, that the response of the inverting circuit continues to roll off at 20dB per decade, and does not flatten out like the first circuit.

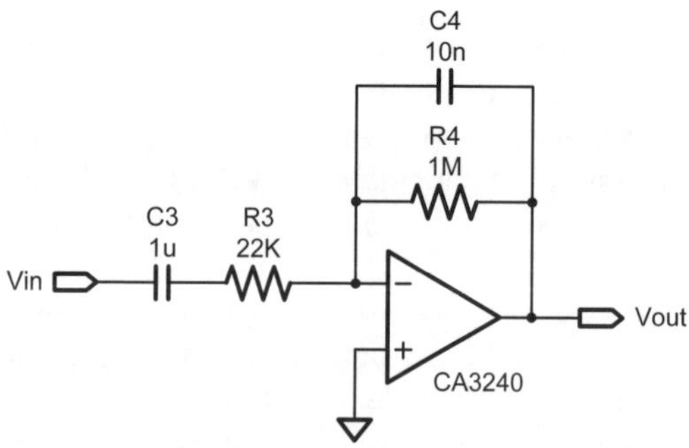

Fig. 9-13: **Second Differentiator Stage**

At this point let's recap where we are. The GEB signal is generated by sampling the RX signal at the zero-crossing point where the signal crosses from negative to positive. If either a ferrous or non-ferrous target enters the electromagnetic field of the search coil, the signal phase-shifts to the left, relative to the TX signal. Thus the sample pulse sees a positive voltage for all targets. Non-ferrous target detection is boosted somewhat, because the amplitude also increases, and since ferrous targets cause a decrease in magnitude, this tends to desensitize the detector to small iron fragments.

Considering only the GEB channel at this time, it should be immediately obvious that this is the "all-metal" mode of operation. The ground matrix is ignored (as explained earlier) because there is only a magnitude change with no associated phase shift. In this case, the zero-crossing point remains fixed, and the result is still zero. Finally the audio circuit requires a signal from the GEB channel that will increase in amplitude when a target is present in order to produce an audio tone.

Now - if you've been paying attention, you're probably thinking that there's a mistake here somewhere. Let's think about this …

The output from the GEB analog switch increases for all targets, and the audio circuit requires a signal that increases. So what's the problem? Simply that we have placed a second differentiator stage in the GEB channel that produces an inversion of the signal. So doesn't that mean that this channel now produces a signal that decreases in amplitude for all targets, rather than the other way round? Let's try and explain what's going on.

In order to understand this better, we need to do a little calculation, then we'll run a simulation to make things clearer. We're going to make some estimates and assumptions based on a best guess, but the results will be close enough to prove a point.

First let's imagine we are swinging a detector through an arc as if we are searching the ground outside. If the distance from the centre of the arc is 1 metre and we swing the search coil through an angle of 90 degrees, then the distance covered in one movement from left-to-right will be 1.57m. This could also be from right-to-left, but there's no difference for these calculations. If we make one pass with the coil in 1 second, then the velocity of the swing will be 1.57m/s.

Next, let's assume that the search coil we are using is a concentric type with a TX coil of 200mm (approx. 8") diameter and an RX coil of 60mm. If you monitor the RX signal with an oscilloscope it is easy to see that a coin starts to produce a signal that increases in amplitude as it crosses the edge of the TX coil. This signal continues to increase until it reaches the center of the receive coil, and then starts to decrease again. The increase and decrease is not linear across the face of the coil, so this experiment is just an approximation. Although any difference in the results between this simple model and reality is virtually undetectable.

In order to demonstrate the secret, a SPICE simulator was used to emulate what happens when a coin crosses the face of the coil. Both the full GEB and DISC channels were simulated, and the input to each channel was driven by an idealised target response. Referring to Figure 9-15, you will see that the top trace consists of two signals - GEB input and DISC input, which show the idealized input signals. These represent the modulation extracted by the synchronous demodulators (in other words, the analog switches).

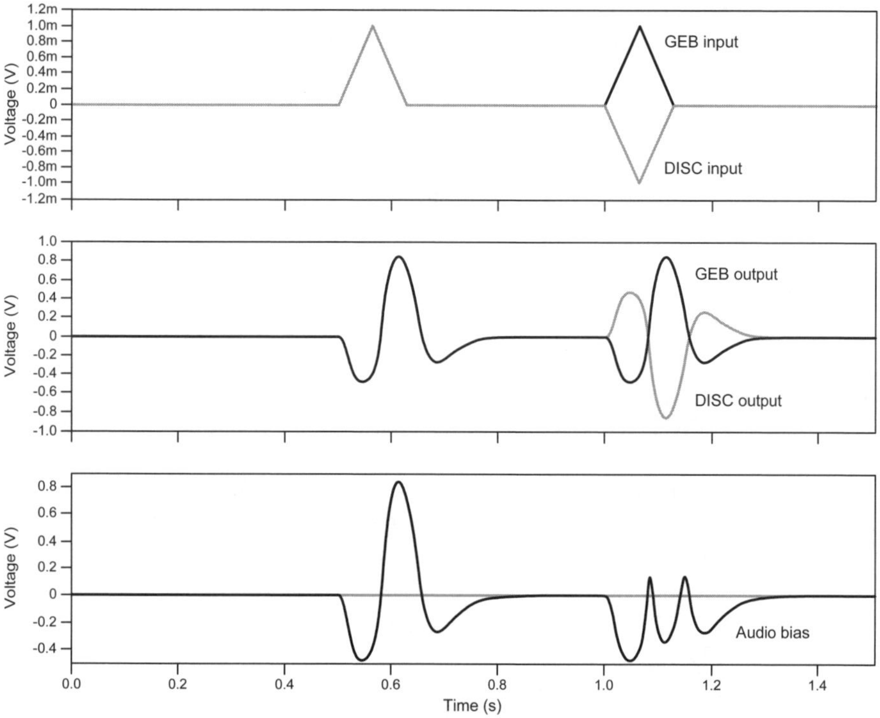

Fig. 9-15: **Response of the Complete Channel**

In this example, it was assumed that a coin would pass from the edge of the coil to the center in a time of 63.7ms, calculated from the search velocity of 1.57m/s and the radius of the coil, which is 100mm. Likewise, it would take the same time to cross the other half of the coil. There are two triangular pulses shown which represent two separate targets. The first pulse is a non-ferrous target, and the second one is ferrous. For simplicity they are both returning a maximum amplitude of 1mV. For the GEB input, the two targets give a positive response, and the DISC input gives a positive response for non-ferrous and a negative response for ferrous, as explained in some detail earlier in this chapter.

The second trace shows the output waveforms obtained from both the GEB and DISC channels in response to these input signals. Due to the differentiating action of the two channels, there are both positive and negative pulses generated as the circuit tracks the rate-of-change of the signals, and not the DC levels of the signals themselves. In order to further simplify the experiment, and to allow the GEB and DISC output waveforms to be overlaid, the DC level that would normally be present at the input to the DISC channel was removed. This makes absolutely no difference to the results.

The final (bottom) waveform is the result of cross-coupling the outputs of the GEB and DISC channels. Please see Figure 9-16 to see how this is achieved.

What is happening here is that the non-ferrous (good) signal is being allowed to

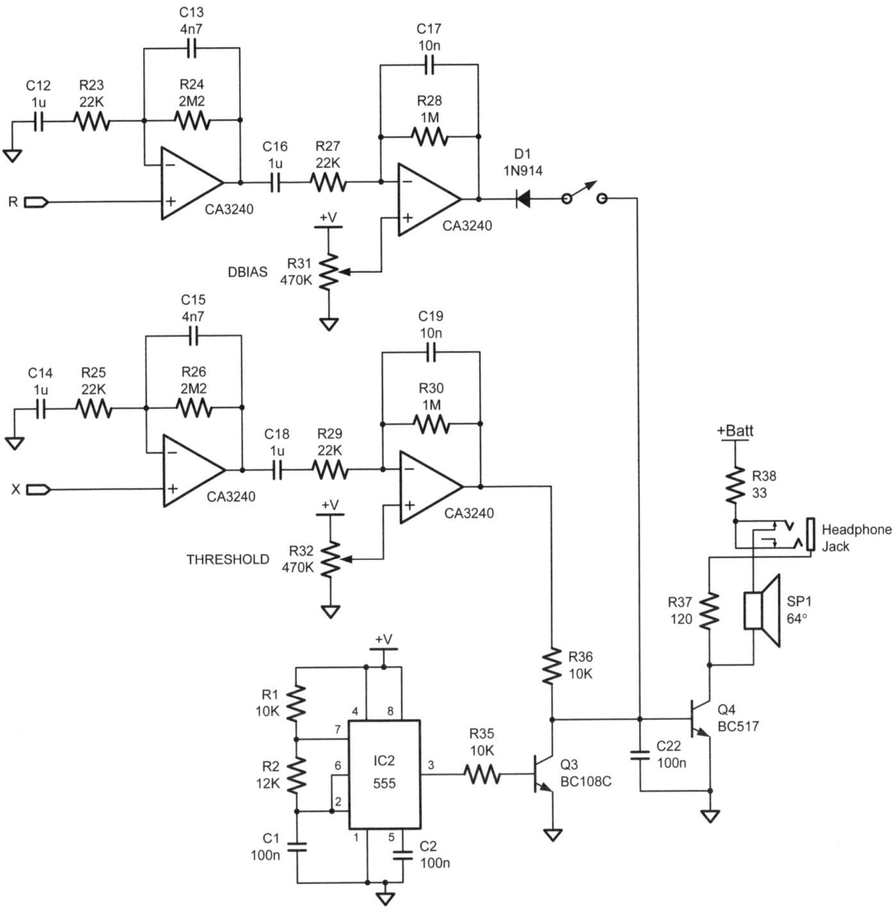

Fig. 9-16: **GEB, DISC and Audio Circuits**

pass through and bias on the audio amplifier, and consequently produce an audio tone in the headphones or speaker. Whereas the ferrous signal is disabled by the DISC channel via the diode D1. Here are a couple of important points that should be noted from this exercise:

1. It is clear from the simulation results that the second differentiator stage needs to include an inversion of the signal.
2. The audio bias signal shows that some chatter could occur during discrimination when the unwanted target is close to the coil. This chatter takes the form of two small pulses as the rejected target crosses the search coil. Although, in practice, this does not result in two audio tones being heard by the user. Instead it sounds more like a click.

Now - this brings us to another interesting question. Is it possible to determine the optimum sweep speed using the same SPICE simulation? Actually the answer is yes. By defining the time for a target object to cross the radius of the coil (let's call this the transition time) as a parameter, it is then possible to sweep this parameter over a range of values. Examining the results shows that a transition time of 35ms yields the highest amplitude signal at the output of the GEB channel. This represents a coil sweep speed of 2.86m/s, and an increase of 34% in signal amplitude over the previous sweep speed of 1.57m/s. Many commercial designs must be swept very slowly to avoid missing targets, whereas this design can tolerate a higher sweep speed. However, we must also consider that going too fast may result in a louder audio response, but it will also be a much shorter duration signal. So there may be some benefits in taking it a little slower. Less signal strength, but a longer duration signal. It was found in practice that this detector can actually tolerate a wide range of sweep speeds.

A Little More Info On Differentiation

Before we move on and complete our design, let's explore the subject of differentiation a little more, in order to clear up any remaining misunderstanding concerning its use in the GEB and DISC channels.

Firstly, what is differentiation?

Differentiation is a part of a branch of mathematics called calculus, and the process of differentiation produces something called the derivative. The derivative is a measurement of how much a quantity is changing at a given time. As an example, the derivative of the position of a car at a specified time is its velocity. In other words, the "rate-of-change" of position with time. If you were to then find the derivative of the car's velocity, you would have a value for the acceleration. In this case you have calculated the "rate-of-change" of velocity with time. So the process of differentiating the car's position twice (known as the second derivative of the function) yields the acceleration. This is exactly the same process that is happening in both the GEB and DISC channels.

In the previous investigation we used a SPICE simulator to visualize the signals in the GEB and DISC channels. This was achieved by emulating, in an idealized way, what happens to the target signal as it undergoes the process of double-differentiation. One thing that is important to understand is that both channels are only capable of amplifying AC signals, as can be seen from the frequency response curves shown earlier. So what would happen if we were to input a sine wave to the differentiation circuits?

Mathematically we can represent the sine wave input as:

$$X = A\sin(\omega t) \qquad \text{Eq 9-6}$$

Taking the derivative of this signal gives:

$$\frac{dX}{dt} = A\cos(\omega t) \qquad \text{Eq 9-7}$$

Then taking the derivative again gives:

$$\frac{d^2X}{dt^2} = -A\sin(\omega t) \qquad \text{Eq 9-8}$$

Note that the sign of the second derivative is negative. This is the reason why we need to have an inversion in the second stage filter. Of course, using a sine wave as the input is a special case, as it enters and exits the filters as a sine wave, albeit inverted and with a larger amplitude. You could also look at this another way, by considering that the two channels are monitoring the "acceleration" of the target signal as it crosses the coil.

This is probably one of the least understood areas of motion detector design, especially by hobbyists who would like to construct their own detectors. Misunderstandings concerning the underlying mathematics can lead to a non-optimal design, and quite possibly to incorrect coil construction and alignment.

One more point concerning discrimination (not to be confused with differentiation) is required before we proceed further. By advancing the DISC control on the front panel clockwise, it is possible to increase the rejection of some of the more conductive objects, such as foil and pulltabs. So how does this work? We stated earlier that ferrous target rejection was achieved by sampling at the positive peak of the received waveform. Since non-ferrous objects cause an increase in amplitude and ferrous objects do the opposite, then it is clear that a distinction can easily be made. In order to understand how this works for say a pulltab, you need to remember that all targets also exhibit a phase shift. The GEB channel uses this knowledge to distinguish between targets that have a phase shift (both non-ferrous and ferrous) and those that don't (neutral soil and ferrite). In order for the DISC channel to reject some non-ferrous targets that exhibit a different phase-shift than coins for example, the sample pulse must be moved to the right, away from the peak of the received waveform. If you wave a pulltab a few inches in front of the coil, then adjust the DISC control clockwise until the pulltab is rejected, you will have found a sample point where the effect of the phase-shift dominates the amplitude change. What happens is that the sample pulse is measuring more of the negative part of the received signal as it is left-shifted.

Admittedly this is one of the most difficult parts of detector technology to comprehend, and an examination of the U.S. Gifford patent (4,486,713) may make this clearer. But be prepared that you may end up even more confused than before. However, let's not worry about this for the moment. Just remember that turning the DISC control clockwise will increasingly reject non-ferrous (good) targets. Most importantly, your motto should be "too much discrimination is bad". Only use the minimum that you need for the conditions.

One final comment - note that the 1uF capacitors used in the differentiating circuits are polyester types. This was an intentional part of the design, with the resistor values being calculated to allow a relatively low capacitance value to be used. Many commercial designs use 4uF capacitors in the non-inverting differentiating stage, which necessitates the use of bipolar electrolytics, although some hobbyists have replaced these with two electrolytics back-to-back. Using electrolytics in filter circuits is a really bad idea, as they can introduce large amounts of unwanted noise. The PCB layout is designed to use 2 x 470nF polyester capacitors in parallel, as the equivalent 1uF capacitor is physically larger and more difficult to obtain.

Automatic Battery Check Circuit

So far we have used a total of 5 opamps in this design. The CA3240 is a dual-opamp package, which means that we have one opamp spare. The 4066 is a quad package which leaves us with 2 spare analog switches.

A simple automatic battery check circuit can be implemented using the spare opamp and one of the analog switches, as shown in Figure 9-17.

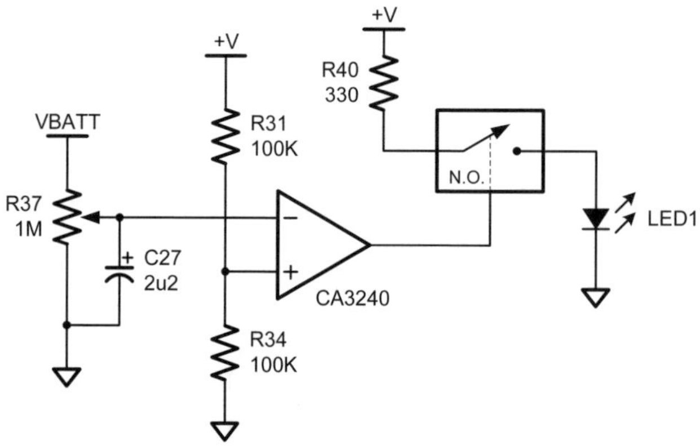

Fig. 9-17: **Automatic battery Check Circuit**

When the detector is switched on, the inverting input of the opamp gradually charges up to the voltage level defined by the Batt Adj preset adjustment. At the same time the non-inverting input rapidly settles at 2.5V as defined by the potential divider resistor network across the +5V power supply. During the time that the capacitor is charging via the battery pack (+12V) the opamp output turns on the analog switch, causing the LED to illuminate. The Batt Adj preset should be adjusted by replacing the battery pack with a +10.5V supply, and setting the voltage at the inverting input of the opamp to +2.5V. Probably the easiest way to achieve this is to use a bench power supply. Assuming that the battery supply voltage is above +10.5V, the LED will illuminate for just under one second. This check is not only performed at switch on, as the circuit constantly monitors the state of the battery voltage during use, and if it becomes illuminated, the batteries need to be replaced.

Battery Pack

The completed design draws only 64mA quiescent current, and never peaks above 70mA. In this case it was decided to simply use alkaline batteries, and dispense with rechargeable cells, such as NiCd or NiMH. It was also felt that a 12V sealed lead-acid battery would make the detector too heavy to use for extended periods. Of course, there is no technical reason why rechargeable cells cannot be used. But if you plan to use either NiCd or NiMH, then you will need to use 10 cells to achieve the required level of 12V.

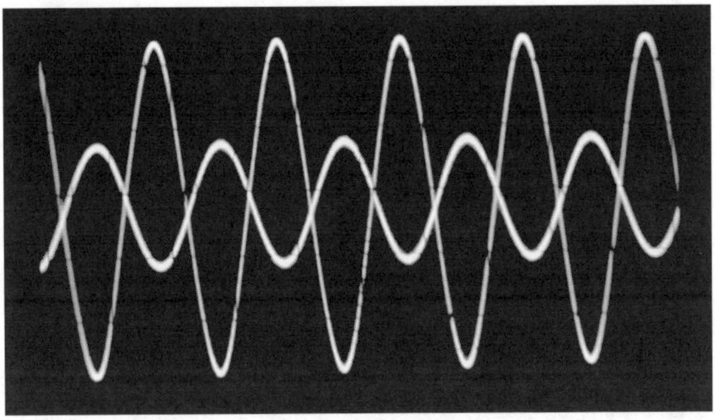

Fig. 9-18: **RX Signal (lower amplitude) Compared to TX Output**

The Completed Prototype

The setup procedure for the automatic battery check circuit has already been mentioned, and we will now cover the remaining "factory" adjustments that need to be made before we test the detector in the field.

Initial Setup Procedure

You will need a 2-channel oscilloscope to make these measurements and adjustments.

1. Check the offset of the received (RX) signal at the output of the pre-amp, when compared to the transmit (TX) signal, is approximately 200 degrees. See Figure 9-18. If the RX signal appears to be inverted, then you probably have the RX coil connections swapped over.
2. The RX amplitude at the output of the pre-amp should be between 250mV and 300mV, which equates to approximately 12mV to 15mV at the coil input. This value may vary from coil to coil.
3. Monitor the RX signal at the output of the pre-amp (with channel 1) and the DISC sample pulse (with channel 2). Adjust the DISC OFFSET preset until the DISC sample pulse is located over the positive section of the RX waveform. Make sure the DISC pot on the front panel is set fully anti-clockwise. This will make sure that ferrous targets are eliminated when the DISC pot is set to minimum discrimination. This is the coarse DISC adjustment.
4. Then slightly re-adjust the DISC OFFSET preset to obtain a maximum DC level at the output of the DISC sample gate.
5. Monitor the RX signal at the output of the pre-amp (with channel 1) and the GEB sample pulse (with channel 2). Adjust the GEB OFFSET preset until the GEB sample pulse is located at the positive-going zero-crossing of the RX waveform. The GEB preset should first be set to mid position. This is the coarse GEB adjustment.
6. Then adjust the GEB preset to obtain zero volts (or as close as possible) at the output of the GEB sample gate.

7. Switch to all-metal mode and adjust the THRESHOLD pot (on rear panel) until a faint audio tone can be heard. Then switch to DISC mode, and adjust the DBIAS preset until the audio tone just starts to reduce in volume. As you switch between the two modes, the audio level should change very slightly, higher for all-metal, and lower for DISC.

Be sure to make these adjustments with the coil attached to the stem and inclined at the correct angle for normal use. This will insure that any metallic parts, such as the stem or fixing bolts do not cause a problem when the detector is in use.

Real World Testing

Many hobbyists have designed and built their own detectors over the years, only to be disappointed by a less than startling performance when it was taken out for some real world testing. So how does this detector really perform, or is it just a "workbench wonder"?

The first thing you must do, after turning on the detector, is to switch to All Metal mode and adjust the Threshold control so that an audio tone can only just be heard in the headphones. On the prototype detector, the Threshold was on the rear of the control box, and rarely needed adjustment unless accidentally moved.

This detector has been designed with a preset ground balance control, which makes it a simple to use "turn on and go" detector. If you wish to detect in an area with mineralized soil, then it may be advisable to mount an external GEB control on the front panel. In this case, simply use a 100k potentiometer, and connect it in place of the GEB preset using flying leads.

Next, switch to Discriminate mode, and initially set the DISC control fully anticlockwise, to the position where ferrous targets start to be rejected, then start searching.

On the first outing with this detector, it was soon clear that its ability to detect non-ferrous targets in an iron infested area was outstanding. The area that was chosen for the initial test was a riverside site. This area has been visited many times over the years, and is pretty much hunted out. So it was particularly surprising that, within a period of 3 hours, 10 coins: 5 fishing weights, 1 pulltab, 1 large iron ring, 1 tent peg, and 4 miscellaneous non-ferrous items were recovered. The first two finds were actually large pieces of foil, but after adjusting the DISC control to eliminate these unwanted items, only one other foil target was recovered very close to the surface. A number of tin cans (either beer or soft drinks) were also detected, but these are not shown in the photograph. See Figure 9-19.

You may have spotted that this detector does not have a pinpoint capability. In practice this was by no means a disadvantage. Pinpointing was easily achieved by marking an X with the coil. At no time was the target found to be offset from the expected position. This capability is a combination of two factors. The Tesoro concentric coils have a small central receive coil that enhances the ability to locate small objects, but the main reason is the double-differentiating architecture, which results in the detector responding to the target as it leaves the center of the coil, and not as it enters.

But Can We Do Better?

This design was intended as a learning tool to help understand how a modern analog VLF-GEB-IB detector works. It was designed to be as flexible as possible, very stable, and easy to construct, even using stripboard. It is also possible to use a number of common substitutes, if the specified components are not available. The target separation on this detector is excellent, with fast recovery, enabling searching in areas saturated with nails or small ferrous trash

However, referring to the quotation at the start of the previous chapter, we should ask ourselves if this good enough, or can we do better?

One problem with the current prototype is that deep targets give a weak audio response, and it is therefore almost a necessity to wear headphones. Is it possible to boost the response of these deep targets and consequently improve the sensitivity of the design? It is this problem that we will look at next.

Improving the Design

One simple modification would be to add a single comparator (with hysteresis) at the input of the audio amplifier to improve the response to weak targets. The hystere-

Fig. 9-19: **First Finds With The Prototype IB Detector**

sis is required in order to avoid any instability due to noise. However, one side-effect of the comparator hysteresis in this configuration is that unwanted chatter is also louder and longer in duration. The result is that target rejection close to the coil is compromised.

An alternative approach would be to place a separate comparator directly at the output of each of the GEB and DISC channels, and to cross-couple the comparator outputs in a similar manner as before. There are also two possible approaches to driving the audio from the comparators. The audio can either be silent by default and enabled by a target response, or alternatively the audio can turned on by default and disabled when no target is present. After several experiments it was determined that the latter approach gave the best results. The final comparator circuit can be seen in Figure 9-20.

Comparator Operation

The simplest way to explain how the comparator circuit functions is to refer to a truth table:

DISC	GEB	AUDIO
0	0	0
0	1	0
1	0	0
1	1	1

A '1' in the table indicates that the signal is high (+5V) and '0' indicates that the signal is low (-5V). It is therefore clear that the audio is only enabled when both the DISC and GEB channels are high, which indicates the presence of a non-ferrous target.

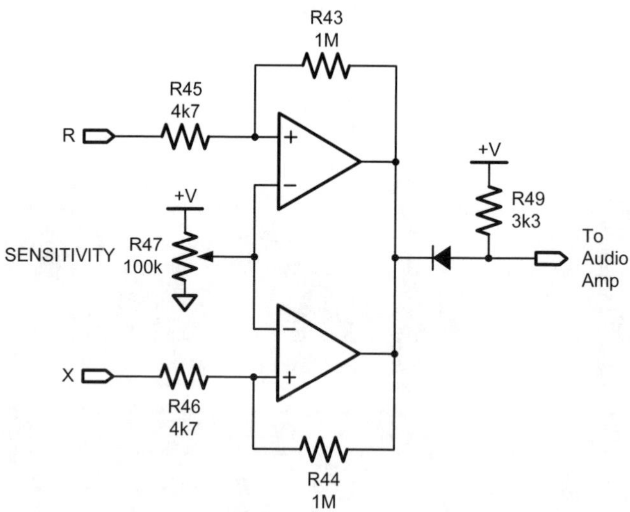

Fig. 9-20: **Comparator Circuit**

Let's examine each line of the table:

1. Both DISC and GEB are low, indicating no targets present. The result is no audio.
2. Remember that the GEB channel goes high for all targets, but ignores any signal from the ground matrix. Since the DISC channel is low, this indicates that the target is ferrous and is consequently ignored.
3. Although the DISC channel is high, the GEB channel is low. In this case there is no target present, and no audio signal is generated.
4. Both the GEB and DISC channels are high, which indicates a non-ferrous target. This is the only combination that produces an audio response.

The only confusion here could be the 3rd line in the table. Since the GEB channel is not indicating a target, how can the DISC channel ever go high, especially since the ground produces a decrease? Or is this a state that can never occur in reality? To answer this question we need to recall that a metal target produces both positive and negative pulses as the circuit tracks the rate-of-change of the signals. For example, if the target is ferrous, the GEB channel will first produce a high followed by a low pulse. However, the DISC channel will first go low, followed by a high. The situation with the secondary pulse (DISC = 1 and GEB = 0) is being shown in line 3. Likewise, line 1 could also occur during the secondary low pulse produced from a non-ferrous target. Essentially the cross-coupled comparator circuit, in association with the diode and pullup resistor, acts like a logic AND gate.

Hysteresis is implemented in the comparator circuit by feeding back a small fraction of the output voltage to the non-inverting input. In order to keep the existing circuit configuration of the detector, it was necessary to use the comparators in the non-inverting mode.

The amount of hysteresis can be calculated as follows:

$$\begin{aligned} Hysteresis &= \frac{R1}{R2}(VOH - VOL) \\ &= \frac{4k7}{1M}(5-(-5)) \\ &= 47mV \end{aligned} \qquad \text{Eq 9-9}$$

The result is a detector that functions as a true "silent search". Although there is some chatter in the presence of ferrous targets, which is reassuring to the operator, without having an annoying background threshold tone.

Some analog IB detectors also have a third channel, known as the All Metal (AM) channel, that is used during pinpointing. This AM channel is actually a non-motion T/R detector with autotune. These detectors often have a RESET button on the front panel or a trigger switch on the detector stem close to the operator's forefinger. As we found in the previous chapter, the T/R detector provides very good iron rejection, and consequently the AM channel can provide additional discrimination against ferrous targets. In the three channel detectors it provides some DC bias to the audio stage which stops the unwanted chatter caused by ferrous objects close to the coil. This iron blanking is very good in an air test, but is less impressive in practice.

The main advantage of the AM channel is its usefulness as a non-motion pinpointing tool. Our own design does not have a pinpointing capability, but as mentioned previously this was not found to be a disadvantage.

A modification was also made to remove both the THRESHOLD and DBIAS presets, as these are now unnecessary, and replace these with a potential divider. Also, the All-Metal (AM) / DISC switch was removed in the final design.

The final Raptor circuit can be seen in Figure 9-22.

Note that both the DISC and SENS potentiometers were replaced by a switched network of resistors, as shown in Figures 9-23 and 9-24.

The SENS control is split between a preset on the PCB and a set of switched resistors on the front panel. The resistors are soldered to the contacts on the rear of the switch. The table below shows the effect of the SENS adjustment on detection distance of a Tesoro 9x8 web coil for a Victorian penny. Since the voltage at the inverting inputs of the comparators versus the detection distance has a non-linear relationship, this switched resistor network was found to be an excellent solution.

SENS setting	Detection distance
1 (minimum)	7" (17.8cm)
2	8" (20.3cm)
3	9" (22.9cm)
4	10.5" (26.7cm)
5 (maximum)	12" (30.5cm)

One drawback of the more flexible sample pulse generator, based around the LM339, is that the DISC control produces some noise at the audio output when it is adjusted. Therefore the DISC pot was also replaced with a switched resistor network. By careful calculation of resistor values, it was then possible to have selected switch settings for predefined unwanted targets. Please see table below:

DISC setting	Unwanted target
1	Ferrous (iron)
2	Foil
3	Pulltabs
4	Non-ferrous #1
5	Non-ferrous #2

DISC switch positions #4 and #5 are not allocated to a specific target, but provide increasingly more discrimination such that non-ferrous targets start to be eliminated. The Victorian penny was still able to be detected, even in switch setting #5, but some lower value modern coinage was ignored.

Detector Calibration

The GEB and DISC setup procedure has already been documented, and the following refers to the SENSITIVITY preset adjustment.

First set the SENSITIVITY pot on the front panel to maximum sensitivity (position #5). Then adjust the SENSITIVITY preset on the PCB until the detector audio is turned on. If you wish, you can monitor the centre pin of the preset with a voltmeter during this adjustment. Re-adjust the preset until this voltage reaches 0.45V. This is the most sensitive setting, and the audio should be remain quiet even if left over a period of several minutes If you are not using a voltmeter, then slowly turn the preset until the audio just turns off. At this point a 9x8 Tesoro web coil will give the following results:

Small hammered silver coin	7.5”	(19cm)
1 Euro coin	10.5”	(26.7cm)
Victorian penny	12”	(29.2cm)
Aluminium drinks can	19.5”	(49.5cm)
Aluminium sheet (5.5” x 3.6”)	21”	(53.3cm)
Non-etched PCB (10” x 6”)	25.5”	(64.8cm)

Detailed construction information and a synopsis of the calibration procedure for the final circuit are provided later in this chapter.

Real World Testing (again)

With the latest comparator modifications the detector worked very well in the field. Faint signals were now much clearer than before, allowing identification of targets at greater depth. In heavily iron-infested areas it was a simple matter to reduce the sensitivity using the front panel control, and non-ferrous targets were easily picked out amongst the trash.

In Conclusion

A block diagram of the overall detector can be seen in Figure 9-21. The transmit oscillator is a 2-transistor circuit designed to operate at 10KHz using a standard Tes-

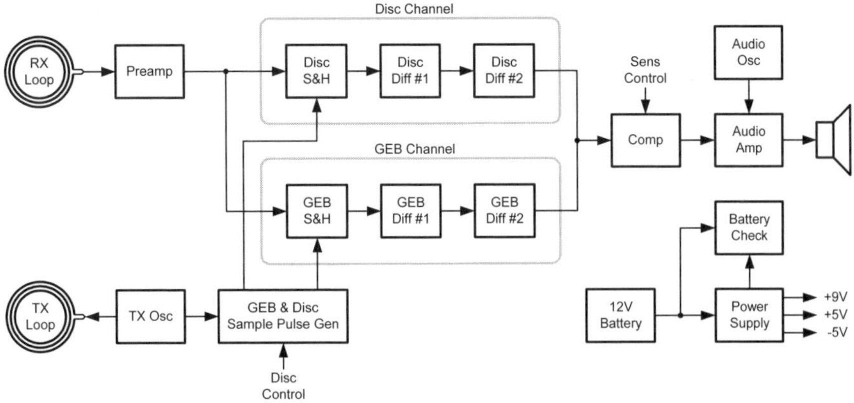

Fig. 9-21: **Block Diagram**

oro coil with an inductance of 5.9mH. The GEB and DISC sample pulse generator is adjustable to allow correct alignment of the DISC control on the front panel, and for variations that may occur if the home constructor wishes to make his own coil. The GEB adjustment is internal and is preset at the "factory". The power supply is stabilized, and provides 9V for the TX and audio circuits, and +/-5V for everything else. The result is a 2-channel switch-on-and-go silent-search detector with a double-differentiating topology that works well in the field.

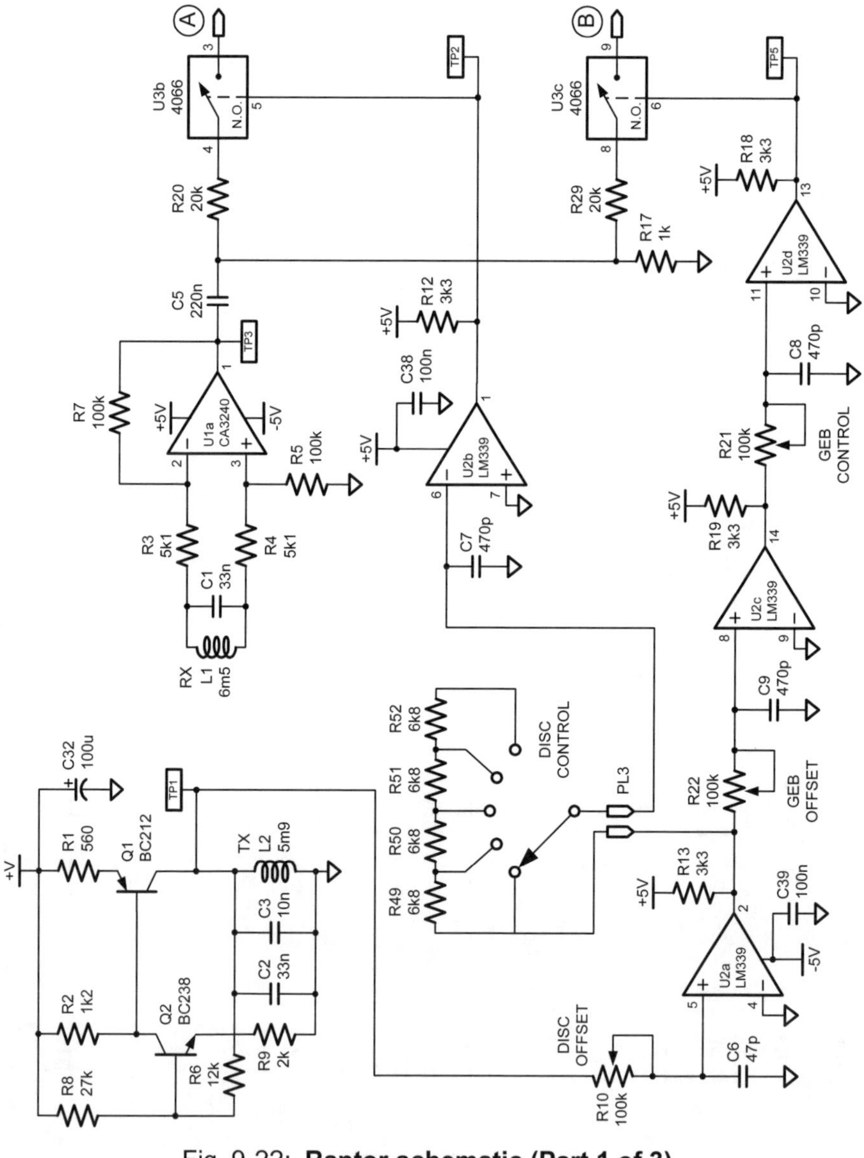

Fig. 9-22: **Raptor schematic (Part 1 of 3)**

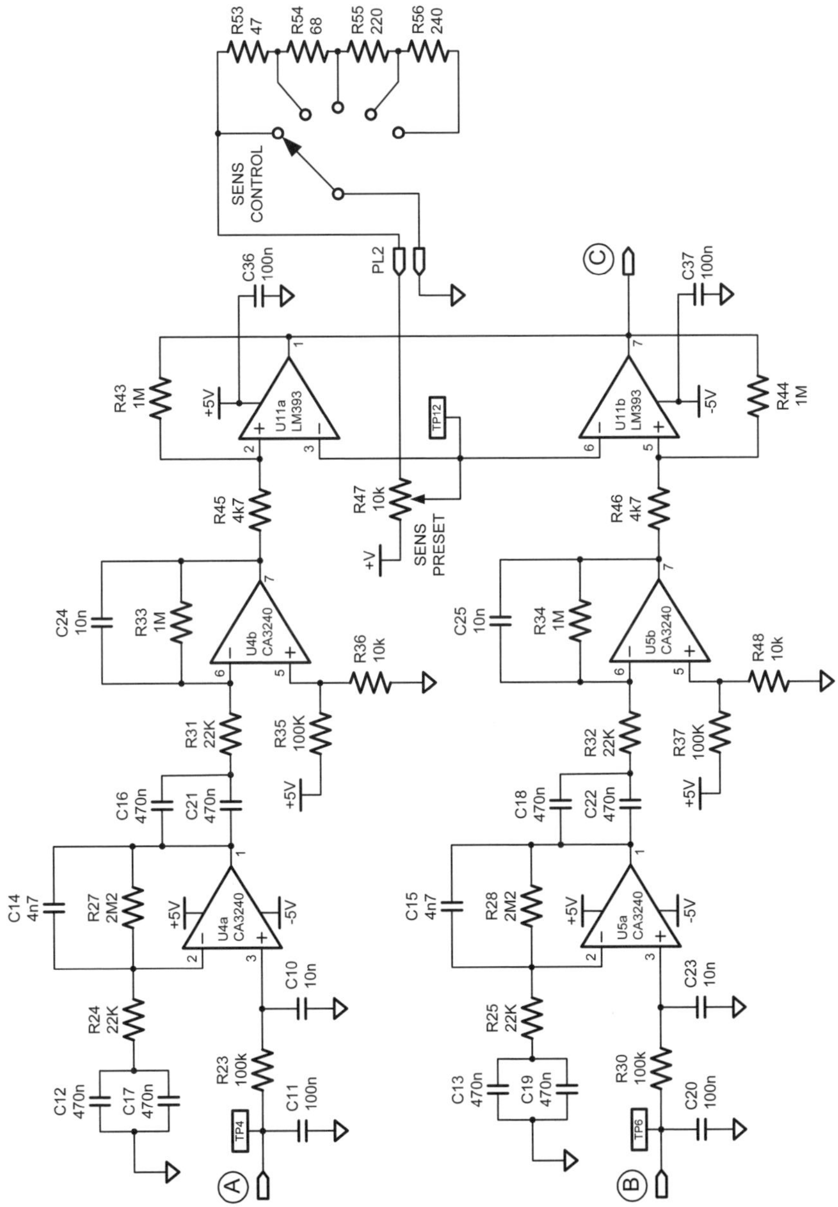

Fig. 9-23: **Raptor schematic (Part 2 of 3)**

The main goal of the Rapter project was to design and build a VLF metal detector using an inductively balanced coil, with the capability to exclude the ground effect which plagues T/R detectors. In addition the design was required to be relatively simple to construct, stable, and easy to calibrate. The Raptor project was also successful at producing a design that can easily compete with many commercially available analog metal detectors, and in some cases even exceed their capabilities.

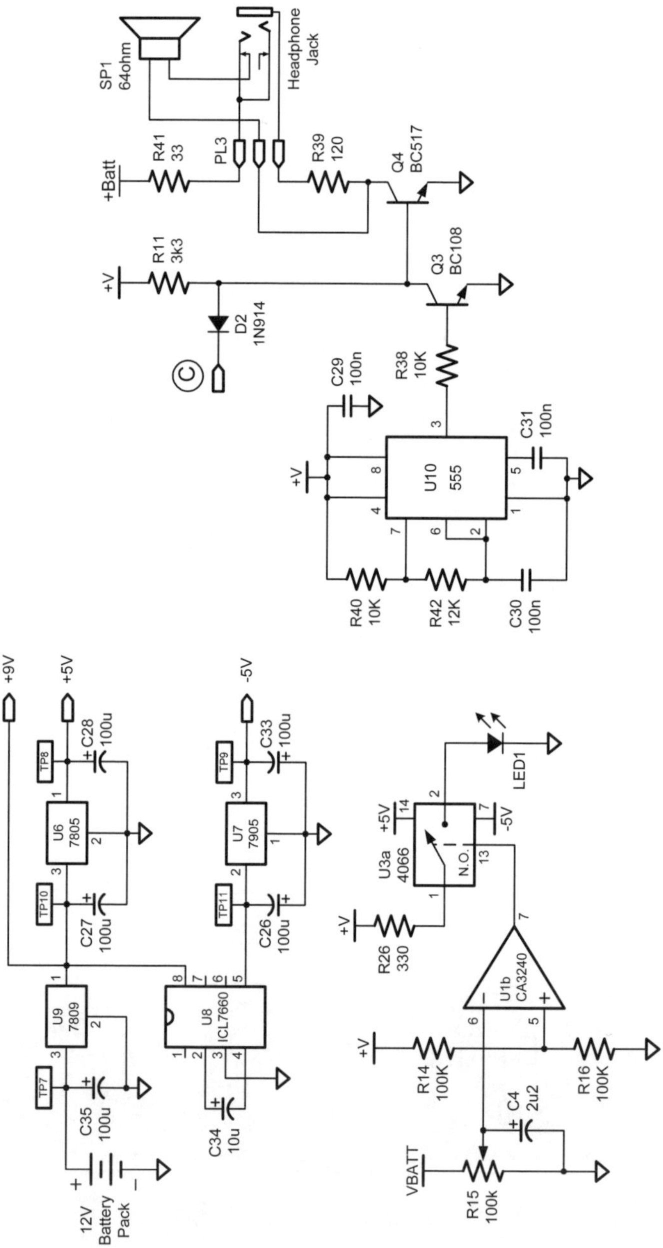

Fig. 9-24: **Raptor schematic (Part 3 of 3)**

Construction Details

The component placement and PCB layout are for illustrative purposes only. You may need to adjust the layout to accommodate components available in your area. In particular, please note that transistor pinouts can vary depending on the country of manufacture, even for what appears to be an identical part number. See Chapter 15 for more details.

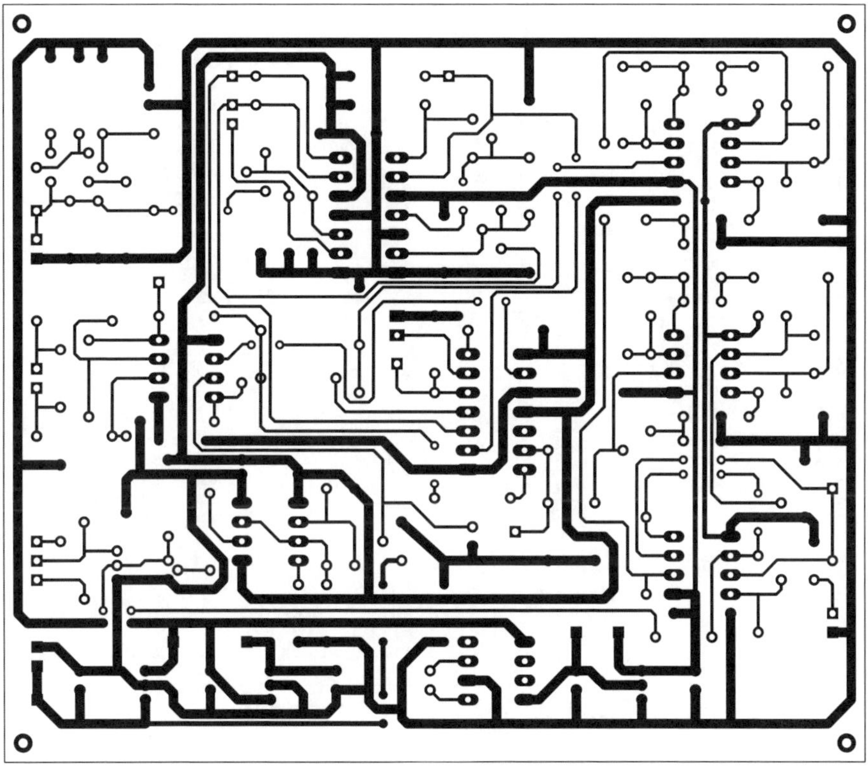

Fig. 9-25: **PCB Layout (single-sided)**

The PCB layout[1] view (Figure 9-25) is from the topside of the board (looking through). The actual size of the PCB is 4.6" x 3.95" (11.7cm x 10cm). Note that there are 20 jumpers required in this layout. The horizontal lines at the top and bottom of the component layout define the keep-out area for both components and PCB tracks. This is to allow the PCB to slide into the aluminum runners inside the Hammond enclosure. The parts placement (Figure 9-26) is shown from the top-side. The 3-dimensional view (Figure 9-27) provides an idea of what the final product will look like in real life. The connectors are designated as follows:

PL1 = Battery

PL2 = SENS control

1. Artwork should be at 1x scale.

PL3 = Headphone socket / speaker

PL4 = DISC control

However, in the prototype there were no connectors used. The wiring was soldered directly to the PCB.

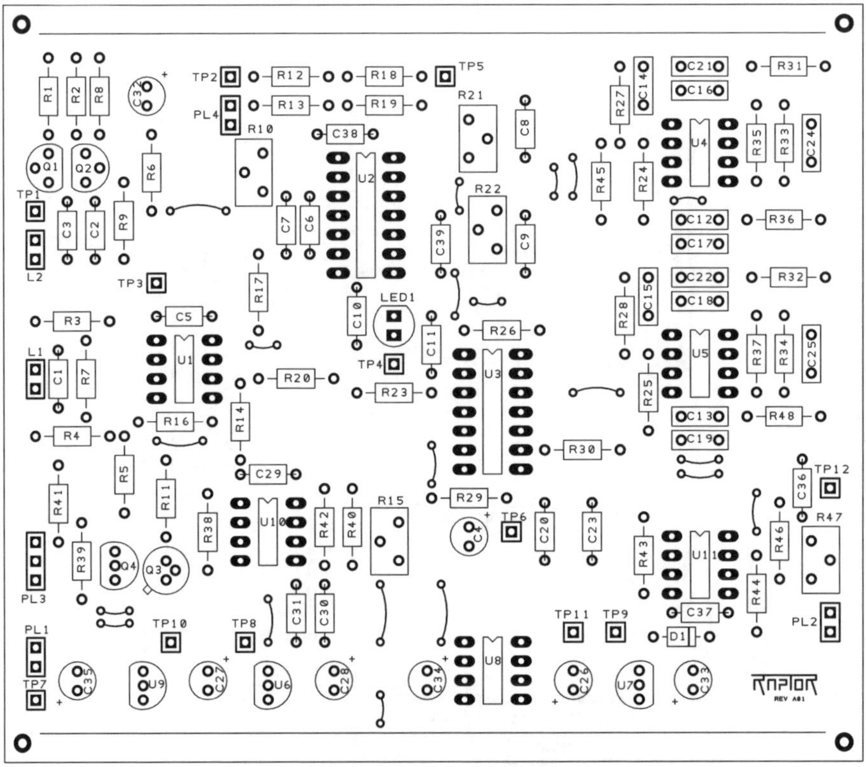

Fig. 9-26: **PCB Parts Placement**

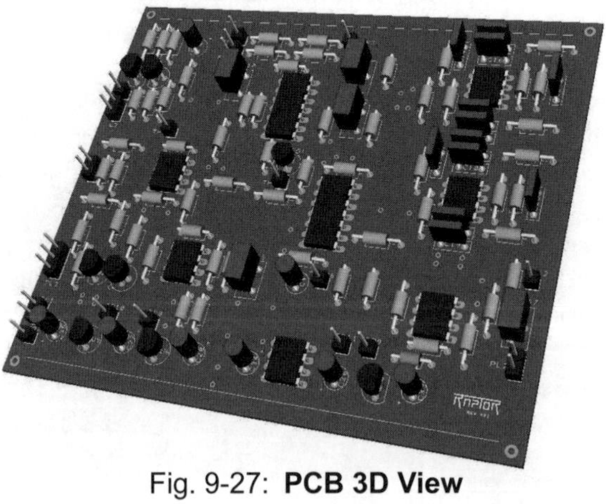

Fig. 9-27: **PCB 3D View**

Parts List

Resistors: (5% 1/4W)

R1	560
R2	1k2
R3, R4	5k1
R5, R7, R14, R16, R23, R30, R35, R37	100k
R6, R42	12k
R8	27k
R9	2k
R10, R15, R21, R22	100k preset (22-turn)
R11, R12, R13, R18, R19	3k3
R17	1k
R20, R29	20k
R24. R25, R31, R32	22k
R26	330
R27, R28	2M2
R33, R34, R43, R44	1M
R36, R38, R40, R48	10k
R39	120
R41	33
R45, R46	4k7
R47	10k preset (22-turn)
R49, R50, R51, R52	6k8
R53	47
R54	68
R55	220
R56	240

Capacitors:

C1, C2	33n
C3, C10, C23. C24, C25	10n
C4	2u2 elect., 16v
C5	220n
C6	47p
C7, C8, C9	470p
C12, C13, C16, C17,C18, C19, C21, C22	470n
C11, C20, C29, C30, C31, C36,C37, C38, C39	100n
C14, C15	4n7
C26, C27, C28, C32, C33, C35	100u elect., 16v
C34	10u elect., 10v

Inductors:

L1, L2	Search coil (see chapter 9 for details)

Transistors:

Q1	BC212 PNP

Q2	BC238 NPN
Q3	BC108 NPN
Q4	BC517 NPN (Darlington)

Diodes:

D1	1N4001
LED1	Red LED (with holder)

ICs:

U1, U4, U5	CA3240
U2	LM339
U3	CD4066
U6	LM7805
U7	LM7905
U8	ICL7660
U9	LM7809
U10	LM555
U11	LM393

Switches:

S1	6-way 2-pole (SENS / On-Off)
S2	12-way 1-pole (DISC)

(Note: Both S1 and S2 must be "make-before-break")

Misc:

Speaker	64 ohm (see text below)
Headphone socket	1/4" jack - plastic bezel (see text below)
IC sockets	(2) 14-pin, (6) 8-pin
Coil plug (control box)	5-pin locking chassis plug
Coil socket (from coil)	5-pin locking line socket
Control box	Hammond - extruded aluminium[2] with die cast aluminum[2] panels Approx. (6.3" x 4" x 2")
Control knobs	(2) 20mm ridged
Battery holder	To hold 8x AA alkaline batteries
Battery box	Suitable ABS enclosure
Test points	(12) single pins

Making the PCB

To make the PCB, first print the image onto toner transfer paper (iron-on) using a laser printer. This will not work with an inkjet printer.

Sand and degrease the copper board. This must be a bare board that is not covered with photoresist. Iron the paper onto the board. This takes around 3 to 5 minutes Put the board in a bowl of hot water and peel off the paper.

Then drill all the holes using a 0.8mm drill bit. You can use a 1mm drill bit for the plug, test point and battery connections. Repair any problem areas using a PCB

2. Depending on which side of the Pond you reside.

etch pen.

Finally put the board into ferric chloride solution. Once etching is complete, wash the board in water and use acetone to remove the toner.

Populating the Board

In order to minimise errors and mistakes during construction, please follow these instructions:

1. Fit and solder all the IC sockets and the 20 wire links.
2. Build the power supply, making sure the voltage regulators are fitted correctly, and observing the correct polarity for the electrolytic capacitors.
3. Connect the battery wires to PL1, also observing the correct polarity. The LM7809 regulator may be destroyed if you incorrectly connect the battery.
4. Measure the voltages at the following test points and check that they are correct:

   TP7 = +12V (battery voltage)
   TP10 = +9V
   TP8 = +5V
   TP11 = -9V (approx.)
   TP9 = -5V

5. Build TX circuit and connect coil. Connect scope to TP1 and confirm that the frequency is approximately 10KHz with a peak-to-peak amplitude of around 12V. Be very careful with the pinout of the transistors in the TX circuit, especially the BC212. The silkscreen shows a format of CBE, but some versions have a format of ECB. It would be wise to check each device in a transistor checker before soldering. If the TX oscillator refuses to oscillate, the fault is more than likely an incorrectly fitted transistor. See Chapter 14 for more details.
6. Insert the components in the pre-amp stage and battery check circuit, but leave U1 (CA3240) and U3 (4066) until last.
7. Monitor the signal at TP3 with an oscilloscope and confirm the amplitude is between 120mV and 250mV, depending on the coil.
8. Next test the battery check circuit by connecting the PCB to a bench power supply. Set the voltage to +12V, then monitor the TX output using an oscilloscope. Gradually decrease the supply voltage until the TX amplitude just starts to drop. At this point increase the supply voltage slightly and adjust the Batt Adj preset until the LED is illuminated. Returning the power supply voltage to +12V will turn off the LED.
9. Build the sampling circuits (based around U2 - LM339) and insert the input resistors and output capacitors to the sample gates (based around U3 - 4066). Leave U2 and U3 until last.
10. Temporarily short the two pins that connect to the DISC_CONTROL pot. This will be the same as turning the DISC_CONTROL pot fully anti-clockwise.

11. Monitor both the DISC_SAMPLE signal at TP2 and the RX pre-amp output at TP3 using an oscilloscope, and adjust the DISC_OFFSET preset until the pulse is positioned centrally over the positive part of the RX signal.
12. Set the GEB_CONTROL preset to its mid position. Then monitor both the GEB_SAMPLE signal at TP5 and the RX pre-amp output at TP3, and adjust the GEB_OFFSET preset until the pulse is positioned centrally over the positive-going zero-crossing of the RX signal.
13. Insert components for the differentiator stages that form the GEB and DISC channels, but leave U4(CA3240) and U5 (CA3240) until last.
14. Insert components for the comparator output stage. Fit U11 (LM393) last.
15. Temporarily short the two pins that go to the SENS control (max sensitivity). Then adjust R47 (SENS preset) so that the voltage at TP12 is 0.46V.
16. Monitor pin 1 of U11 while passing a non-ferrous object across the front of the coil at a distance of approximately 5cm (voltage will rise briefly to +5V). Repeat with a ferrous object (voltage will remain at -5V). This demonstrates the ferrous discrimination is working.
17. Fit the remainder of the components, and temporarily connect the speaker so that it is in series with R41 (33 ohms). A 64 ohm speaker is specified for this design, but these can be difficult to find. In the prototype design an 8 ohm mylar speaker was used, and the volume was found to be adequate. A mylar speaker is a good choice as it is more resistant to moisture. When the headphone jack is not inserted the speaker is placed in series with R41 (33 ohm). When using headphones both R41(33 ohm) and R39 (120 ohm) are connected in series to reduce the volume. However, if you are planning to use headphones with built-in volume controls, you should replace the 120 ohm resistor with a wire link.
18. After successful testing of the complete circuit, connect the headphone socket via flying leads to the PCB. See Figure 9-28 for details. Note that the headphone socket must have a plastic bezel if you are planning to use an aluminum front panel, otherwise it could short the +9V supply to ground via R41.

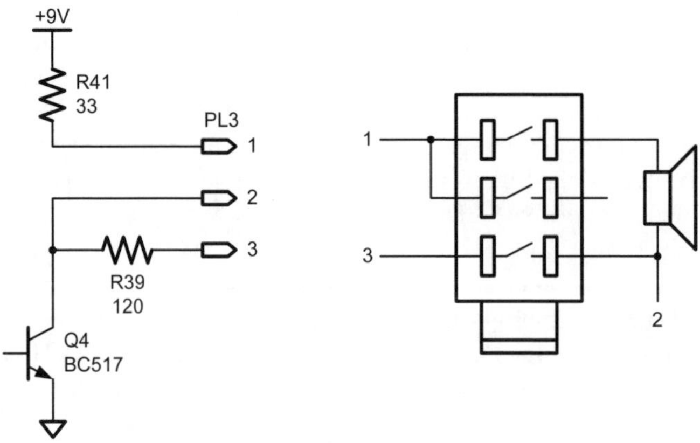

Fig. 9-28: **Headphone Socket Wiring**

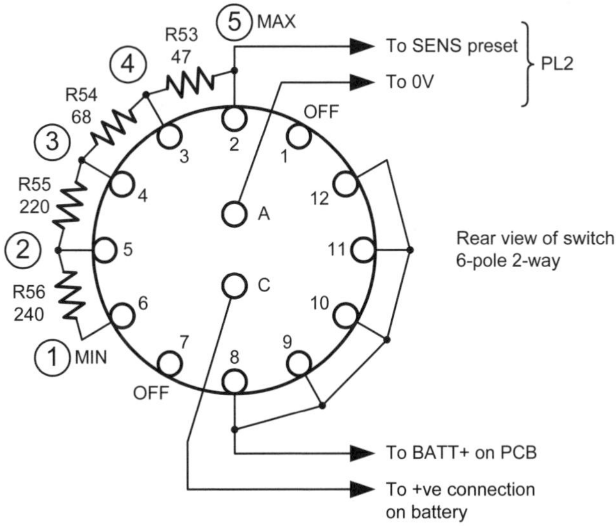

Fig. 9-29: **SENS Switch Wiring**

19. Solder the resistors to the SENS control switch and connect via flying leads to the PCB. Refer to Figure 9-29 for details.

20. Solder the resistors to the DISC control switch and connect via flying leads to the PCB. Refer to Figure 9-30 for details.

21. Assemble the whole detector, and adjust the angle of the search head to its normal operating position. Then repeat the DISC, GEB and SENS adjustments. At this point you must be using a battery and not a bench power supply, and insure that the aluminium enclosure is connected to ground (0V). This will allow you to set the detector to its maximum sensitivity without any unwanted oscillations.

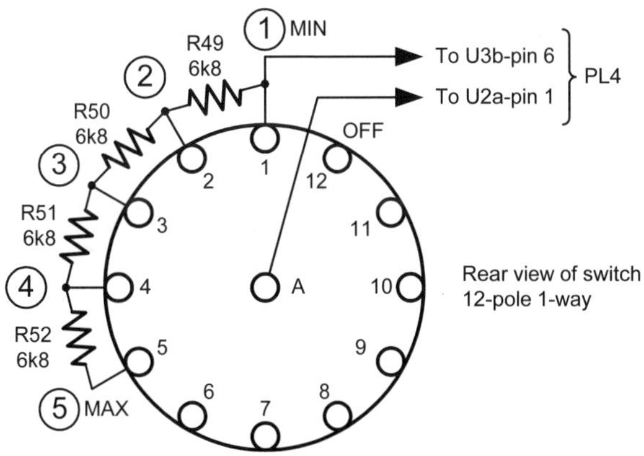

Fig. 9-30: **DISC Switch Wiring**

Notes on Switch Wiring

SENS switch - It is important to use a 6-way 2-pole make-before-break type with the limit stop set to position 6. Contact C is used as an On/Off switch. Contacts 8, 9, 10, 11 & 12 are shorted together. Contacts 1 & 7 are the OFF position. The SENS switch could be replaced with a switched 1K pot if desired.

DISC switch - Again, you need to use a make-before-break type, but this can be either a 12-way 1-pole, or a 6-way 2-pole. The prototype detector used a 12-way, with the limit stop set to position 5. The DISC switch could also be replaced by a pot, but this will result in some audible noise at the output when the control is adjusted.

Front Panel Design

A simple front panel was produced by printing the design shown in Figure 9-31 onto ordinary office paper, and then covering the paper with sticky-backed plastic. The left-hand hole is for the headphone socket. See Figure 9-32 for the final result.

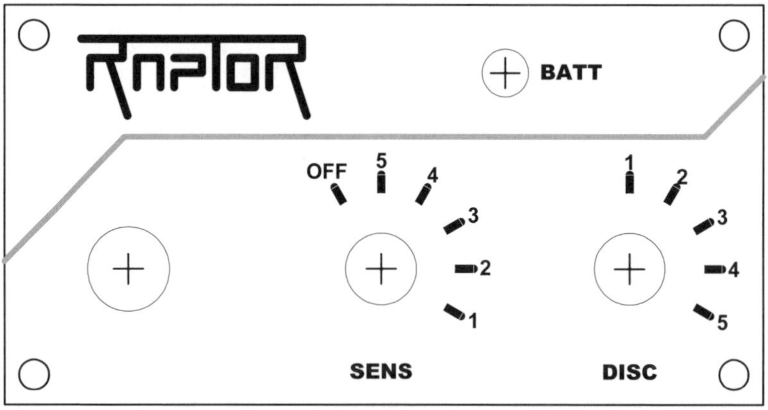

Fig. 9-31: **Front panel layout**

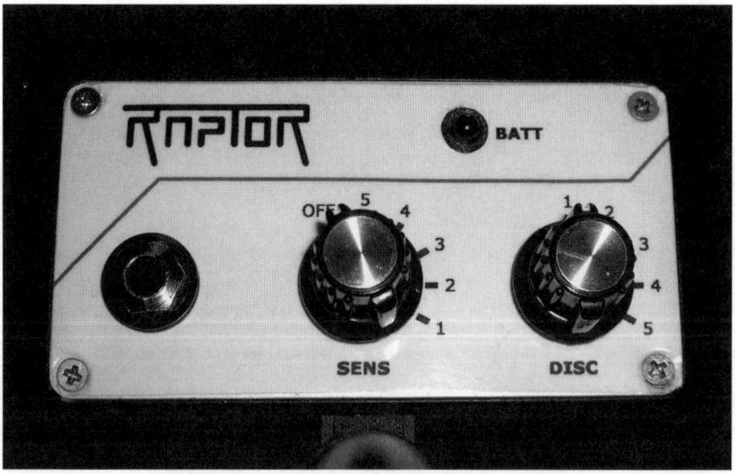

Fig. 9-32: **Completed front panel**

Conclusion

Many commercial detectors will demonstrate well in an air test, showing good detection distance and even good discrimination. However, the real test is out in the field, and especially in trashy or iron-infested areas. In that situation other factors come into play, and you could be missing targets amongst the clutter, or even digging deep holes and finding a large rusty nail. Raptor compares well with most mid-range metal detectors, and even some of the high-end models. It provides excellent stability, fast recovery time, good target separation, iron rejection and reasonable target depth.

The prototype detector (shown in Figure 9-33) was constructed using a Viking

Fig. 9-33: **Completed detector**

stem and arm rest, a Hammond aluminum extruded enclosure for the electronics, an ABS enclosure (located under the arm rest) for the battery pack, and a Tesoro 9x8 web coil. An inside view of the control box is provided in Figure 9-34.

The main purpose of the Raptor project was to act as a learning tool for VLF-IB detectors, and to demonstrate the principles behind the operation of an analog motion ground-cancelling detector. A wealth of information has been presented in Chapters 6 - 10 to enable the home constructor to understand all the relevant parts of the detector, and to allow further modification if so desired to achieve even greater performance. This could be achieved by either circuit modification and/or the creation of different coils. The choice is up to you.

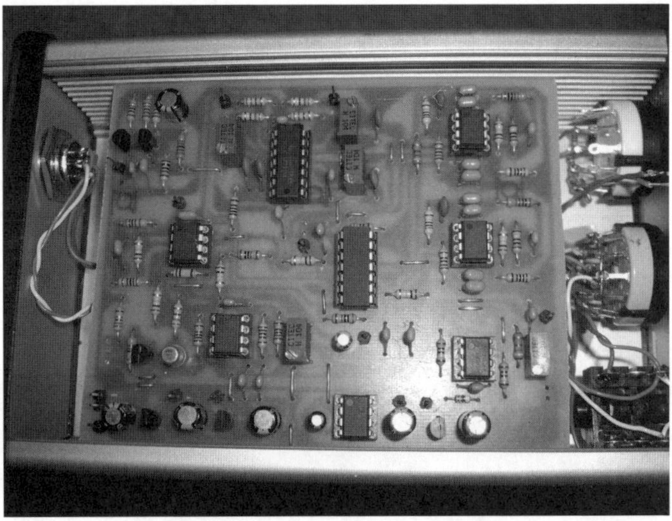

Fig. 9-34: **Inside view**

CHAPTER 10

IB Coil Experiments

"What I am going to tell you about is what we teach our physics students in the third and fourth year of graduate school... It is my task to convince you not to turn away because you don't understand it. You see my physics students don't understand it... That is because I don't understand it. Nobody does."

— Richard P. Feynman

There are two main parts to this chapter. In the first section we will investigate whether it is possible to use coils from manufacturers other than Tesoro with the Raptor IB design. In the second section we will perform some experiments with various homemade coils, using a number of different configurations.

Experiment #1: MD-3030 10" Concentric Coil

The MD-3030 is an inexpensive IB metal detector manufactured in China. It possesses a large (and somewhat heavy) 10" diameter concentric spider coil, as shown in Figure 10-1.

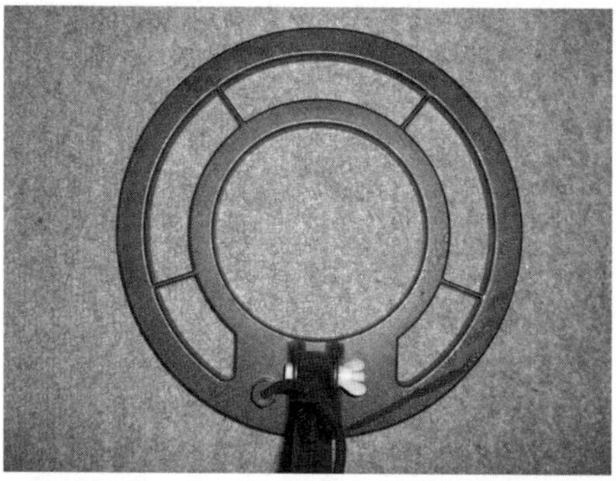

Fig. 10-1: **MD-3030 10" Diameter Concentric Coil**

The TX coil has an inductance of 2.6mH, which is much lower than the Tesoro's 5.9mH; and the RX coil has an inductance of 11mH, which is much higher than the Tesoro's 6.5mH. This will be a good test of the flexibility of the Raptor design, if it can utilize a coil with such widely different parameters. A diagram of the plug wiring is shown in Figure 10-2.

Note that there is a connection between the Green wire of the RX coil to the Screen. This connection is made in the socket of the MD-3030 located at the back of the control box.

The operating frequency of the MD-3030 is 5.7kHz, but we will attempt to use this coil at our preferred frequency of 10kHz. To achieve this, it will be necessary to increase the value of the TX tuning capacitor to 100nF. By now you should be able to

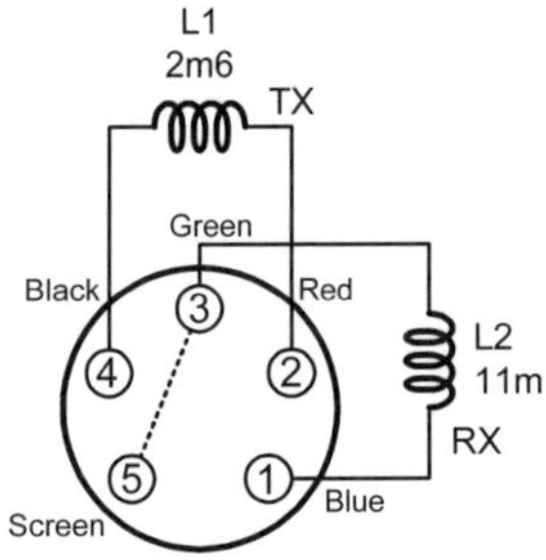

Fig. 10-2: **MD-3030 Plug Wiring (inside view of plug)**

confirm this is correct by using the equation given in Chapter 9. For the RX coil to also be tuned to 10kHz, the tuning capacitor needs to be 23nF. The nearest preferred value is 22nF, and this is the value that we will use in the initial experiment.

Unlike the Tesoro coil, there is no internal connection within the search head that connects one side of the RX coil to zero volts (screen). This is because the MD-3030 uses a forced oscillator to drive the coil, and this is connected in series with a small transformer, which provides the quadrature signals to the GEB and DISC sampling circuits. In this particular design, the TX coil is not connected directly to the zero volt line, and therefore it would not be possible to provide such a connection in the search head. However, in the Raptor design, the TX coil must be connected to zero volts, and consequently the screen of the connecting cable. So, for correct operation, we need to connect the screen to 0V, and this can be most easily achieved by connecting pins 4 and 5 together on the back of the socket. The original connection between pins 3 and 5 (shown in Figure 10-2 as a dotted line) is not present once the coil had been unplugged from the MD-3030 control box. See Figure 10-3 for the modified plug wiring.

Everything was connected as described, and the TX and RX caps replaced with their new values. The TX frequency was measured as 10kHz, but the behaviour of the RX signal in the presence of metal was found to be completely the reverse of the Tesoro coil. In this case, all metal targets caused a phase-shift to the right, non-ferrous targets gave a decrease in amplitude, and ferrous targets gave an increase. For the detector to work correctly with this coil arrangement, it then becomes necessary to do the GEB sampling on the negative-going zero-crossing of the RX signal, and the

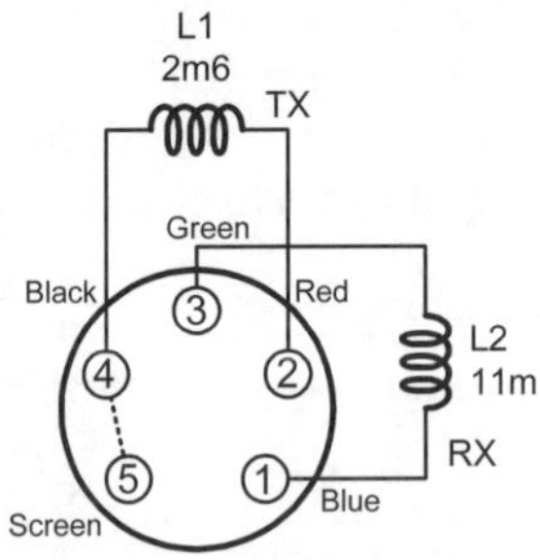

Fig. 10-3: **Correct wiring for Raptor (inside view of plug)**

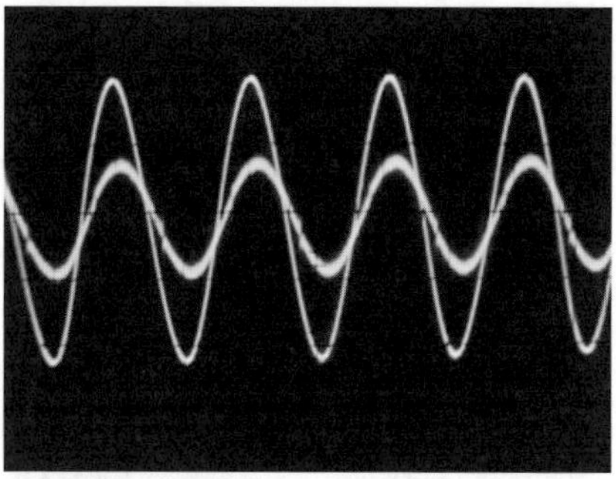

Fig. 10-4: **TX and RX Waveforms for MD-3030 Coil**

DISC sampling on the negative peak.

The reason the MD-3030 behaves differently is due to the way it was balanced at the factory. It is also likely that the nulling coil is in series with the RX, rather than the TX coil. The result is that pin 3 (green) must be connected to RX1 and pin1 (blue) must be connected to RX2. The desired TX and RX waveform relationship is shown in Figure 10-4. Note that the RX waveform is now inverted, when compared to Figure 9-18 in the previous chapter.

With the tuning capacitor we calculated earlier, the RX signal was found to have a phase offset, relative to the TX signal, that is within the range of the sampling circuit.

So how does inverting the RX signal, sampling on the negative-going zero-crossing, and the negative peak, produce the same results as before?

When a metal target crosses the face of the coil, there is a phase-shift to the right. Therefore it is necessary to sample at the negative-going zero-crossing in order to obtain a positive amplitude. Take a look at Figure 10-4, and note the position of this zero-crossing. Next imagine that the RX waveform moves to the right. You can then clearly see that there will be a positive voltage at the sampled position.

Since the non-ferrous targets now give a decrease in amplitude, it is necessary to sample the negative peak, because (when the amplitude decreases) this creates a less negative signal. i.e. a positive increase. Which is what we need for the discrimination to work correctly. You can follow a similar argument for ferrous targets.

The main advantage of using this particular coil is the price. It is possible to purchase a complete MD-3030 metal detector for less money than a Tesoro 9x8 web coil alone. On the disadvantage side, since the RX amplitude increases for ferrous targets, it does not have the same inherent rejection to small iron fragments as the Tesoro coil. In addition, the MD-3030 coil is much heavier.

However, it does work well, and represents a cheap alternative. Not forgetting that purchasing a complete detector would also supply all the required donor parts, such as an extendable stem, arm rest, plastic bolts, etc.

Experiment #2: Fisher M-scope 11" Concentric Coil

Although the Fisher search coils have inductance values that are reasonably close to the Tesoro coil (see Figure 10-5) there is an added complication that one of the

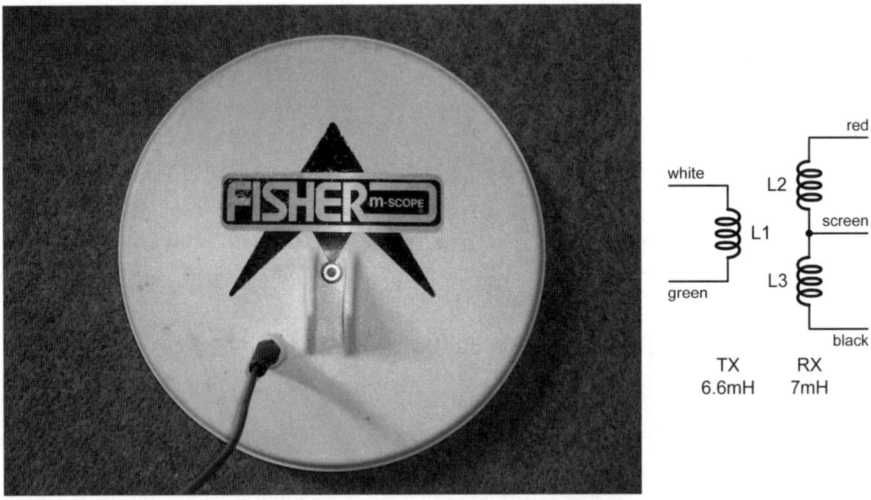

Fig. 10-5: **Fisher M-Scope 11" Concentric Coil & Details**

coils has a center tap that is connected to the screen.

In order to use this coil successfully we must connect the green wire of the TX coil to 0V, which will also be connected to the screen.

For a transmit frequency of 10kHz, we require a tuning capacitor of 38.4nF. The nearest preferred value is 33nF. This gives a transmit frequency of 10.8kHz, which is close enough. However, with an RX tuning capacitor of 33nF, the pre-amp is saturated, and the output is severely clipped. Even reducing the capacitor to 22nF, the pre-amp output is still 2.5Vpp, and sensitivity is very poor. Unlike the MD-3030, the Fisher coil is not well balanced at this higher frequency.

The m-SCOPE detectors usually have a TX frequency of between 5.5kHz and 5.9kHz, and we can easily get close to this value by using a TX capacitor of 100nF, giving a frequency of 6kHz. See Figure 10-6 for details of the final coil connections and capacitor values. The plug details are not provided in this instance.

Initially an RX capacitor of 100nF was tried, but the phase-shift between TX and RX was too large. By using two 47nF capacitors in parallel (to give 94nF) the phase-shift was correct.

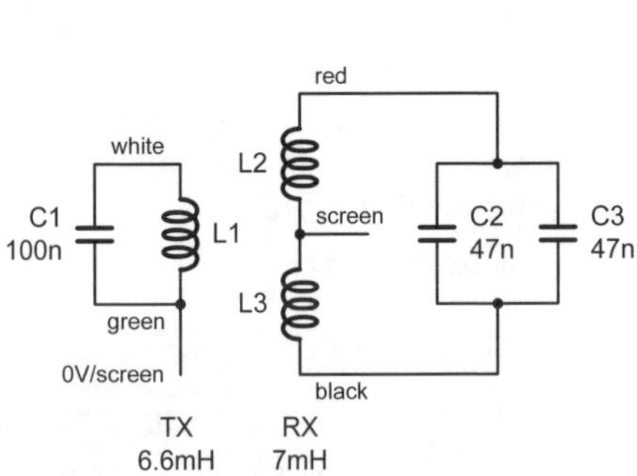

Fig. 10-6: **Fisher M-Scope Coil Connection Details**

This coil has the same characteristics as the Tesoro coil: non-ferrous targets cause an increase in amplitude, ferrous targets cause a decrease, and all targets produce a phase-shift to the left.

Although this 11" coil will work at 6kHz with the Raptor circuit, it is less sensitive than the 9x8 Tesoro web coil, and is much heavier. This also confirms our earlier findings from the experiment in Chapter 6, where we investigated the effect of different transmit frequencies versus various targets. With this coil there was a definite lack of depth in tests performed with small hammered silver coins.

Experiment #3: Garrett Crossfire 8" Concentric Coil

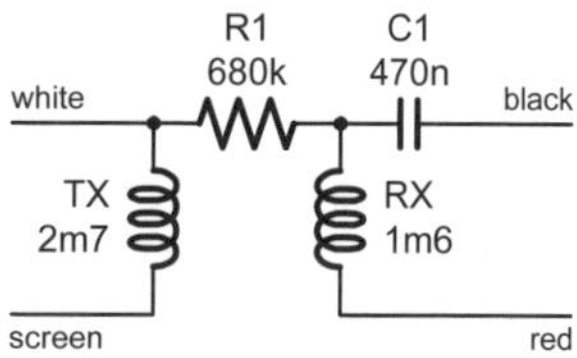

Fig. 10-7: **Garrett Crossfire 8" Concentric Coil & Schematic**

This coil is more complicated than the two coils we have already investigated, as it contains some passive components in the search head; see Figure 10-7.

It is not possible to use this coil, as supplied, with the Raptor circuit. The main problem is the presence of the capacitor in series with the RX coil. It is interesting to note that the Crossfire search head has the nulling coil in series with the RX. This is different to the Tesoro coil which has the nulling coil in series with the TX.

To continue with the experiment, it was first necessary to dismantle the search head and remove the passive components. In this case, the schematic was reduced to that shown in Figure 10-8. Dismantling the search head was easier said than done. This required cutting away the bottom cover with a sharp knife to expose the innards. An effective repair can then be made by gluing a coil cover in place over the original

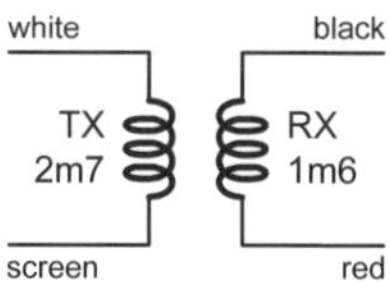

Fig. 10-8: **Garrett Crossfire Coil (with passive components removed)**

bottom cover.

Although much time was expended testing this coil, trying various frequencies, the balancing was such that it was not possible to achieve the required phase relationship between the TX and RX signals. Even when the phase offset was close to the desired value, it was found that non-ferrous and ferrous targets were producing phase-shifts in opposite directions. The result was that both types of target were either accepted or rejected, depending on which way round the RX coil was connected.

This was quite disappointing, as the Crossfire coil is well made, and very lightweight. Unfortunately the conclusion was that the Crossfire coil is totally incompatible with the Raptor circuit, and had to be abandoned.

Experiment #4: Troy 9" Concentric Coil

Fig. 10-9: **Troy 9" Concentric Coil & Schematic**

At first sight this coil could easily be mistaken for a Tesoro coil, but a quick test with the inductance meter shows that the coil values are considerably different. Of course, this should not be so surprising, as the Troy X3 is designed to operate at 19kHz.

As you can readily see, from Figure 10-9, the inductance of the TX coil is very low when compared to the Tesoro. Again, this evidence suggests that the nulling coil is in series with the RX coil. Unfortunately this means that the Raptor's TX oscillator fails to start when this coil is connected.

Which leads us to an interesting experiment. Let's swap over the coil connections, so that the 6.9mH coil acts as the TX, and the 1.1mH coil acts as the RX. In this case we can use the following values of tuning capacitor:

TX capacitor = 10nF, which results in a frequency of 19.2kHz.

RX capacitor = 47nF, 10nF and 4.7nF in parallel (61.7nF), so that RX coil is tuned to 19.3kHz.

This configuration (perhaps surprisingly) works reasonably well. Except that (like the MD-3030 coil) non-ferrous targets give a *decrease* in amplitude, ferrous targets give an *increase*, and all targets cause a phase-shift to the *right*. Consequently it does not have the same sensitivity to small silver items, or the inherent small ferrous trash rejection of the Tesoro coil.

The unusual discovery from this experiment is that it is possible to place the Tx coil inside the RX, and still obtain reasonable sensitivity.

Experiment #5: Viking 6DX Double-D Coil

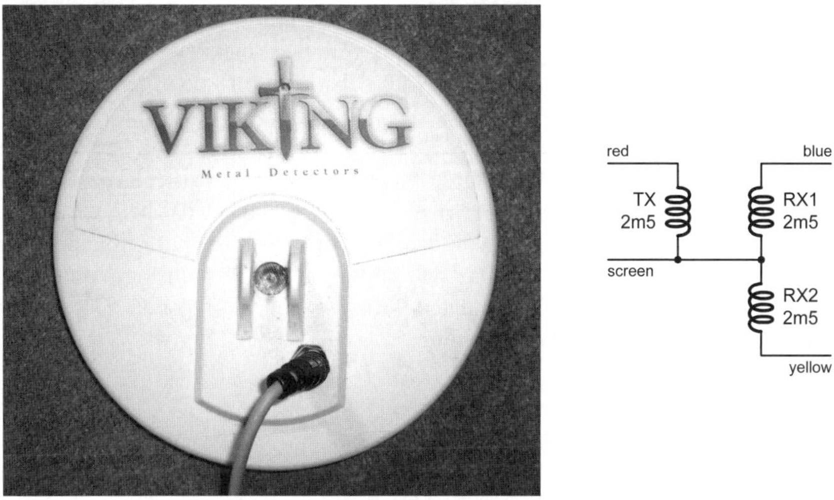

Fig. 10-10: **Viking 6DX Double-D Coil & Wiring (tuning capacitors not shown)**

This is the only double-D coil we will be experimenting with in this section. It is a complicated coil to decipher from measurement alone, without dismantling the search head. The RX coil is center-tapped, with the tap connected to the screen within the search head, in the same way as the Fisher m-SCOPE. The difference here is that one side of the TX coil is also internally connected to the screen. The other important point is that the tuning capacitors are built-in. There were no tuning capacitors present on the PCB, and unfortunately this means the inductance measurements may be inaccurate. Examination of the PCB also assisted in determining the correct coil configuration. We did not dismantle the coil during this experiment.

The TX frequency was found to be 22.6kHz. Non-ferrous targets gave an amplitude decrease, with a phase-shift to the right. Ferrous targets gave an amplitude increase, with a phase-shift to the left. This setup is not compatible with the Raptor circuit. If additional capacitance was added across the RX coil, the signal amplitude decreased, and the RX waveform was shifted further away in phase, relative to the TX, from its ideal position. However, inserting 20nF of additional capacitance across the TX coil moved the RX waveform closer to the ideal position, resulting in a TX frequency of 18.6kHz. The phase relationship was then such that all targets caused a shift to the right, non-ferrous targets caused an amplitude decrease, while ferrous targets caused an increase. With this setup the coil worked ok, but lacked sensitivity when compared to either the Tesoro or the MD-3030 coils.

If you wish to perform your own coil experiments, similar to those shown above, then you may sometimes notice that a target will give a double signal as it crosses the face of the coil, even at a distance of 4 inches (10cm). This is a clear indication that the RX connections need to be swapped over, although this does depend on the screen connection. In cases where both the TX and RX coils share a common screen connection, and the RX coil is not center-tapped, then there could be a problem.

It is also clear that it is more important to understand the correct relationship of the coil to the detector circuit, then it is to simply follow a "monkey see, monkey do" set of instructions. In other words, there is no substitute for knowing what you're doing.

It can be readily determined from these experiments, that not all coils perform the same, and there is a lot more to coil construction than meets the eye. There are several possible configurations, but only one of these leads to optimum performance when used with the Raptor design. Of course, one important question still remains, "Is it possible to create a homemade IB coil that will outperform the Tesoro?" This will be the subject of the second part of this chapter.

Read on ….

Constructing Homemade IB Coils

This is undoubtedly the most difficult and daunting part of building your own metal detector. It is also poorly understand by amateur constructors, and even some professionals. Hopefully you have already absorbed many interesting and important facts concerning coil construction, and will have a pretty good idea of the necessary requirements.

Before we get started, let's recap on the optimum setup required for the Raptor design:

1. All metal targets must produce an RX phase-shift to the *left*, relative to the TX signal.
2. Non-ferrous targets must give an *increase* in amplitude.
3. Ferrous targets must give a *decrease* in amplitude.
4. The RX signal (with no metal targets present) must be offset by approximately 200 degrees relative to the TX signal.

Depending on the design of a particular detector, these parameters can vary considerably.

Double-D Coil Experiment

For this experiment we used a 14" x 10" open-loop DD coil housing from Hays Electronics.

In order to determine the length of wire required for each loop, it was first necessary to lay a single length of wire inside the coil housing. This wire was removed and

Fig. 10-12: **14"x10" Open-loop DD Coil Housing from Hays Electronics**

the length measured and found to be 83cm (32.7"). If this was reformed into a circle of circumference X = 83cm, then the radius can calculated as 132mm. The reason for this is simply to enable easier winding of the coil.

Entering the following values into the coil calculator (Appendix A):

Inner radius = 132mm
Wire thickness = 0.2mm
Number of turns = 89

Which results in:

Mean radius = 132.94mm
Coil thickness = 1.89mm
Inductance = 6.8mH

After tying the coil with thread at a number of points, the inductance increased to 6.8mH. The coil was then placed in the housing, and reformed into an elliptical shape, which resulted in the inductance being reduced to 6.2mH. Once it had been determined that the inductance value was within tolerance and it would actually fit within the coil housing, the next step was to tightly bind the loop with insulation tape, and add a layer of aluminum tape, making sure to leave a small gap. Without this gap the electrostatic shield will look like a shorted turn, and drastically reduce the sensitivity of the detector. The final value of inductance was measured as 6.5mH.

A second identical coil was then wound and placed in the other section of the coil housing.

Balancing procedure:

1. Both coils were fixed in position within the housing, but with the two parallel sections in the center left free to move.
2. TX capacitor was calculated as 37.7nF. i.e., 33nF and 4.7nF in parallel, and RX capacitor calculated as 33nF.
3. The parallel sections were carefully moved together until they lay on top of each other. Then gradually moved apart, while the TX and RX signals were monitored on the oscilloscope. The windings were carefully adjusted until the RX signal had an offset of about 200 degrees relative to the TX.

4. The coils were fixed in this position.
5. Both GEB and DISC phase-sampling circuits were adjusted, as detailed in Chapter 9, so that the GEB was sampling on the negative-going edge of the RX signal, and the DISC was sampling on the negative peak.

Although we previously stated that sampling on negative-going zero-crossing gives inferior results to sampling on the positive-going zero-crossing, there is no other option available for the DD coil. However, there was found to be one positive benefit from this arrangement, which appears to be a quirk of the DD configuration. That is, non-ferrous targets give a larger phase-shift than ferrous. This means that the DD configuration is less sensitive to ferrous targets, even when used in all-metal mode.

Note: If you have performed the adjustments above, but are getting a double signal as the target crosses the coil, then you can be certain you have the RX connections swapped over.

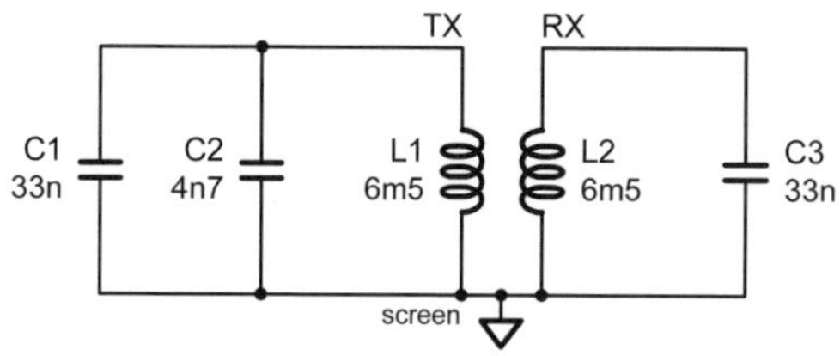

Fig. 10-13: **Wiring for Homemade DD Coil**

All targets should produce a phase-shift right. Non-ferrous targets give a *decrease* in amplitude, and ferrous give an *increase*.

When you first start to balance the coils it is quite difficult to find the correct position. You do need to persevere and have a lot of patience, but with careful and methodical work it is possible to produce a coil with similar sensitivity to the commercial Tesoro coil.

Unlike a concentric, the DD has its sensitive area along a central strip that runs between the front and the back of coil. This particular example had reduced sensitivity to small targets, probably due the larger diameter. The advantage of an elliptical DD is improved ground coverage and wider scan, without the necessity to overlap sweeps.

Hays Electronics supply plastic off-cuts with their coil housings that can be broken into small pieces and mixed with methyl ethyl ketone (MEK). This can be left to set for a couple of days and will result in a paste that can be used to seal the housing. It should be pointed out the MEK is a strong solvent, and can result in irritation to the eyes, nose and throat. So be very careful when using this technique. An alternative sealing technique is to use foam gap filler that can be sprayed from a can. This can be incredibly messy, and you should wear gloves when working with this material. It is also possible to use epoxy, but be aware that this takes a long time to set, and the coil will be much heavier than most commercial coils.

Apart from the obvious problems of aligning the coils correctly, and mechanical stability of the coil assembly, there is also the problem of making a search head that looks professional. Amateur constructors have made their own coils using many weird and wonderful techniques. A great number are simply made on a plywood base, sometimes sandwiched between two wooden boards, or even just covered in epoxy. These coils may work ok, but they lack that finished look. When using plywood, a more professional finish can be achieved by covering the coil assembly with a commercial coil cover. See the example at the end of this chapter.

Some people actually enjoy making their own homemade coils. If you are one of these individuals, then read on, as we explore the OO, concentric and omega coil configurations.

Double-O Coil Experiment

These coils are either referred to as double-O or OO coils. For this experiment we created two identical coils of 132mm radius, as in the previous DD experiment. However, in this case, there was no requirement to reform the coils into a different shape.

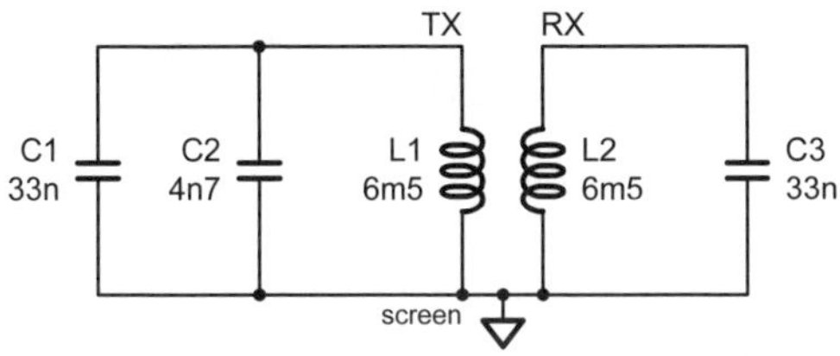

Fig. 10-14: **Wiring for Homemade OO Coil**

It was found to be relatively difficult to find the correct balance point between these coils, but with some patience it was possible, albeit somewhat tricky. As you might expect, the mode of operation was the same as for the DD. That is, all targets phase-shifted *right*, non-ferrous caused a *decrease* in amplitude, and ferrous caused an *increase*.

This type of coil is more difficult to make because of the non-availability of coil housings. It is claimed to have better ground exclusion than other types of coil, and in practice it did appear to provide better ferrous rejection. The only commercial metal detector company known to employ this coil configuration is Nexus Metal Detectors.

Fig. 10-15: **Example of Commercial OO Coil From Nexus**

Concentric Coil Experiment

The concentric coil configuration is the most difficult for the home constructor. Although there are coil housings available from Hays Electronics, it is the nulling coil that causes the most confusion. The outer coil is the transmit, and the inner coil is the receive. There is also a nulling coil that lies on top of the receive coil, which is electrically connected in series with the transmit coil, but in anti-phase.

Fig. 10-16: **8" Concentric (Spider) Coil Housing from Hays Electronics**

The Hays coil has two circular channels. One for the outer TX coil, and one for the inner RX and nulling coils. Each channel is approximately 0.5" wide. In terms of metric measurements, a TX coil with an inner diameter of 184mm, and an RX coil of 80mm will fit comfortably within the coil housing, allowing for the thickness of each set of windings.

First we need to determine the number of turns for the TX and RX. Again we can make use of the coil calculator, as follows:

For the TX:

Inner radius = 92mm
Wire thickness = 0.2mm
Number of turns = 105

Which results in:

Mean radius = 93.02mm
Coil thickness = 2.05mm
Inductance = 6.055mH

And for the RX:

Inner radius = 40mm
Wire thickness = 0.2mm
Number of turns = 186

Which results in:

Mean radius = 41.36mm
Coil thickness = 2.73mm
Inductance = 6.479mH

So far so good, but it is at this point that most home constructors encounter problems. The trick to determining the correct number of turns for the nulling coil will now be revealed. You need to first find the ratio between the TX inner-radius-squared and RX outer-radius-squared. In this particular case:

ratio = $92^2 / 40.2^2 = 5.24$

Therefore the number of turns for the nulling coil is 105 / 5.24 = 20 turns. By using the calculator it is easy to determine that the inductance of the nulling coil will be 103μH.

Of course, the TX and nulling coils are in series, so this makes a total inductance of 5.9mH. Note that the nulling coil inductance must be subtracted from the TX inductance, as they are connected in anti-phase. Also, you need to use the outer radius of the RX coil for the nulling coil calculation, because the nulling coil is wound on the outside of the RX. Finally, make sure to leave enough extra wire to form a movable loop to assist with the coil balancing procedure.

All that is left to do is balance the coil, as described previously, and noting that all targets phase-shift *left*, non-ferrous items give an *increase* in amplitude, and ferrous items give a *decrease*.

The concentric coil works very well with the Raptor design, and provides accurate pinpointing, and improves the ability to hunt in iron-infested areas.

Omega Coil Experiment

The last coil configuration we will examine is the omega coil, which is sometimes referred to as the 4B.

For this type of coil you must use a closed coil housing, such as shown in Figure 10-17. It performs in a similar way to the concentric, but the sensitive area is slightly offset away from the center. Please also refer to Figure 5-18 for a visual description.

This configuration was popular with some Radio Shack and Heathkit detectors. With the Heathkit designs, the RX coil was wound onto a movable disk, that allowed it to be easily positioned during the coil setup procedure. You can see this arrangement in Figure 10-18, which shows the inner workings of a Heathkit GD-348 metal detector search head.

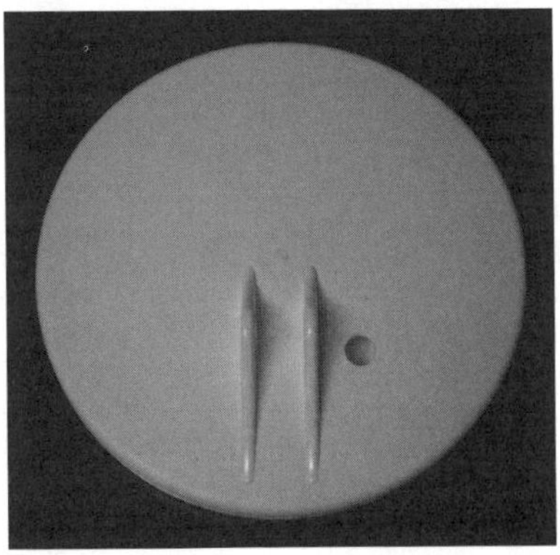

Fig. 10-17: **8" Closed Coil Housing from Hays Electronics**

In some ways the omega coil is similar to the concentric. The difference lies with the nulling coil. In the omega the nulling coil is offset from the center and has the same number of windings as the TX. This can make it tricky to determine the correct number of windings required. If, for example, we use the same TX coil as for the concentric, this would result in a nulling coil with an inductance of 2.1mH, which means a total inductance of 8mH. So how can we solve this?

In practice it is not easy to perform this calculation, and unfortunately it comes down more to trial and error. However, the omega is definitely easier to balance than the concentric coil, and can be a useful alternative for the amateur constructor.

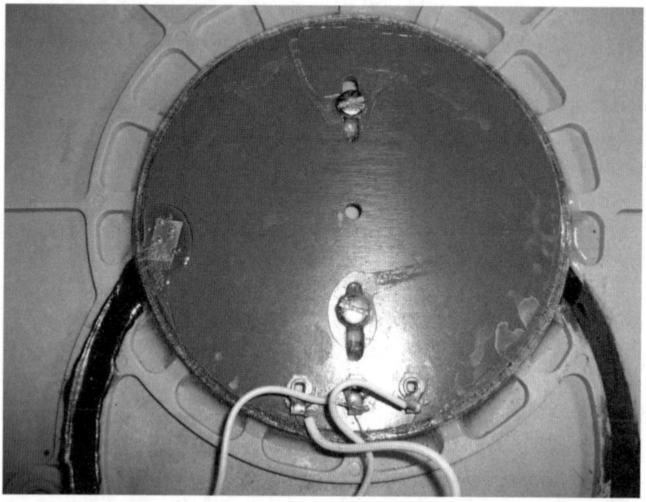

Fig. 10-18: **Adjustable Nulling Coil in Heathkit GD-348 Search Head**

This coil responds in the same way as the concentric. That is, all targets phase-shift *left*, non-ferrous targets give a *increase* in amplitude, and ferrous give a *decrease*.

Conclusion

If you enjoy a challenge, then you should certainly enjoy building your own coils. However, it is not for the faint hearted, impatient, or non-methodical. Second-hand Tesoro coils can be purchased at a reasonable price, either from a metal detecting shop, or on the Internet. If you are at all uncertain about constructing and testing electronic equipment, you should most certainly take this approach. Otherwise it will be difficult to fault find the problem if you are not confident about the coil.

Step-by-Step Instructions For Constructing a DD-Coil

There will inevitably be those people who still insist on building their own coils, despite the warnings given previously. This may simply be because Tesoro coils are not available locally, coils prices are outside the project budget, or they just want to see how it's done. Whatever the reason, here is a set of step-by-step instructions that you can follow to build a DD-coil from scratch using low-cost materials.

First acquire a solid coil cover. Either purchase one new, or go to your local metal detector shop, as they will often have a "bucket of bits" where you can find one for a cheap price. You will need one that is capable of holding a 20cm diameter (slightly under 8") coil. The one used in this example was 21cm (see Figure 10-19). The manufacturer is unknown.

Fig. 10-19: **Coil Cover**

Next cut a wooden circle from a sheet of 5mm thick 3-ply that will fit tightly inside the coil cover. This will form the top section of the search coil. You could substitute 5mm thick MDF (medium density fiberboard), if you prefer, as it may be less susceptible to twisting.

Using a separate wooden base, draw a circle of 20cm diameter, and mark out the D-shape according to Figure 10-20. Using either nails, right-angled cup hooks or wooden doweling, wind the coil into the D-shape using 114 turns of 0.2mm enamelled wire. This first coil will be the TX.

Secure the windings tightly in several places using electrical insulation tape. See Figure 10-21.

With reference to Figure 10-22, wrap the wire tightly with small strips of self-adhesive aluminum tape. Make sure to leave a gap, otherwise the aluminum tape will act as a shorted turn instead of an electrostatic shield.

Loosely spiral wind the coil (see Figure 10-23) with tinned copper wire. The wire used in this example was 22SWG (0.71mm). Again, make sure there are no loops created during the winding process. Leave a 2cm tail of wire at the end, as it will eventually need to be connected to the screen (0V). This is known as the drain wire.

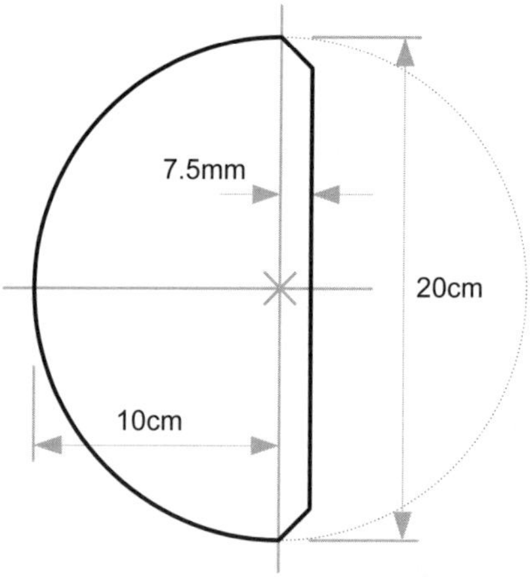

Fig. 10-20: **Coil Details**

Finally wrap the entire coil with Gaffa (Gaffer) / Duct tape. This is a strong, tough, fabric-based sticky tape. See Figure 10-24.

Repeat this operation for the RX coil, using 119 turns of 0.2mm enamelled wire.

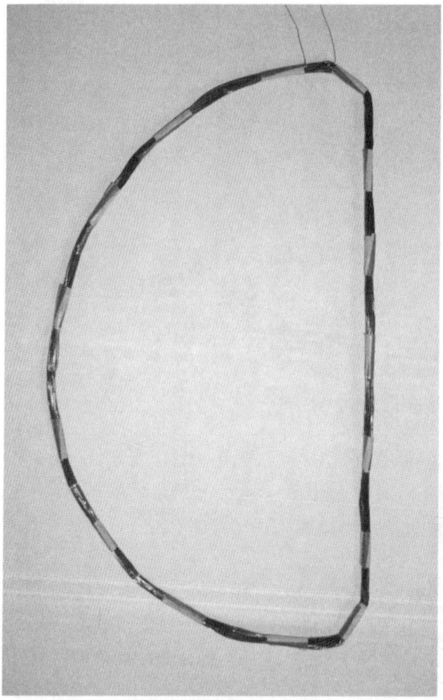

Fig. 10-21: **Coil wound with electrical tape**

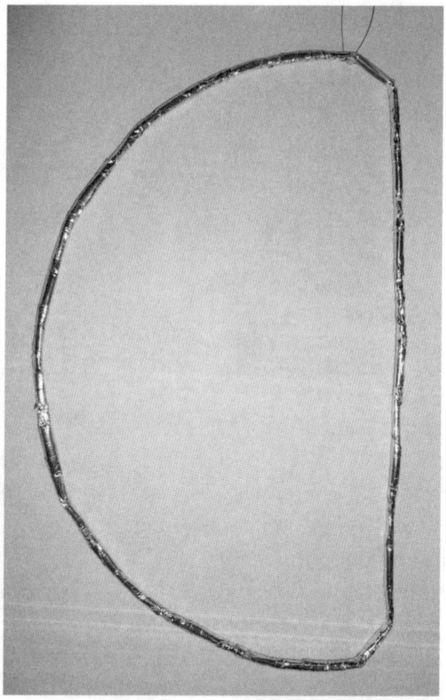

Fig. 10-22: **Coil wound with electrostatic shield**

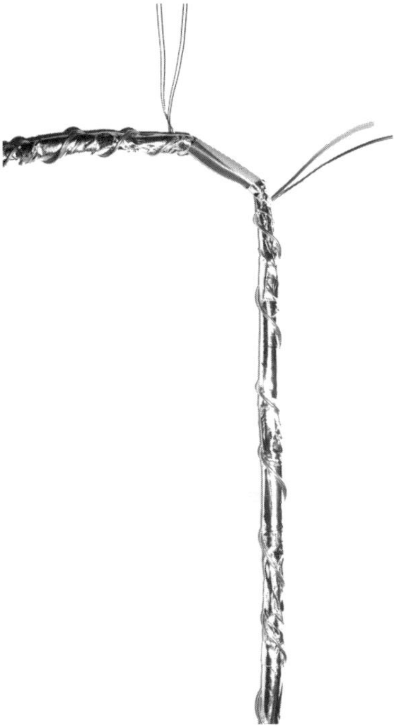

Fig. 10-23: **Detail of drain wire**

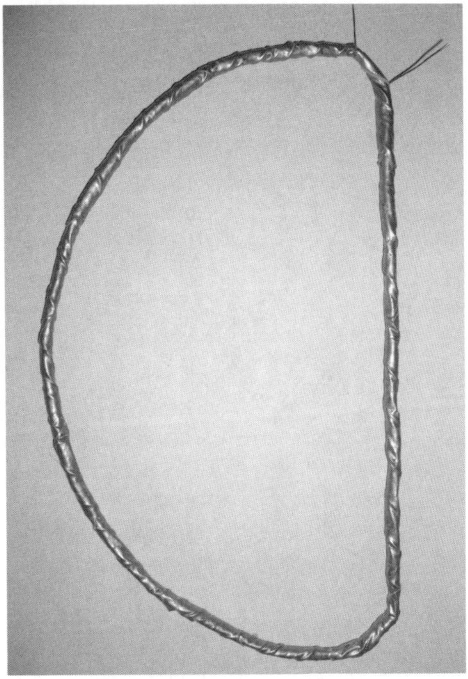

Fig. 10-24: **Coil wound with final layer of Gaffa/Duct tape**

Hopefully you should now have two D-shaped coils. Mark each one so that you can identify which one is which. Then construct the mounting bracket, that connects to the detector stem, according to Figure 10-25. These should be made from 5mm thick MDF.

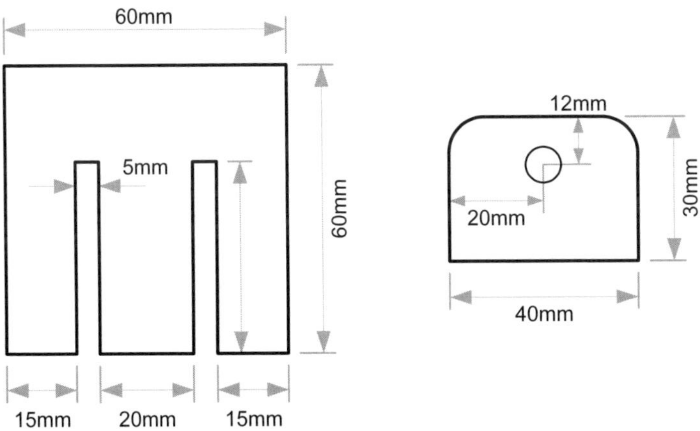

Fig. 10-25: **Mounting bracket base and lug (x2), 5mm MDF**

Fig. 10-26: **Mechanical parts required for search head**

Other mechanical parts required are (see Figure 10-26):

- 1x plastic bolt, washer and wing nut
- 1x cable entry sleeve
- 2x rubber washers
- 4x coil cover spacers

Depending on the internal depth of the coil cover used, either find or make 4 spacers to support the plywood top. For the example coil shown here, it was found that a small plastic shelf support was the exact size required. These were glued into the base of the coil cover. See Figure 10-27.

Assemble the mounting bracket and glue it onto the plywood circle, as shown in Figure 10-27. Finally, paint the whole assembly to match the coil cover.

Now the mechanical construction is complete, it's time to fix the TX and RX coils in position. Using a hot glue gun, fix the TX coil first. Once the glue has set, solder the connecting cable to both coils and connect to the detector. Leave a small loop of wire at the end of the RX coil to allow some final adjustment prior to sealing the search head. Remember that both the TX and RX coils share a common connection, as described in Chapter 9, and shown in Figure 9-2. Remember to connect the drain wires to the screen.

Then follow this procedure:

1. With the TX and RX coils slightly overlapping, monitor both the TX signal (TP1) and the RX pre-amp output (TP3) using an oscilloscope.

Fig. 10-27: **Coil cover with spacers**

Fig. 10-27: **Mounting bracket complete**

2. Gradually move the RX coil towards the TX coil until the RX signal reduces to a minimum.
3. Continue to move the coils together until the RX signal has an initial phase-shift of 20 degrees (approximately 6μs). This is clearly shown in Figure 10-4. If you cannot achieve a 20 degree phase-shift try reversing the RX leads.
4. Fix the RX coil in this position using hot glue.
5. Solder the drain wires to the screen, as shown in Figure 10-28. At the 20 degree position the residual voltage should be somewhere between 200mV and 400mV peak-to-peak when measured at TP3, but it may go as high as 1V, depending the actual coil shapes.
6. Follow the directions near the end of Chapter 9 to calibrate the detector for use with your coil.
7. If you wish, you can fix the coils more firmly in position by using epoxy glue, but hot glue was found to be adequate for the purpose. Figure 10-29 shows both the TX and RX coils fixed in their final positions.
8. Once you have successfully calibrated the detector, attach the coil cover and make it waterproof using a silicon sealer.

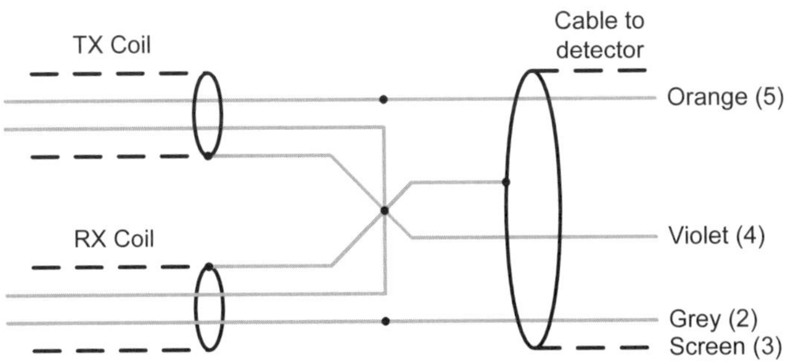

Fig. 10-28: **Search head wiring**

The fully assembled search head is shown in Figure 10-29, attached to the detector stem.

The internal glued coils might not look very pretty, but the result was a stable search head that provided some reasonable results. It was found that our DD-coil had a slightly reduced performance when compared to the Tesoro 9x8 web coil. It was also noted that the Tesoro concentric was better at pinpointing than the DD, and was the preferred coil to use in trashy and iron-infested areas.

You may also like to experiment by changing the shape of the overlap between the TX and RX coils. With the parallel overlap, shown in this example, the sensitive area is a thin strip between the front and rear of the search head. One advantage of this approach is that you do not need to overlap consecutive sweeps as you must do with the concentric. The main disadvantage is a less than perfect pinpointing capability.

It takes a lot of time, energy and nerves to construct your own coils, but the result can be quite satisfying. As we mentioned earlier, the problems usually start when both the coil and electronics are unproven. If you still feel up to the challenge, then good luck with your experiments.

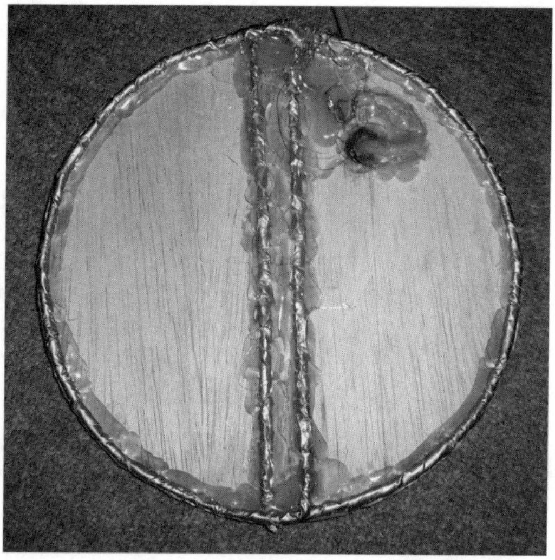

Fig. 10-29: **Coils glued in position**

Fig. 10-29: **Search head mounted on stem**

CHAPTER 11

Pulse Induction Principles

"And now I see with eye serene,
the very pulse of the machine."
— William Wordsworth

So far, we have investigated two methods of designing a metal detector: BFO, and IB. Both of these generally employ a continuous wave signal—usually a sinusoid—and both methods look for shifts in the signal: a frequency shift for BFO, and an amplitude or phase shift for IB. Since frequency, amplitude, and phase are all characteristics of the frequency domain, these are considered to be *frequency domain* methods[1]. There is a third popular metal detector design, one in which the detection takes place in the *time domain*, using pulses.

PI Transmit

As we saw in Chapter 5, a changing magnetic field produced by the search coil induces eddy currents in metal targets, and the counter-magnetic field of the eddy currents can be detected. BFO and IB detectors commonly use a sinusoid to generate the changing magnetic field, but any time-changing waveform will do[2]. In fact, the faster the magnetic field changes, the greater will be the induced eddy currents:

$$i = -\frac{1}{R} \cdot \frac{dB}{dt} \quad \text{Eq 11-1}$$

It stands to reason, therefore, that a fast-slewing signal, such as a pulse, will produce a stronger changing magnetic field than, say, a sinusoid.

The pulse induction, or PI, detector does just that. The coil is briefly energized with a DC current which produces a static magnetic field, then the current is quickly shut off which collapses the magnetic field. The higher the coil current and the faster it is shut off, the higher the magnetic field transient dB/dt will be. Figure 11-1 shows a circuit which creates a high coil current and a fast turn off.

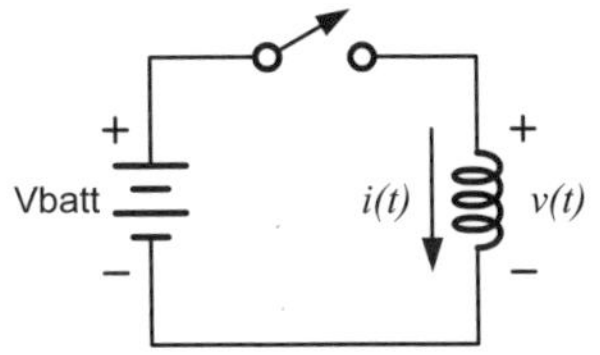

Fig. 11-1: **Switched Coil**

1. Single-frequency IB can also be considered a *phase-domain* technique.
2. For example, Fisher's CZ-series detectors use a 5kHz square wave. A square wave has strong odd-order harmonics, so the 5kHz fundamental and the 15kHz 3rd harmonic are used in the phase demodulators, effectively producing a 2-frequency detector.

When the switch is closed, the battery is connected directly across the coil. An ideal battery has zero resistance, and an ideal coil has zero resistance, so in the ideal world this would produce an infinite current through the circuit. However, real batteries and real coils have some amount of resistance[3], so the current will be limited to perhaps a few amps. That's still a lot of current, and you only want to apply a battery directly across a coil for a very, very short time.

As the switch closes the coil current does not immediately jump up to a DC level, but exponentially builds up to a peak value, following a classic RL response determined by the inductance of the coil, and the non-ideal series resistances of the coil, the switch, and the battery. At the same time, the coil voltage initially jumps up to the same value as the battery, but then exponentially decays to some lower level determined by the resistances of the coil, switch, and battery. Figure 11-2 illustrates what happens at turn-on.

With the switch closed, the current through the coil generates a magnetic field around the coil, just like the electromagnet experiment from grade school science. When the switch is opened, the current suddenly drops to zero (the turn-on time constant does not apply), and collapses the magnetic field creating a high dB/dt. It so happens that, because the coil is a simple inductor and inductors resist sudden changes in current, the high-slewing current at turn-off also generates a large voltage spike across the coil[4]:

$$v = L \cdot \frac{dI}{dt} \qquad \text{Eq 11-2}$$

This is called the "flyback" voltage and, depending on the charging current and the

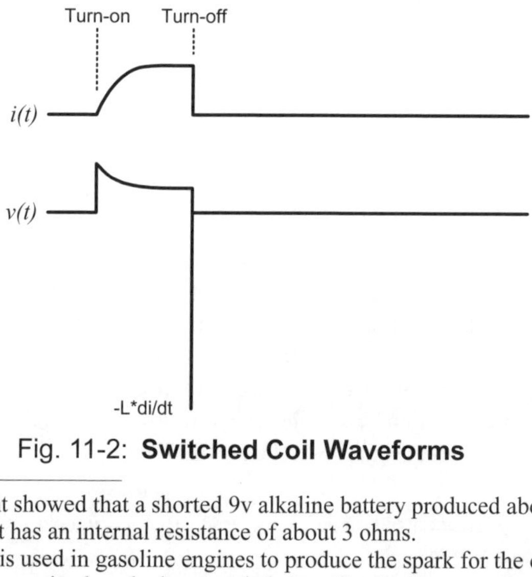

Fig. 11-2: **Switched Coil Waveforms**

3. A quick experiment showed that a shorted 9v alkaline battery produced about 3 amps, which means that it has an internal resistance of about 3 ohms.
4. This same method is used in gasoline engines to produce the spark for the combustion chamber. A coil is energized, and when a switch (e.g., "points") opens, the resulting high voltage (boosted by a transformer) creates a spark across the gap of the spark plug.

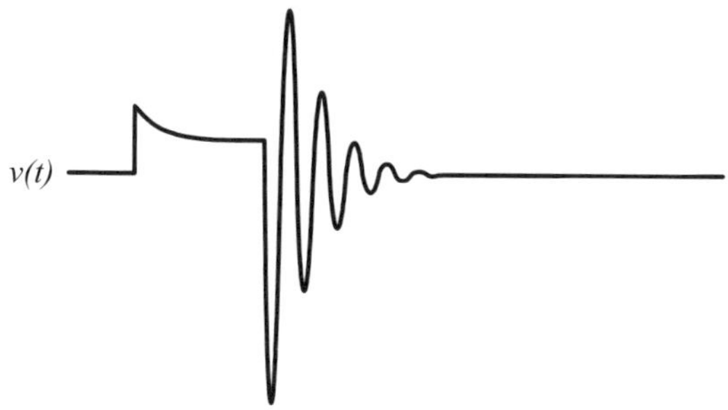

Fig. 11-3: **Undamped Waveform**

speed at which it is turned off, it can be as large as a few hundred volts[5]. For an ideal inductor, the flyback voltage is an instantaneous spike that ends abruptly when the coil current reaches zero. Again, Figure 11-2 illustrates the turn-off reactions.

Because of the way the circuit (Figure 11-1) has been drawn, the flyback voltage has a negative value; in some designs the waveform is inverted so that the flyback is positive. Although an ideal inductor has an instantaneous spike, a real inductor is far more complicated, and has series wire resistance, plus capacitance between windings. If you were to build the circuit in Figure 11-1 and look at the flyback on an oscilloscope, you would see a waveform like that in Figure 11-3. The ringing caused by the coil's distributed inductance and capacitance will follow an exponential decay determined also by coil resistance.

While it might be possible that this decayed ringing is useful for detecting metal, it is not used in traditional pulse induction designs. Instead, a damping resistor is added in parallel with the coil (Figure 11-4) to get rid of the ringing and provide a nice, smooth exponential response. The value of Rd is chosen (usually by trial-and-error) to give a response that is close to critically damped. While the charging current follows an exponential curve set by the inductance and (total) series resistance, the

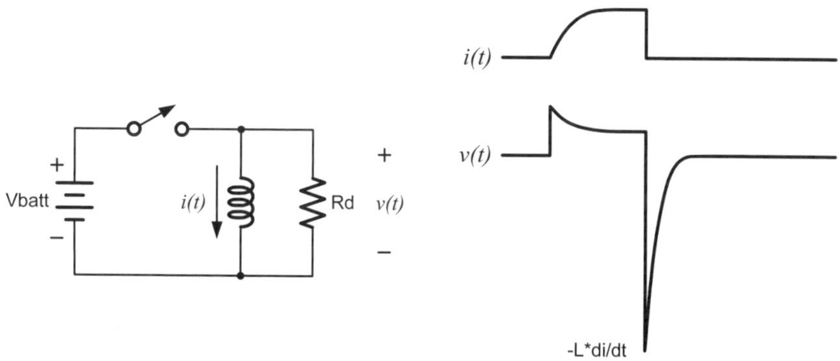

Fig. 11-4: **Damped Coil & Waveform**

5. Or, in the case of a gasoline engine, a few thousand volts. Which explains why it doesn't take long to let go of a spark plug wire.

flyback exponential follows a exponential response determined by the inductance and parallel damping resistor. Thus there are two different time constants: one for turn-on and one for turn-off.

What is the purpose of getting a nicely damped flyback decay? Recall that when the coil is turned off, the magnetic field collapses in a fast transient dB/dt; this magnetic transient induces eddy currents in a metal target. From the sudden dB/dt, the induced eddy currents jump to some value, and then they also decay exponentially, with a time constant determined by the electrical characteristics of the target. The counter-magnetic field of the eddy currents follows this decay, and this counter-magnetic field alters the flyback decay as shown in Figure 11-5. In other words, in the absence of a target, the flyback decay is determined by only the coil and damping resistor. But when a target is present, the flyback decay is altered by the counter-magnetic field, which decays with the tau of the target. So we can monitor the flyback decay and determine the presence of a target, and possibly even discriminate based on the shape of the altered response.

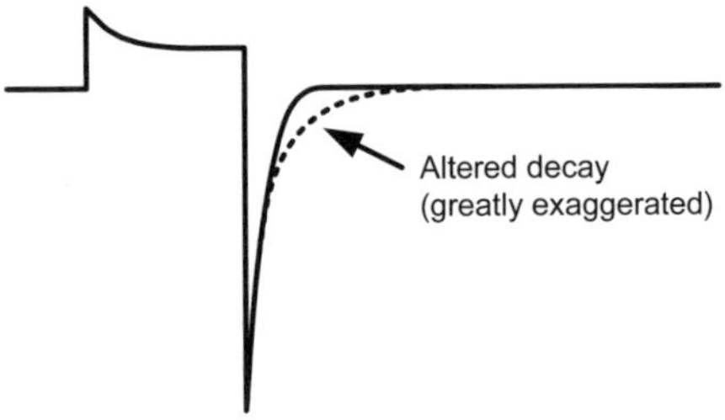

Fig. 11-5: **Target Response**

The altered decay shown in Figure 11-5 is greatly exaggerated; in reality, the decay shift caused by targets will be in the microvolts to millivolts range. Keep in mind that this small perturbation is riding on top of a waveform that peaks at perhaps a few hundred volts. Fortunately, we can design the coil so that its flyback response is faster than a typical target decay, which allows us to look for the target perturbations in a region where the coil flyback is very close to its final settled (zero) value.

The maximum speed of the coil's response is largely determined by its inductance and interwinding capacitance, and the resulting damping resistor required to prevent ringing. Both inductance and interwinding capacitance increase with the number of turns of wire, but so does the strength of the transmitted B-field (for a given peak transmit current) and the sensitivity to a target secondary field in receive mode. Therefore, the coil should be designed with the greatest number of turns possible for best depth, while still achieving the desired settling speed[6]. Typically, PI coils range from 200µH - 500µH[7]; we will use 300µH for a nominal coil value in the next chapter.

6. More windings for a given transmit current produces a stronger B-field but, in practice, more windings also have a higher series resistance which reduces the transmit current. The reality is that a lower inductance transmit coil produces a stronger B-field for a given drive voltage. However, the same coil is also used as a receiver, and a higher turn coil improves receive sensitivity. There is no free lunch.

7. This is for mono coils; PI detectors can also be designed to use separate TX/RX coils, in which case the RX coils are often designed to be a higher inductance, especially if the coil is inductively balanced.

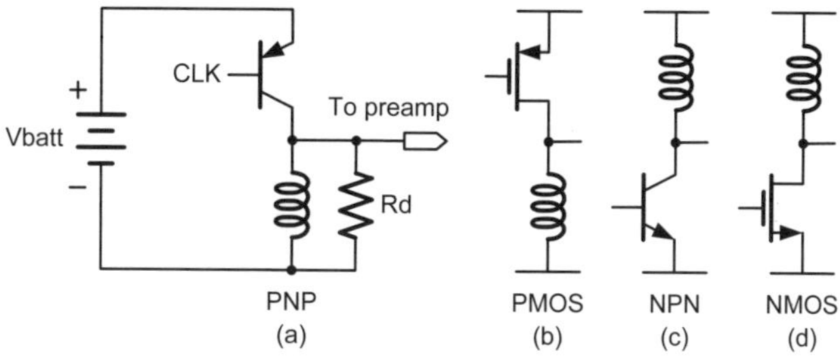

Fig. 11-6: **Coil Switches**

Before we leave this section, let's take a closer look at the coil switch. So far, it has been presented as a simple ideal switch. In a real design, we would avoid a mechanical switch and opt for a solid-state device. Because the target's eddy current response depends on a fast-changing magnetic field (Equation 11-1), the time it takes the coil to switch off is important. So the switch must be able to handle large (albeit short-duration) DC currents, and have fast switching. Typically, the switch is simply a single transistor, either a bipolar or a MOSFET. Figure 11-6 shows a PNP transistor directly replacing our ideal switch. We could also use a PMOS device, or move the coil to the positive rail and use an NPN or NMOS transistor. The NMOS transistor tends to be the most popular choice because of its lower on-resistance, lower capacitance, and faster switching speed.

PI Front End

Now that we know how to create the necessary transmit pulse, we need to design a circuit to look at the resulting decay for those small target perturbations. Because the target signal is so small, an amplifier stage is needed before any other kind of processing is attempted. This amplifier is the first gain stage in the receiver, so it is commonly called the *preamp*. Fortunately, by designing a fast-settling coil, the receive circuitry will only need to deal with the final portion of the decay curve as it approaches zero[8]. But the high-voltage flyback is still present, so the coil response needs to be clamped before it gets to the amplifier. This is easily done with simple diodes, as shown in Figure 11-7.

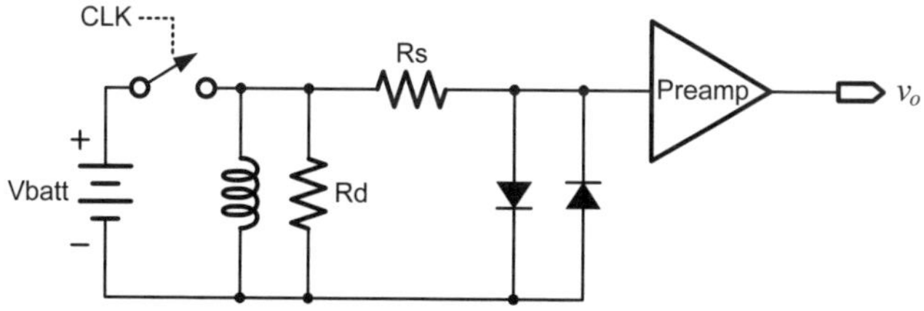

Fig. 11-7: **Preamp Clamping**

8. Depending on the switch arrangement in Figure 11-6, "zero" can mean the positive rail or the negative rail.

The parallel diodes, along with Rs, form a crude clamping circuit that clips the high-voltage part of the flyback voltage (as well as the charging voltage, Vbatt) and limits the input to the opamp to roughly ±0.7V. It's important to note that as long as the clamp is active, Rs is effectively in parallel with Rd. This means there are now two damping time constants: a slower time constant (L/Rd||Rs) when the flyback exceeds about 1 volt, and a faster time constant (L/Rd) when the flyback decays below 1 volt.

Clamping allows the use of an ordinary opamp for the preamp stage. Typical gains for the preamp are 100-1000, with the higher gains often split between two gain stages. Although the diode clamps prevent the high-voltage flyback from frying the opamp[9], the clipped levels will still overdrive the opamp output, even for low gains. Therefore, the opamp used in the preamp stage must not only have a fairly high gain-bandwidth product (15MHz or more), but also needs to have fast overdrive recovery.

Thus far, we have a circuit that will switch a current through a coil, clip the flyback voltage, and amplify the remaining signal. A simple front-end PI circuit that does all this is shown in Figure 11-8. In this example, the opamp is configured in a

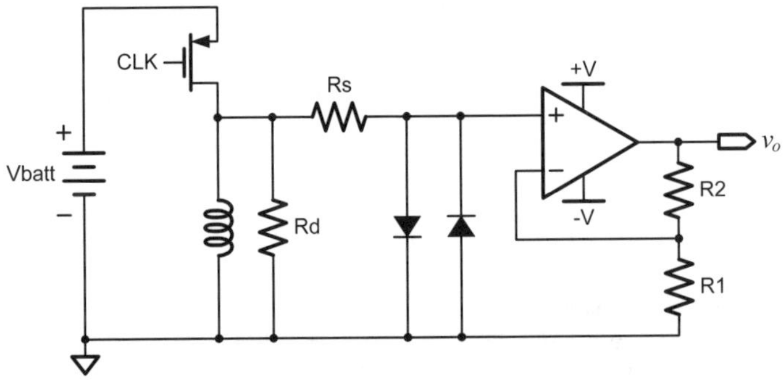

Fig. 11-8: **Simplified Front-end Schematic**

non-inverting mode with a gain of (R2/R1+1). Notice that circuit "ground" is defined to be at the battery's negative terminal. This is also the level to which the coil will decay, and therefore the input to the opamp will "see" input voltages that are (roughly) ±1V about ground. What this means is that the opamp's supply rails (+V, -V) need to be referenced to ground, which further means that the -V supply will be below the voltage range of the battery. This requires either a second battery or, as is more often the case, an inverting charge pump to create the lower rail. This issue exists regardless of the type of coil switch used (Figure 11-6).

The output waveform of the opamp is shown in Figure 11-9. Note that the output is overdriven during the coil turn-on time, and also for part of the flyback decay. At some point, when the flyback has decayed close to zero, the opamp output will come out of its overdriven state and show the final portion of the decay. If the opamp gain is 1000, then this represents the last millivolts of the coil decay. This is the region where we will look for variations due to metal targets.

9. And without the diode clamps, this *will* happen.

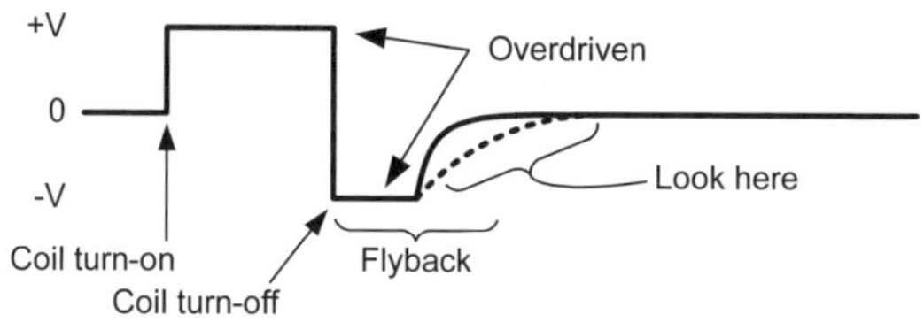

Fig. 11-9: **Opamp Output**

Before we move on, it should be noted that the opamp in Figure 11-8 can also be connected in an inverting fashion. For the inverting opamp, Rs can be eliminated and R1 (the feedforward resistor) can serve as the clamping resistor. Figure 11-11 shows not only the inverting opamp, but also replaces the P-mode switch with the more widely used N-mode switch, and flips the coil and damping resistor to the high side of the battery. But the opamp input still needs to see a decay that ends up at 0V, so 'ground' is now defined to be the positive terminal of the battery. It is now the +V supply that extends beyond the voltage range of the battery, and will need to be supplied by a second battery or a charge pump. The coil voltage and the opamp output for this configuration are shown in Figure 11-10.

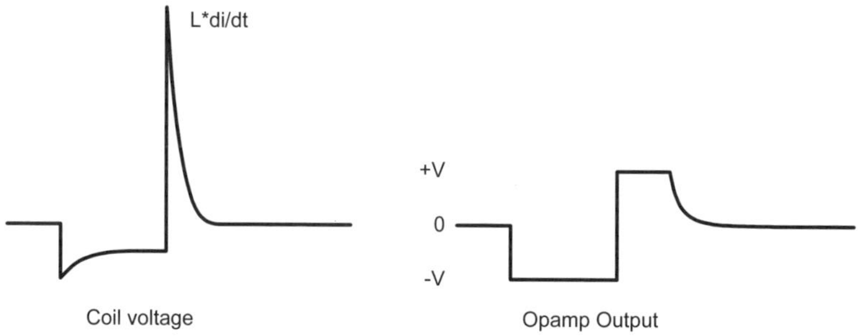

Fig. 11-10: **Inverting Opamp Waveforms**

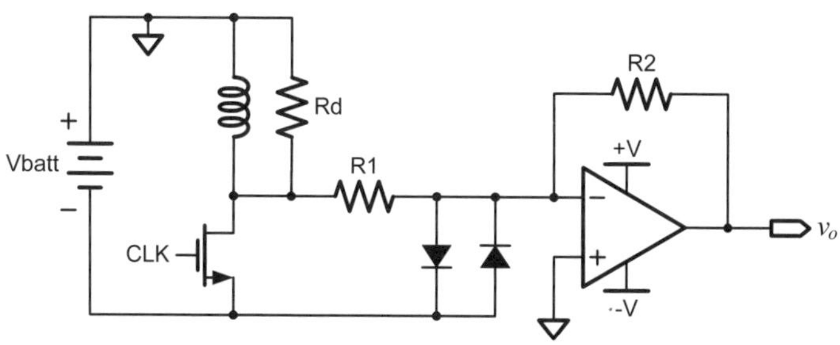

Fig. 11-11: **Inverting Opamp**

Sampling

With the flyback voltage limited and the rest of the decay amplified, we can now look for subtle changes in the decay due to metal targets. This is still problematic because the output signal of the preamp is decaying by a few volts, and we want to find a much smaller change on top of this. One solution to this is to use an amplifier whose gain *increases* logarithmically with time, which will flatten the decay and reduce the range of the changing voltage. While such a solution is sometimes used in other applications like ultrasound imaging, it is difficult to implement.

A method which is commonly used in commercial detectors is to *sample* a short portion of the decay curve and look for variations in the sampled level. The sampled portion of the decay will have some nominal voltage level — depending on where the sample is taken along the decay — and a target will cause the level to change. Figure 11-12 shows the waveforms for this process; the pulse that drives the coil switch and the coil voltage are included for completeness. The preamp output swing is limited to the supply rails of the opamp (±V). The sample pulse "looks" at a small portion of the preamp output somewhere between the point at which the opamp comes out of overdrive and the point where the decay has settled to zero. Usually the sampled point is close to the final settled value, where the coil response is out of the way but a target response is still decaying. Figure 11-13 shows what happens when the decay is disturbed by a target (dotted line). The falling edge of the sample switch determines the value of the sampled voltage, and in the presence of a target it will be consistently different than the nominal (no-target) value.

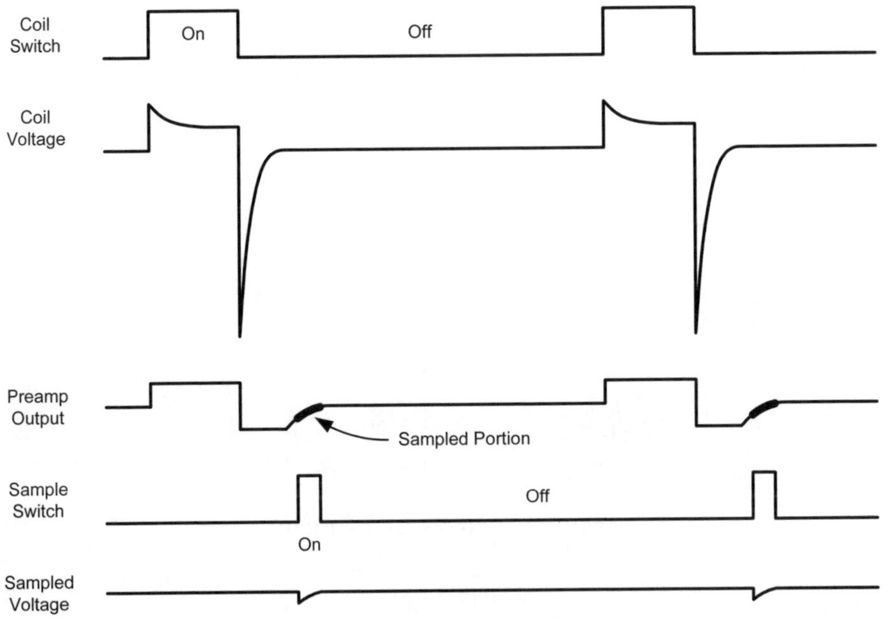

Fig. 11-12: **Sampling Waveforms**

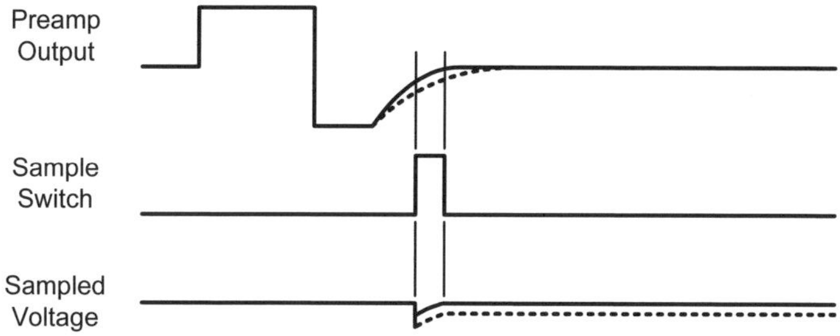

Fig. 11-13: **Target Response**

Figure 11-14 shows a circuit that samples the output of the preamp with a simple switch and capacitor. A second opamp stage buffers the sampled voltage on C1 and can provide additional gain if the range of sampled voltages is close enough to zero. Switch SW2 is the sample switch and is controlled by a clock signal that is timed off the falling edge of the coil switch. When the switch opens, the voltage is held on the capacitor, and a level detector can be used to determine if the voltage has increased or decreased due to a target.

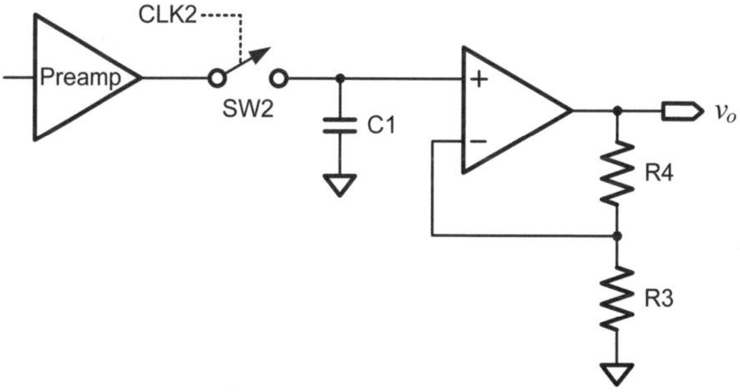

Fig. 11-14: **Sampling Circuit**

But random noise — either from outside the detector, or from the detector's circuitry — can also cause momentary jumps in the sampled voltage. Usually, the pulse rate of a PI detector is at least a few hundred pulses-per-second, so as the coil is swept over a target we would expect quite a few of the sampled portions to have a level that is deviated from the normal level. A noise event, on the other hand, is a random and relatively quick transient that will probably affect only a single sample. So to improve the ability to distinguish random noise fluctuations from a real target, we can look at a running average of the sampled decay (Figure 11-15). Correlated target samples are reinforced over many samples, while random noise averages to zero.

A simple method of averaging the signal is to add a series resistor to the switch. This resistor along with capacitor C1 will form a low-pass filter which averages each new sample with the voltage already held on the capacitor. However, if we use an inverting opamp for the second gain stage, we can just add the capacitor as part of the feedback and get both averaging and gain in one circuit. This is known as a *sampling integrator* and is shown Figure 11-16.

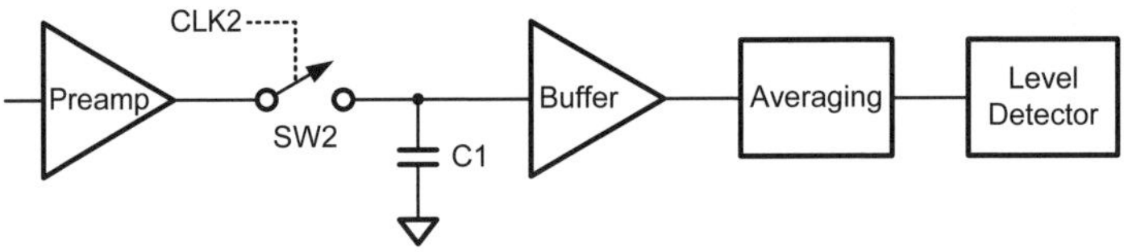

Fig. 11-15: **Averaging**

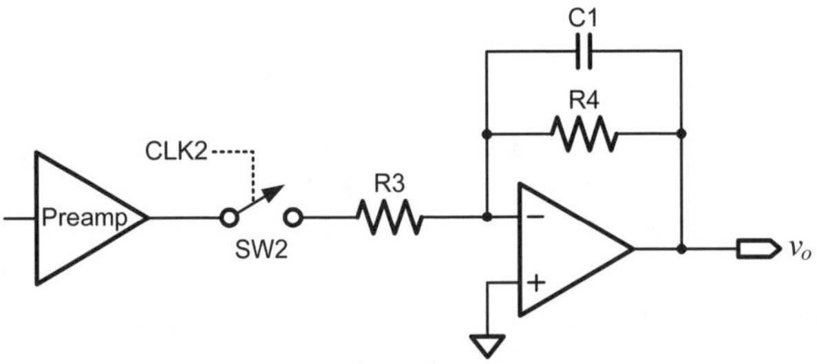

Fig. 11-16: **Sampling Integrator**

We now have a conceptual circuit that transmits a pulsed magnetic wave, samples the pulse decay, and averages the samples to look for variations in the decay level. The only thing lacking is a method to signal the presence of metal, such as an audio beep. The output of the integrator is a DC voltage which rises or falls[10] when metal is detected, so an indicator is easily added. If your preference is for a BFO-like pitch response, the integrator voltage can be used to drive a VCO circuit. For a loudness response, the voltage can be used to control the amplitude of a fixed tone.

The PI concept discussed thus far is fairly simple, yet is the basis for many commercial detectors such as the Tesoro Sandshark and the White's Surfmaster PI and TDI. The details may vary, such as whether discrete logic or a microcontroller is used to generate timing pulses, or whether a PMOS or an NPN is used to switch the coil current[11], but this basic design has been widely used for many years.

Digging in Deeper

In the presence of a target, the BFO detector produces a frequency shift, and the IB detector produces a phase shift. The PI detector has no frequency shift or phase shift, just the pulse decay. Therefore, the important operating parameters of the PI detector center on the timing of the coil and sampling pulses. Figure 11-17 again shows the various waveforms and adds some timing parameters.

The *pulse frequency* (the inverse of the pulse period) is the rate at which the coil is pulsed, and for a basic design, this parameter is not critical. Recall that we normally

10. Depending on the design of the front end.

11. The Sandshark uses a PMOS switch and a micro; the Surf PI uses a PMOS switch and discrete logic; the TDI uses an NMOS switch and has (in different revisions) used both discrete logic and a micro.

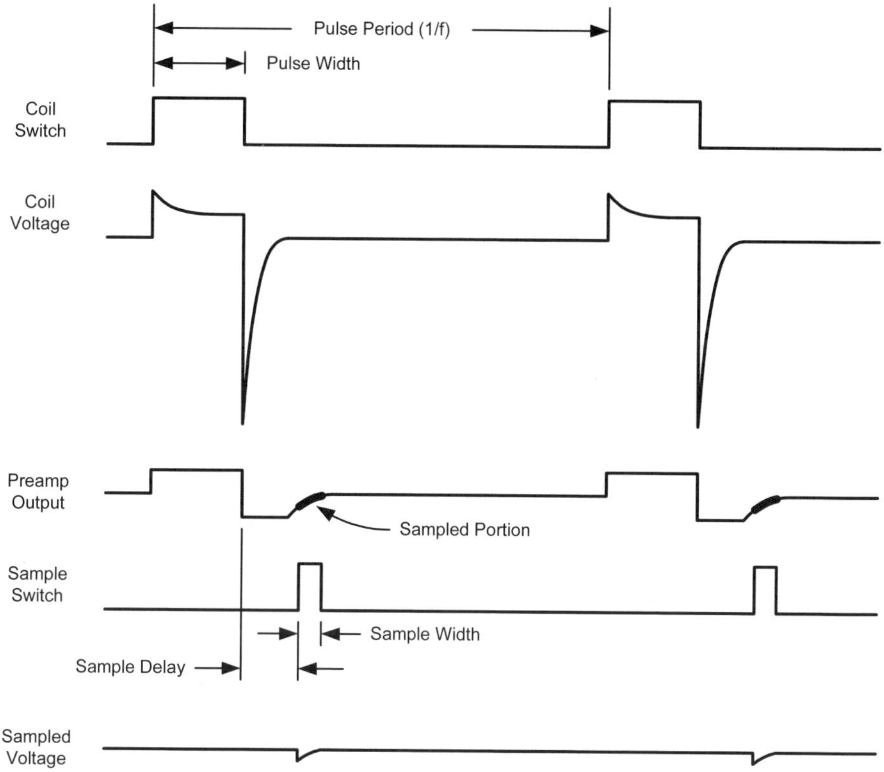

Fig. 11-17: **Switched Coil Waveforms**

look at a running average of the sampled pulse decay, therefore you want to run the frequency high enough so that an average speed of swinging the coil results in a number of sequential "positive hits" for a target. Too low a frequency will result in fewer target responses, but a higher pulse frequency uses more power, everything else being equal. Typical PI frequencies are a few hundred Hz to a few kHz.

The *pulse width* is the amount of time that the coil is energized. In some cases, this controls the strength of the transmitted magnetic field, and therefore detection depth. Recall from Figure 11-2 that the coil current exponentially rises to a peak value, due to non-ideal series resistances. Figure 11-18 shows the full exponential response of the coil current at turn-on, along with three example pulse widths. If the pulse width is too small, then it's possible that the coil will not develop the peak DC current; that is, the current will be switched off during the exponential part of the curve. This means the resulting magnetic field will not be as strong. Once the coil current has reached its peak value, any additional turn-on time is just wasted power.

Let's consider the decay part of the coil voltage again. With no target near the coil, the decay is roughly exponential with a decay rate primarily due to the inductance of the coil in parallel with the damping resistor Rd[12]. As mentioned earlier, the decay is altered by the magnetic field response of targets. It turns out that the decay is

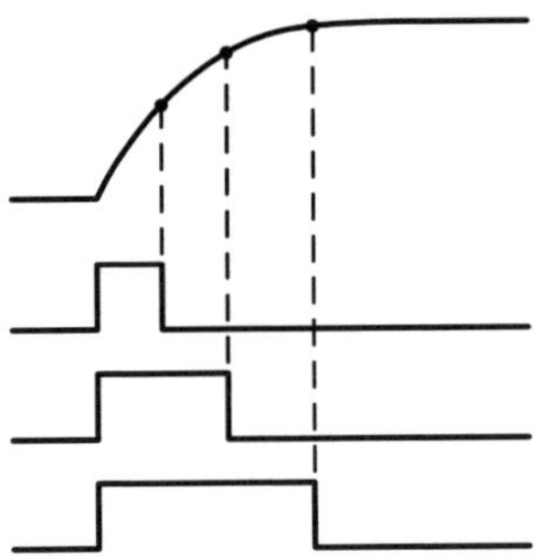

Fig. 11-18: **Coil Turn-on**

also altered by the mineralization of soil. This is not an eddy current response, but a phenomenon known as *magnetic viscosity*. Mineralized soil contains particles (usually magnetite or maghemite) which have magnetic permeability, and can be temporarily magnetized by an external magnetic field. When the field is removed, this temporary magnetization decays in a manner similar to eddy current decay in conductive targets.

Fortunately, the signal component from most soils dies out pretty quickly, while metal targets have a stronger decay. In Figure 11-17, we can see that the preamp sample is taken at some point past the coil turn-off event. The sample should be beyond the point where the ground component (as well as the coil decay) vanishes, but well before a desired target signal vanishes.

With a BFO detector, the frequency shift will often tell you whether the target is ferrous or not. The phase response of an IB detector provides even more information about the target. With a PI detector, it turns out that the electrical characteristics of a target affect the target decay rate, so in theory you could sample several points along the decay curve and roughly identify the metal being detected. This is rather difficult, and beyond the scope of this book.

But we can use this information about conductivity to make the design more sensitive to the most desirable target, gold. Gold has a lower conductivity than silver or copper, so its induced eddy currents die out at a faster rate. Therefore, gold affects the flyback decay at an earlier portion of the curve, and very little in the latter portion. Silver, on the other hand, has a measurable effect well past that of gold, as do other higher conductivity metals[13]. So, in order to make a PI more sensitive to gold, it is important to sample the flyback decay 10-15 μs (microseconds) after the instant the coil is turned off. This is called the *sample delay*. Many commercial detectors set the sample delay at around 15-20μs.

Most commercial PI detectors have no ability to discriminate between different

12. Perhaps with Rs in parallel.

13. Targets of higher conductivity support eddy currents for longer times. In a perfect conductor (superconductor) the eddy currents would never die out, and would produce a static magnetic field. This would be analogous to storing a voltage charge in a battery, except that a current gets stored.

metals, so PI detectors are normally used in areas where trash is tolerable. PI detectors have found two niches: beach detecting and nugget hunting. The saltwater beach is a difficult environment for metal detecting. Wet salt sand is conductive and supports eddy currents, so most VLF detectors[14] are virtually unusable in the surf. Fortunately, the eddy currents in wet salt sand decay pretty quickly, though not as quickly as the magnetic viscosity response of most mineralized soil. For surf hunting, PI detectors will start falsing as the sample delay is reduced much below 15μs, so this is a practical minimum. This delay will still produce a pretty good response to gold jewelry.

With nugget hunting, you want good sensitivity to very small pieces of gold which tend to have very fast signal decays, so nugget PI detectors typically have a faster (sub-15μs) sample delay. Even at 10μs the smallest gold nuggets won't respond, so some PI designs attempt to sample even before 10μs. This brings up two problems. One is that the speed of the coil decay becomes much more critical, and special techniques are required to lower the parasitics and make its response fast enough. The other is that ground mineralization begins to respond strongly, and ground balance techniques are required. These issues and a few others will be briefly looked at in the *Advanced Topics* section of the next chapter.

In the next chapter we will work through a simple PI design, and then look at several variations.

14. Except multi-frequency VLF detectors.

CHAPTER 12

Pulse Induction Designs

"I check my pulse, and if I can find it, I know I've got a chance."
— Paul Newman

Chapter 11 presented the basic concepts and building blocks of pulse induction detectors. In this chapter, we will run through a few simple PI detector designs, and look at some advanced concepts as well.

PI Design 1

The first PI design will follow directly from the simplified circuits presented in Chapter 11. Figure 12-1 shows the front end portion of the circuitry. This is almost identical to the circuit in Figure 11-8. The coil switch is a IRF9630 PMOS transistor which has a breakdown voltage of about 200 volts, so the flyback voltage will get clipped to this level. The preamp is set up with a non-inverting gain of 100, and the input signal is clamped by R3 and D1 & D2, which are common small-signal diodes (e.g. 1N4148).

A major difference between this circuit and Figure 11-8 is the addition of C2 and R4. Practically all PI designs run the coil directly off the battery because of the high transient currents involved. When there is only a single battery, as in this design, the coil will decay to one or the other battery rails, depending on the type of switch used.

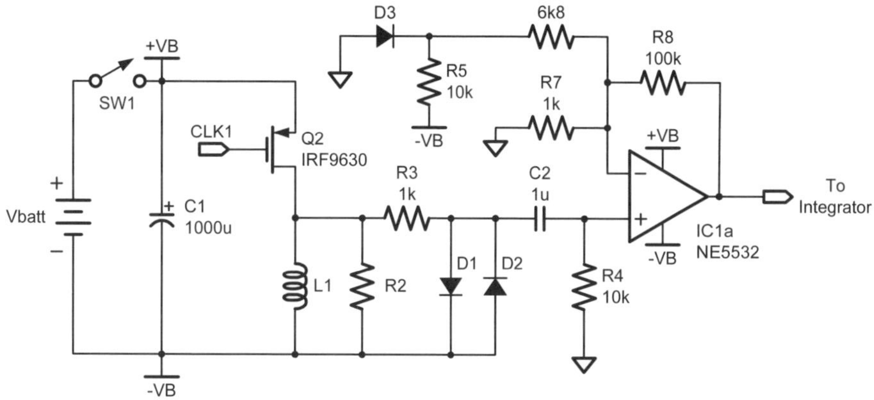

Fig. 12-1: **PI-1 Front End**

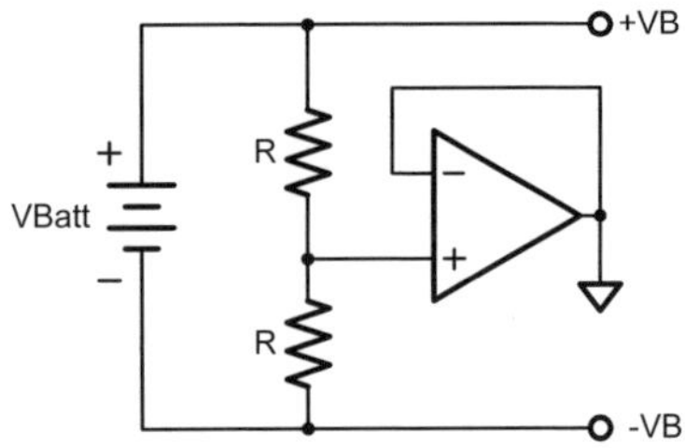

Fig. 12-2: **Creating a "Ground"**

Our circuit uses a PMOS switch so the coil decays to the negative battery rail. But if the preamp is powered by a split (bipolar) supply, also as in this design, it would like to see a signal that decays nominally to its "ground" potential. In Figure 11-8 the opamp is shown with a ±V supply; the +V could come from +VB (the positive side of the battery), but the -V needs to be a *lower* voltage than the negative side of the battery. We glossed over this in Chapter 11, but it's time to tackle it.

There are a few ways to solve this problem. One is to simply add another battery, a common solution in metal detector designs of the 1960's and 70's. An elegant solution found in most modern detector designs is to use a voltage inverter[1] to generate the opposite supply. Although the elegant solution works well, it adds some complexity and we're striving for bland simplicity to start with. We will instead create a "ground," as shown in Figure 12-2, by buffering the mid-voltage point of the battery with an opamp. We simply label the battery terminals +VB and -VB, and now we have the supply arrangement we need.

This still leaves us with the original problem: the input signal from the coil decays all the way to the -VB power rail, which the preamp probably cannot handle. C2 and R4 solve this problem by AC-coupling the coil signal and centering it up to our new "ground." Now the opamp sees a signal swing it can deal with. This method can produce some odd long-term droop with the signal (depending on the opamp used), but that is of no concern as we are looking for delta changes in the signal, and only near the turn-off point of the decay.

In addition to the components for AC coupling, there are also additional components — D3, R5, and R6 — for injecting some offset current into the preamp. Since the preamp has high gain, a small input offset voltage can result in a large output offset voltage. D3 and R5 produce a small (0.7V) battery-independent voltage that creates a current through R6. This DC current flows through R8 to effect an output offset voltage needed to keep the JFET switch in its proper operating region.

The preamp feeds a sampling integrator stage and final gain stage, shown in Figure 12-3. The sampling switch is a depletion-mode JFET transistor, which is "off" when the gate voltage is equal to the source (or drain[2]) voltage, but "on" when the gate is pulled to a lower voltage. The integrator stage is designed with an integration

1. A voltage inverter provides a -V supply from a +V supply (or vice versa) using a capacitive charge pump. A popular chip for this is the "7660" such as the ICL7660.
2. Unlike MOSFETs, drain and source are usually interchangeable on discrete JFETs.

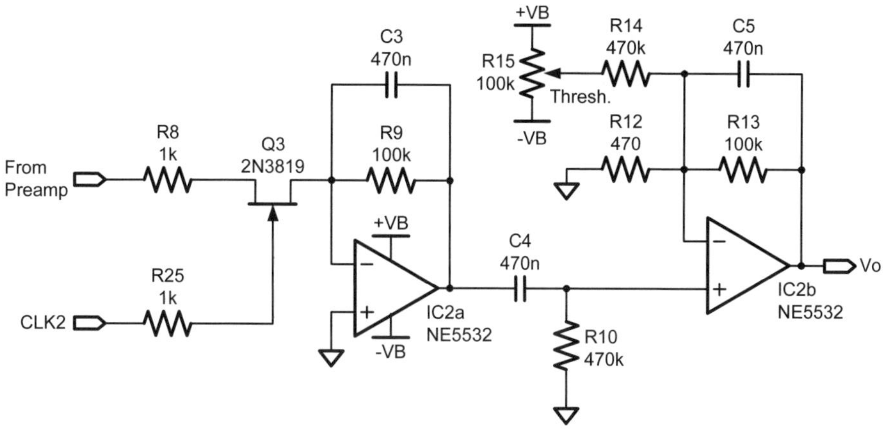

Fig. 12-3: **Sampling Integrator & Gain Stage**

time constant of 47ms. With a sampling rate of 600Hz (or 1.67ms; more on this later), this time constant will provide sufficient averaging while still having a reasonably fast target response.

The output of the integrator stage feeds a final gain stage which, besides gain, serves two other purposes. First, C4 and R10 form a high-pass filter which stabilizes the signal against all the previous gain that has been applied by blocking DC. This is usually called a *self-adjusting threshold* circuit (SAT) and is the reason many PI detectors require some amount of coil motion to produce a target signal. If you hold the coil steady over a target, the integrator stage will continue to produce a target signal but the SAT circuit will quickly tune it out. The speed of the SAT circuit is set by its time constant (R10-C4), 220ms in our case.

Secondly, the gain stage has a potentiometer R15, which can inject a small positive or negative current into the feedback path to decrease or increase the output offset of the opamp voltage. This offset is used to effect an audio threshold level control. The final gain stage has low-pass filtering via C5 for additional noise reduction, and is set to the same time constant (47ms) as the integrator stage.

The final part of the signal path is the audio stage. There are many, many methods to generate an audio signal, from the ultra-simple to the very complex. A popular audio circuit in metal detectors is a voltage-controlled oscillator (VCO), which produces a varying pitch in response to an amplitude change. The BFO circuits from Chapter 3 have a similar pitch response. Although we could implement a simple VCO audio circuit[3], we will bypass this for now and go even simpler. We will just drive the headphones with the output of the gain stage, and add a chopping signal to make it into an AC waveform. See Figure 12-4.

The output of the final gain stage is a DC voltage (ignoring the SAT) with a nominal value (no target) of about zero volts, depending on the threshold setting. When a target is present, the opamp output rises in voltage. Q4 acts as a level shifter so that

3. A 555 timer chip is an easy way.

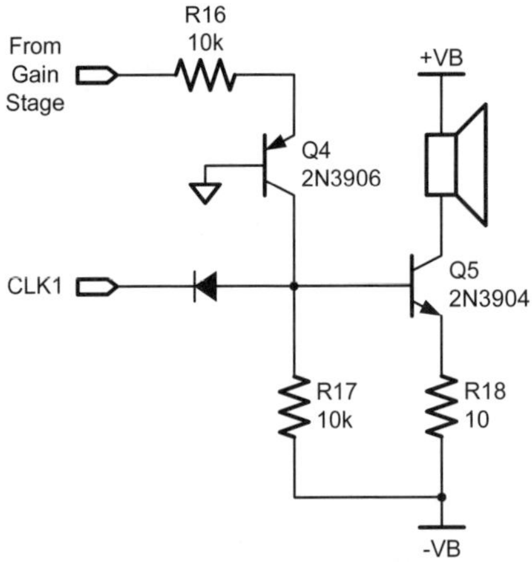

Fig. 12-4: **Audio Circuit**

the collector voltage is nominally close to -VB and rises with target signal. This voltage drives Q5, which is a common-emitter stage that drives the speaker. A clock signal is applied to the voltage drive via a diode; when the clock is low (-VB) it will force Q5 off, regardless of the target signal level. This gating signal gives the audio a fixed pitch tone, and the loudness will rise in response to a target; see Figure 12-5. Many commercial detectors use this approach.

The final piece of our circuit is the clocking. We now need three clocking signals: one to drive the coil switch, one to drive the sampling switch, and one to drive the audio chopper. The first order of business is to determine the frequency we want to use and generate the master clock. As already mentioned in the description of the integrator, we will rather arbitrarily choose 600Hz. A higher frequency allows the integrator to time-average more samples but, everything else being equal, drains the batteries more quickly. A lower frequency saves power, but produces a lower signal-to-noise ratio in the integrator. Common PI frequencies run between a 100Hz or so up

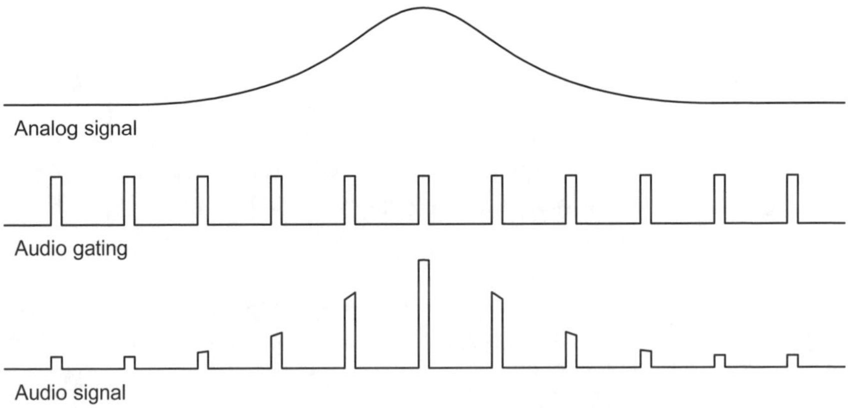

Fig. 12-5: **Audio Response**

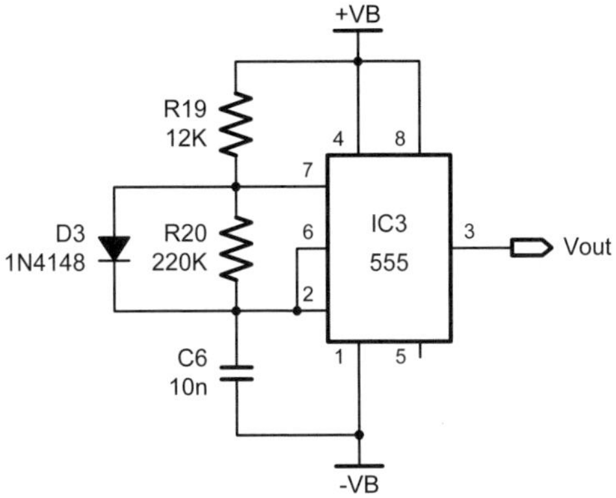

Fig. 12-6: **555 Oscillator**

to many kHz[4], though we need to keep in mind that this is also the frequency of the audio signal in our design. 600Hz will do fine.

As with most subcircuits, there are hundreds of ways to produce a clock signal. Obvious methods are a crystal, a Colpitts or Hartley oscillator[5], a relaxation oscillator, various specialty clock chips, and so forth. A popular circuit with experimenters is the "555" timer connected as an oscillator. It is easy, cheap, and flexible, able to generate a wide range of frequencies and duty cycles. However, its single output pin will satisfy only one of our clock signal needs.

Figure 12-6 shows a simple 555-based oscillator. You might recognize this circuit from Chapter 8, but there is a minor difference. A diode has been added to the R-C components that set the frequency and duty cycle. The diode allows the 555 to produce an output duty cycle that is the opposite of its normal operation, which we happen to need for our circuit. This neat little trick is well-documented in 555 applications sources.

The clock for the sampling switch runs at the same frequency as that for the coil switch. It is delayed by some amount and usually has a different pulse width (see Figure 11-17), so we need a circuit that produces a delayed-and-shortened pulse. Once again, there are many ways to accomplish this; one is known as a "monostable multivibrator." This is a fancy way of saying it's a single-shot pulse circuit. The 7400-series logic family includes a chip with 2 complete monostables, the 74221[6]. We can use the first monostable to set the pulse delay and the second monostable to set the pulse width. Figure 12-7 shows a 74221 configured in a way to give the proper output delay and pulse width.

4. Low pulse rates are often used in large-coil deep-seeking PI detectors where, say, a 1-meter coil is pulsed with a very high current and has a slower response than that of a small fast coil.
5. We've seen these in BFO & IB detectors, where the TX coil forms part of the tank circuit.
6. We will use the 74HC221 which has CMOS (rail-to-rail) logic and can tolerate up to a 15V supply. The 74C221 will also work, but the 74LS221 (or any other TTL variant) will not work.

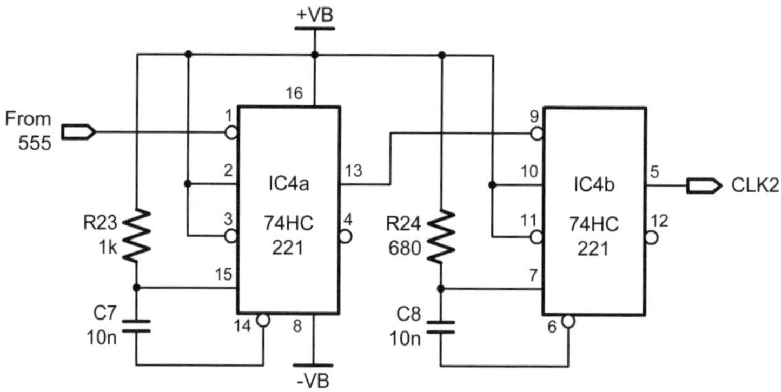

Fig. 12-7: **Monostable Clock Circuit**

Finally, the clock for chopping the audio signal can be any frequency. However, speakers are typically low-impedance and generate large current spikes when they are driven with pulses. These current spikes will usually translate into glitches on the power supply voltages that run the opamps, and these glitches can end up as spurious noise in the signal path. To minimize this problem, the speaker frequency should be the same as the sampling frequency so that glitches correlate in the integrator and become a long-term DC offset. This is another reason why 600Hz was chosen for the master clock frequency. Instead of generating another clock, we will simply use the transmit switch clock (CLK1) to chop the audio signal, and 600Hz is a nice frequency for the audio.

Let's put all the pieces together and make a PI detector. Figure 12-8 shows the complete circuit, and the component list is in the latter part of the chapter. Because we simplified most of the design, there is (hopefully) little to go wrong.

The coil for this circuit is a "mono" loop of about 300µH. This coil can be practically any diameter, but 10 inches (25cm) is a reasonable size to start with. For this, 20 turns of 26 AWG (27 SWG, ~0.4mm) will work. Wire diameter is not extremely critical, so don't worry if you don't have exactly 26 AWG. See Chapter 10 for more information on how to wind and shield coils.

Once the circuit is completely built, power it up, and set the sample delay to minimum, which should correspond to a delay of roughly 15µs. Adjust the threshold until there is a faint hum with the gain set for stable audio. An average silver or cupronickel coin should be detectable at about 6 inches (15cm). Adjust the sample delay and see how detection distances vary for gold, silver, foil, and iron targets.

This design is intended to run on a single 9V transistor radio type battery, although the use of a 6-cell AA battery pack will have lower resistance and last longer. This circuit should run just fine up to 12V although the next few PI circuits may not. You should find that this circuit consumes 90-100mA at a full 9 volts, and the current consumption drops in rough proportion as the battery voltage drops. Performance also decreases slightly with lower voltage, and as it approaches 6.5 volts it becomes unacceptably poor.

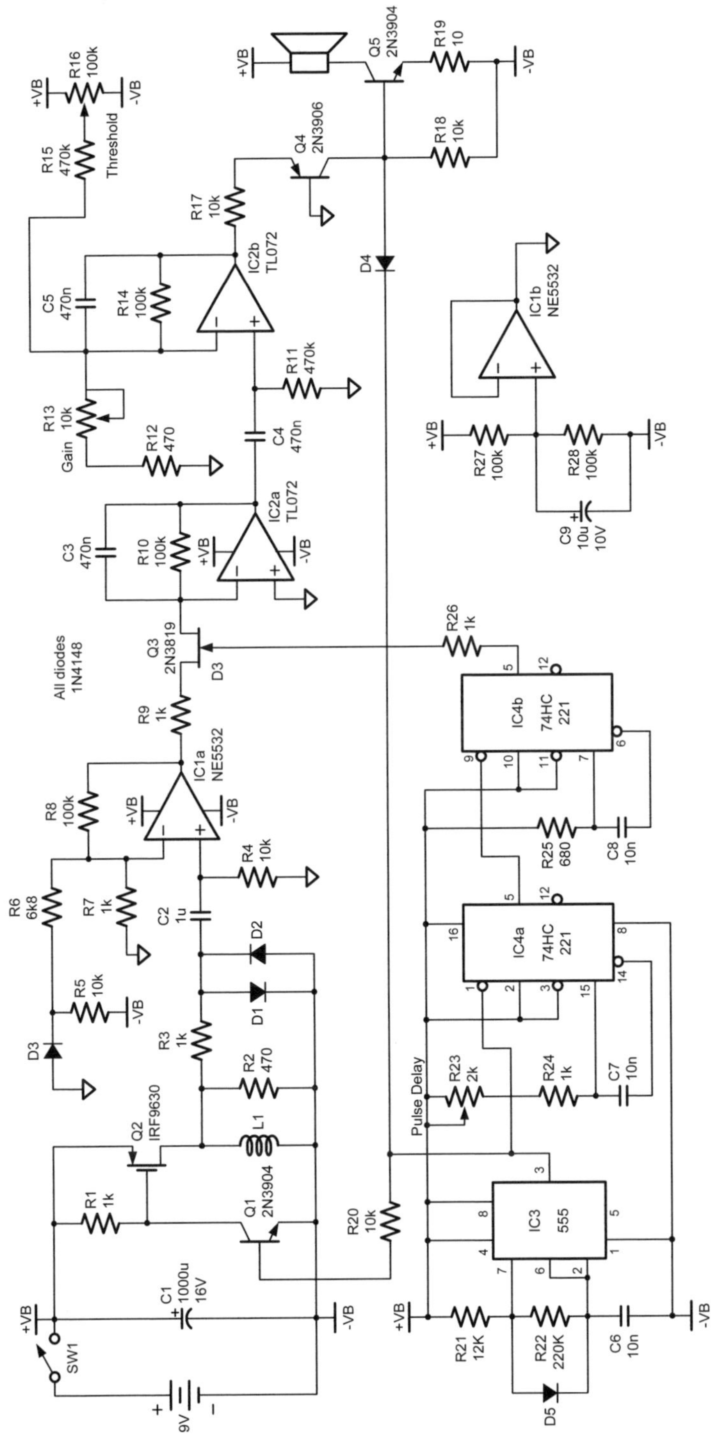

Fig. 12-8: Complete PI-1 Circuit

PI Design 2

In the first PI project we were faced with the need to generate three clock signals. The solution was to use a 555 timer for the master clock and audio gate, and a 74221 for the sampling clock. However, generating low-to-moderate speed clock signals is a task ideally suited to a microcontroller chip. Microchip makes a very popular line of microcontrollers known as "PIC" micros, so let's see how PI-1 can be simplified using a micro.

PICs are available with a wide range of I/O capacity — up to 85 — but we need only 3 output signals so we will choose the PIC12F1840, which comes in an 8-pin package (same as the 555 timer) and has up to six I/O pins. Figure 12-9 shows the pinout for this chip. Immediately you will notice that every pin other than power (VDD & VSS) have multiple functions. Microcontrollers are typically highly programmable and, for example, a given pin can be programmed to be a general purpose logic input, logic output, ADC input, comparator input, PWM output, and so forth. This flexibility has made microcontrollers a very popular choice in processing and control.

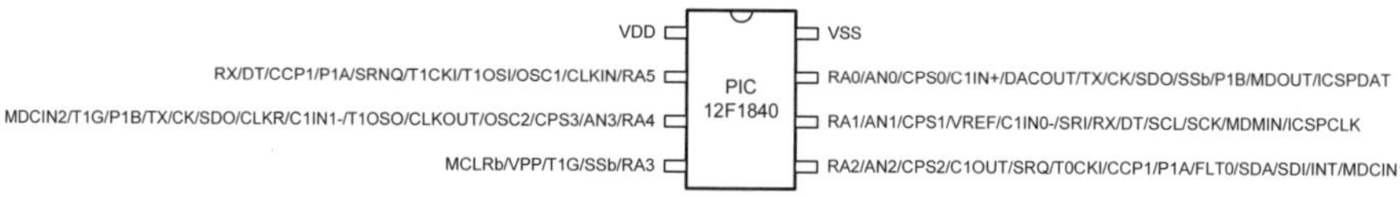

Fig. 12-9: **Microchip PIC12F1840**

Another nice feature of this PIC is that it has a built-in oscillator circuit that runs at up to 32MHz. Code executes at 4 clock cycles per instruction, so we can theoretically achieve timing with a resolution of 0.125μs, which is plenty sufficient for our project. The PIC also offers the option of an external clock source (e.g. crystal), but with no greater speed (though with better precision) than the internal 32MHz oscillator.

Using a PIC does have minor disadvantages. One is that you must write firmware[7] for generating the timing pulses, program the PIC, and possibly debug it. Fortunately, we've done the hard parts for you, and the necessary C code is presented in the construction part of the chapter. The 5-pin header shown in the schematic allows the use of a standard PIC programmer (such as the PICKit-2) for programming the firmware. The second drawback is that timing parameters are set by programmed delays, and normally cannot be varied in the field. In PI-1, the coil pulse width, the sample delay, the sample width, and even the overall frequency were all set by RC time constants, so by using potentiometers instead of fixed resistors any of these parameters could be made adjustable.

It so happens this PIC chip, like most other micros, has a feature that can be used to overcome this limitation. It has a built-in analog-to-digital converter (ADC) with four multiplexed analog inputs (AN0-AN3). We can connect a potentiometer across

7. Firmware is like software, but does not run under an operating system. Instead, it is code that does everything, and the chip has no functionality without it.

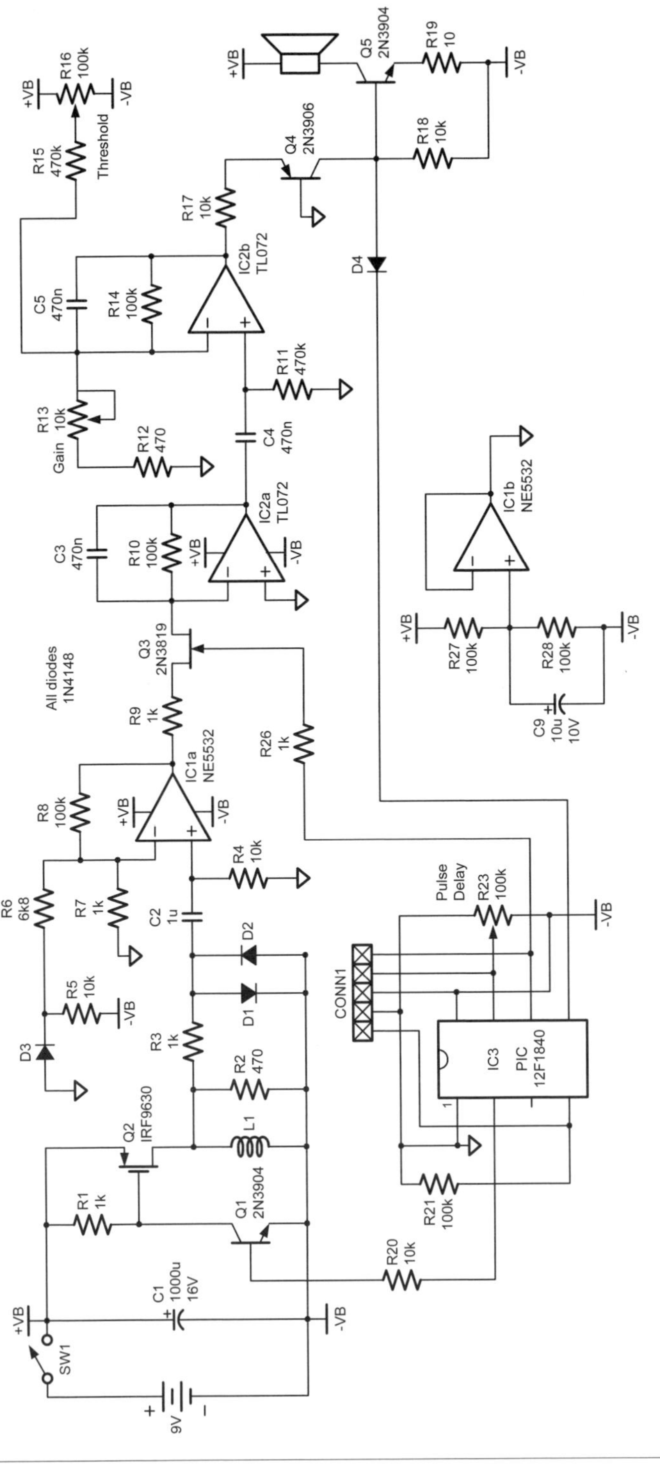

Fig. 12-10: **Complete PI-2 Circuit**

the PIC supply voltage and feed its wiper to one of the ADC inputs, then use the ADC to read the wiper voltage and vary one of the timing parameters on-the-fly. This can be replicated for each of the desired adjustments, up to the limit of the number of available ADC inputs. In the PI-2 design we have only provided for varying the sample delay. This PIC also includes a pulse width modulator (PWM) output which is useful for creating an audio signal, but we will save that for the next project. Instead, we will use this pin to provide the audio gating signal. By making it independent of the transmit pulse we can delay it to a point where it is less likely to cause interference.

Most PIC microcontrollers are limited to a power supply voltage of up to only 5V (nominal). Our design is intended to run on a 9V battery, so we can simply power the PIC off the lower half of the rail splitter, which puts the PIC supply voltage at 4.5V max. This means that trying to run a higher voltage battery (say, 12V) could damage the micro as its half-voltage will exceed the PIC's 5V limit. As the battery voltage diminishes the PIC voltage will drop in proportion, so the voltage at the wiper of the pulse delay pot will also drop. However, we can program the internal ADC to use the PIC supply as the ADC reference voltage, meaning that the resulting value for the pot setting will remain constant as the battery voltage drops.

Figure 12-10 is the full schematic of a PIC-controlled PI detector. Notice that we have greatly simplified the clocking circuitry, and reduced the entire design down to only three chips: two dual opamps, and the PIC. Operation is otherwise the same as PI-1, and uses the same coil. Power consumption is still around 100mA.

PI Design 3

PI-3 is a very minor modification of PI-2. In PI-2, the final opamp stage directly drives the audio stage, and the PIC provides a chopping signal to produce the audio tone. In PI-3 the PIC drives the audio stage from the built-in pulse width modulator (PWM). The PWM can be programmed for a constant frequency output, and varying

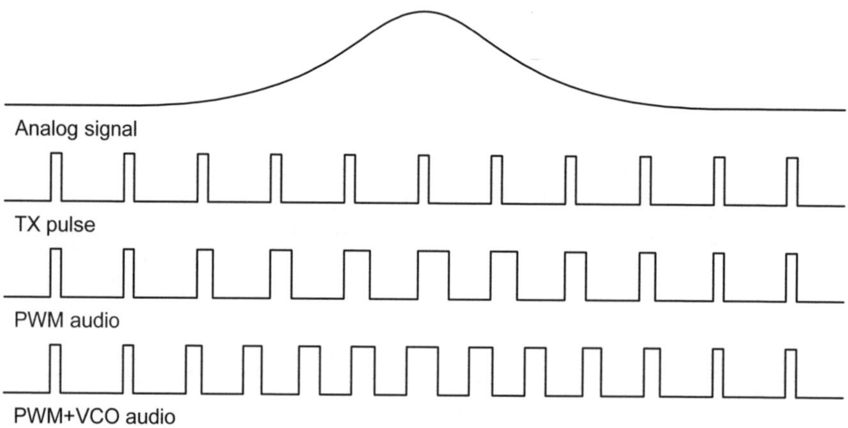

Fig. 12-11: **PWM/VCO Audio Response**

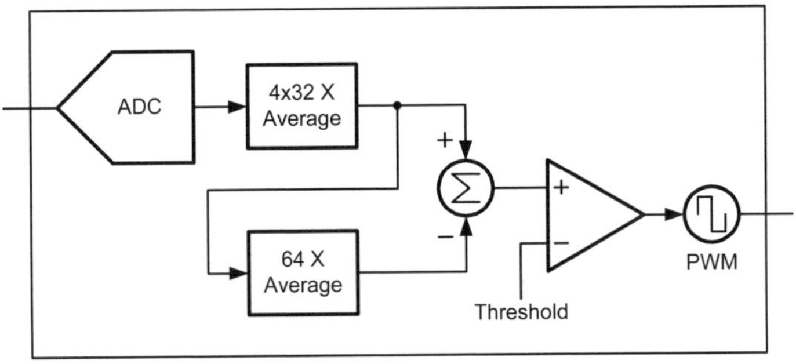

Fig. 12-12: **PIC Firmware Operation**

the duty cycle will produce a loudness change in the audio. But the PWM frequency can also be varied such that the audio response has a change both in loudness *and* in pitch. This gives us the VCO-style response mentioned earlier in the chapter. See Figure 12-11.

To control the PWM, we use one of the PIC ADC channels to sample the output voltage of the final opamp stage. The digitized signal is then used to modulate both the duty cycle and frequency of the PWM. Figure 12-12 shows a conceptual diagram of what happens in the PIC firmware. Once the ADC samples the incoming signal, the digital values are fed to a 2-stage integrator. The first integrator is a simple 128-sample boxcar averaging routine which provides a faster (short-term) average of the samples. The output of this integrator is fed to a second integrator which is a 64-sample boxcar averaging routine which provides a slower (long-term) average of the samples. The long-term average is then subtracted from the short-term average to give a relative target response. The response is compared to a threshold value, and the difference determines the frequency and duty cycle setting of the PWM module. Since the target response is based on the difference of two software integrators there is a built-in self-adjusting threshold function, plus a fixed digital threshold function. Neither of these require hardware or user-adjustment.

Figure 12-13 is the schematic for PI-3; the source code is again at the end of the chapter.

PI Design 4

Let's take PI-3 and simplify it even further, by removing the sampling integrator and final gain stage. We will use the PIC's ADC to sample the output voltage of the preamp at the appropriate time delay, and then use the same firmware-based digital integrator as in PI-3. This is known as *direct sampling* and is becoming widespread in receiver designs (such as radios and cellphones) as ADCs become faster, higher in resolution, and cheaper. Also as in PI-3, the PIC's PWM generates a VCO-style audio response.

Figure 12-14 shows the complete circuit for this project. With only a 10 bit ADC, sensitivity will be somewhat less than in the previous designs. However, this approach might be sufficient for some applications, such as a pinpointer probe (a PI

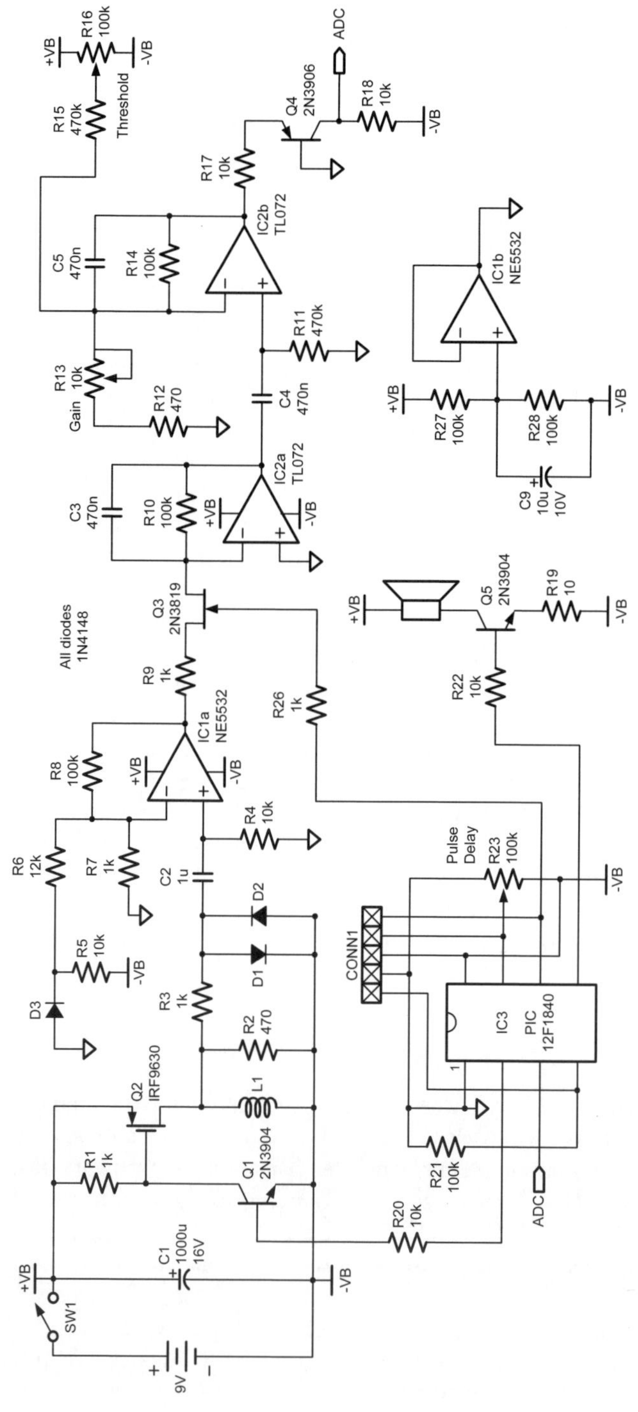

Fig. 12-13: **Complete PI3 Circuit**

version of the Chapter 4 project) or a security wand. A higher resolution ADC, such as 16-24 bits, will significantly improve performance. Micros usually do not have such a high-resolution with their built-in ADC, so you would need an external ADC. Then, with a high resolution ADC, a lowly 8-bit PIC may not have enough processing horsepower and a 16-bit or 32-bit micro may be necessary. The microcontroller market is extremely competitive, with lots of options to choose from at very low prices.

Since the ADC is referenced to the negative half of the battery, it is important that the sampled signal be properly offset so it ends up within the range of the ADC. In this design, a target response results in a decrease in the voltage at the output of IC1a. Like previous designs, R6 provides some offset current injection into the feedback to create a sufficient negative offset at the output. Some adjustment to R6 may be necessary, especially if another type of opamp is used.

As you can see, a lot of the analog circuitry is being replaced by digital circuitry. More precisely, by programmed digital circuitry. The nice part of this is that the characteristics can easily be changed in firmware, without changing any physical parts. In PI-1, changing the integrator time constant meant swapping out a resistor or a capacitor; now we just alter the firmware and reprogram the PIC. Or, better yet, control it with a pot via an ADC channel.

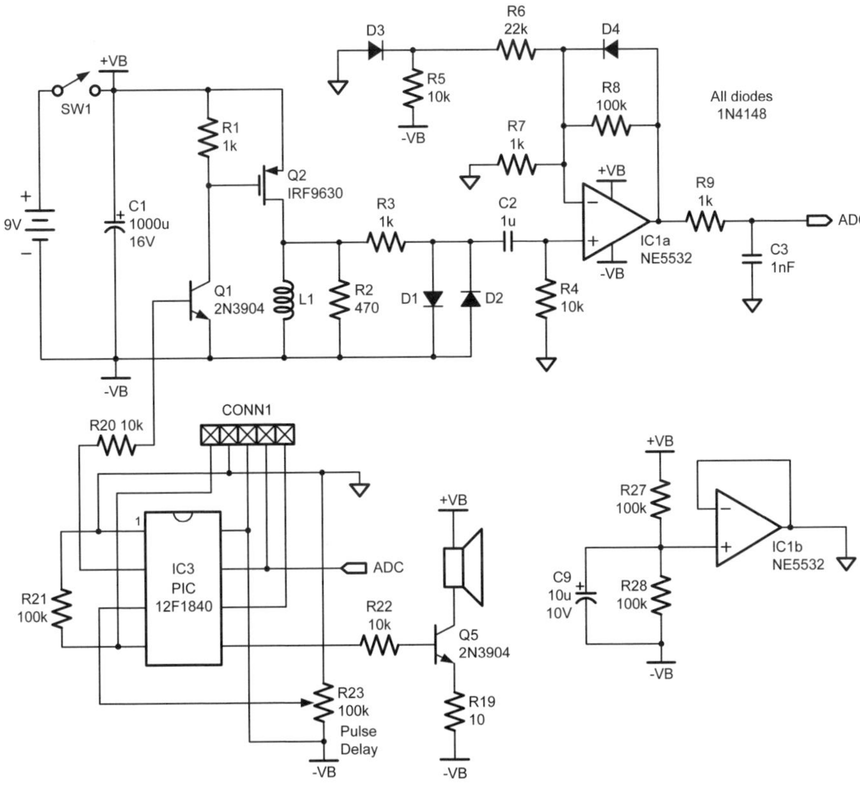

Fig. 12-14: **Complete PI4 Circuit**

The main purpose of this exercise is to demonstrate exactly how metal detector designs, like most other electronic gadgets, are migrating increasingly to the digital domain. Digital designs are more exactly reproducible; that is, there are no component tolerances, no potentiometer tweaking, no drift with age. Every detector will behave exactly alike, no "hot" units or "duds". This also makes them cheaper to build and more reliable.

A second purpose of this project is to show just how small we can make a detector. Since a dual opamp is covering both the preamp and the virtual ground generator[8], we have a design with two ICs and a small number of discrete parts. And this is not a weak design; potentially, with a better ADC[9] and a more powerful processor, this simple approach could have features and performance that exceed high-end commercial detectors. At the end of the chapter are both a normal through-hole PCB layout and a surface-mount PCB that uses SOIC chips and 0805 passive components. While small, these sizes are not overly aggressive and fairly easy to deal with if you have a decent soldering iron. A magnifying lamp is also useful.

Bonus: PI Design 5

The PI projects so far are based on a sampling-integrator design (even if done in software). It is the basis for most commercial PI metal detectors. As we saw with some of the subcircuits, there are many ways to skin a cat[10], so let's try a completely different approach. Instead of sampling and integrating the decay, we'll use the decay to control a *pulse width modulation* circuit. As we saw in Figure 11-5, the presence of a metal target alters the decay of the coil flyback. The previous projects sampled the decay and used the variations in decay amplitude. Figure 12-15 shows that for a given point on the nominal flyback curve, we can view the disturbance by a target either as a variation in amplitude (voltage *dv*) or as a variation in *time* (*dt*). The time variation can be tracked by a pulse width modulation circuit that provides a variable frequency output dependent on the pulse width. This results in a VCO-style audio output.

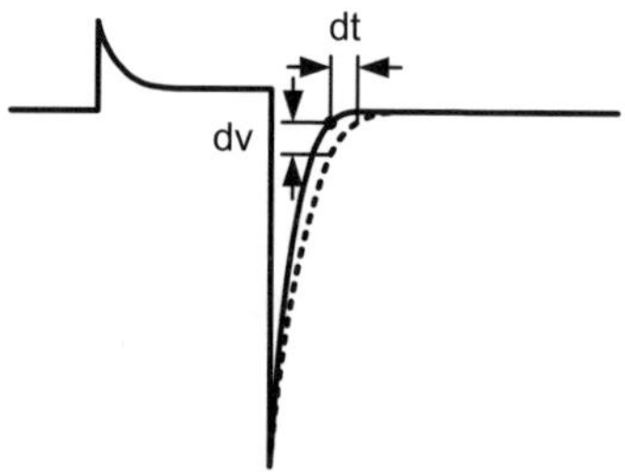

Fig. 12-15: **Time-domain Variation**

Figure 12-16 shows the complete circuit. Note that this design flips the transmitter to use an NMOS switch. The elimination of the sampling integrator and audio

8. Using the NE5532 is way overkill for the virtual ground opamp. If you want to cut power consumption, use a NE5534 for the preamp, and use any micropower opamp for IC1b. Or use the nifty TLE2426 rail-splitter IC.
9. By sampling multiple points on the decay and channelizing them, it could be possible to have not only ground balance but even discrimination. A micro also gives you the ability of doing otherwise difficult mathematical processing.
10. This adage refers to catfish, not house cats.

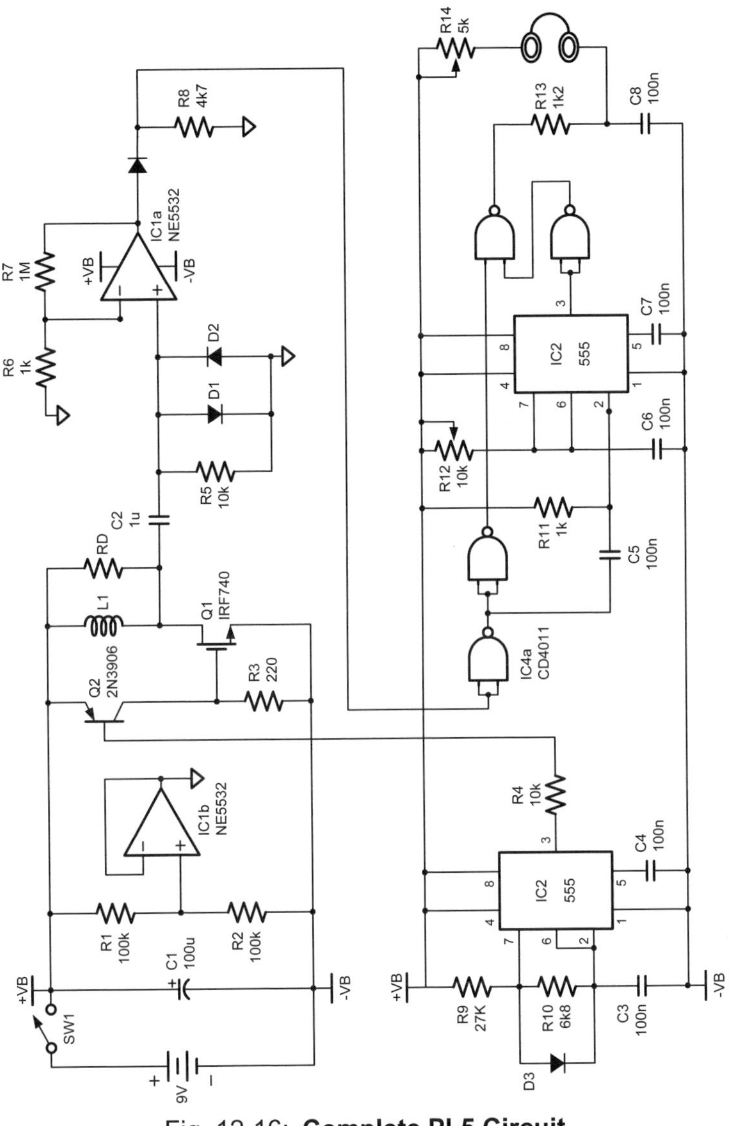

Fig. 12-16: **Complete PI-5 Circuit**

chopping circuit reduces the number of clock signals needed to just the transmit pulse. The output of the preamp is converted to a logic signal; this signal then triggers a 555-based one-shot delay generator. The delayed output is combined with the digital input to produce a signal whose pulse width is modulated by variations in the preamp decay. This bonus circuit is not covered in the construction portion of this chapter.

Advanced topics

The PI projects presented here are very simple in concept, primarily to prevent the basic principles of PI from getting lost in extraneous functions. There are a number of ways to enhance the basic PI design for better performance, so let's look at some of those techniques.

Power Supply

We've already mentioned the possibility of using a charge pump voltage inverter instead of a second battery or a rail splitter. Most commercial designs now use this approach which, of course, only requires a single battery (or battery pack). A charge pump voltage inverter is not difficult to design, but there are chips available specifically for this task. One is the ICL7660; for a positive input, the output is negative at about 80-90% of the input under moderate load. For example, a +12V input might yield a -10V output; these levels could then be regulated down to ±5 volts for opamp power.

One disadvantage of a charge pump is that they are not 100% efficient, some energy is lost in the process. Another potential problem, and one that can be serious in pulse induction, is that the switching action in a charge pump generates a lot of noise, and that noise can feed through the power rails and seriously degrade performance. For the same reason as mentioned for the audio, PI detector charge pumps are often synchronized to the overall pulse rate. Figure 12-17 shows a simple discrete charge pump that can run at most PI pulse rates.

Preamp

PI-1 through PI-4 use a preamp with a gain of 100; PI-5 has a gain of 1000. Many commercial PI designs have a preamp gain of 1000 as well. A high gain requires an opamp with a high gain-bandwidth product, and even then the slower transient response limits how early the target sample can be taken, usually around 15µs. Earlier target sampling almost always yields better sensitivity, but so does a higher preamp

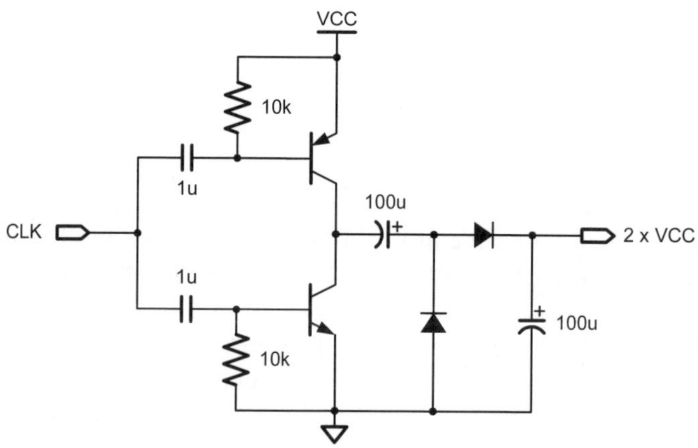

Fig. 12-17: **Charge Pump Circuitry**

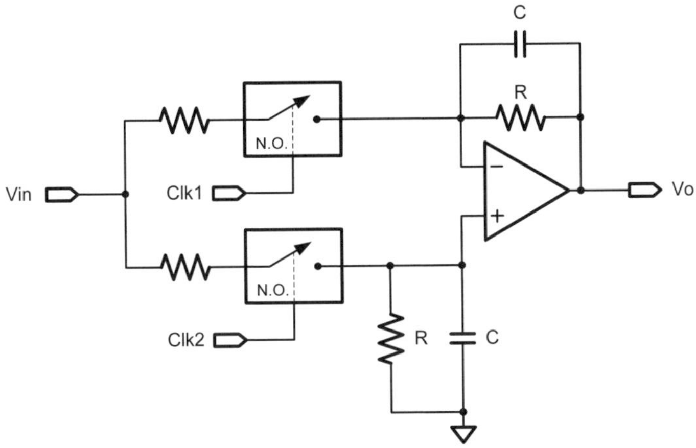

Fig. 12-18: **Differential Sampling Integrator**

gain. A good solution is to split the preamp into 2 stages, each with a lower gain. This provides an overall high gain while maintaining a fast transient response.

Sampling

In the sampling integrator PI designs (PI-1 through PI-3) a JFET was used as the sampling switch. While this makes for an extremely simple switch, it is not the best solution because it has a rather high "on" resistance with a lot of variation, so the gain of the second stage is not very predictable. Also, the JFET is a depletion device, so it has a very limited analog voltage range. A better analog switch can be found in the CD4066[11] logic chip. Actually, this chip contains 4 switches. Each switch is a CMOS transmission gate[12] made up of both a PMOS and NMOS device in parallel, so it works very well over a wide range of analog voltages. It also has a lower and more consistent "on" resistance of about 50Ω.

Also in the sampling designs, a single sample was taken early in the decay curve. This early sample point corresponds to expected variations in the decay due to metal targets, where effects on the decay curve tend to occur in the first 5-50μs after the pulse. After that, the effects from metal targets tend to die out so they have little effect at longer sample delays. Sometimes there are anomalies which persist for much longer times, and affect the decay not only in the region of the sampling but also farther out. One such effect is the induced signal from the Earth's magnetic field as you swing the coil. It appears as an overall offset in the decay and can result in a response that comes and goes as you swing the coil (a "breathing" effect).

A common method of eliminating these unwanted responses is to take a later sample of the decay voltage, and subtract its value from the normal earlier sampled voltage. For long response times (such as Earth field effect) the two samples will largely cancel and there will be little to no differential voltage passed through the

11. Or the CD4016.

12. Also called a *bilateral switch*.

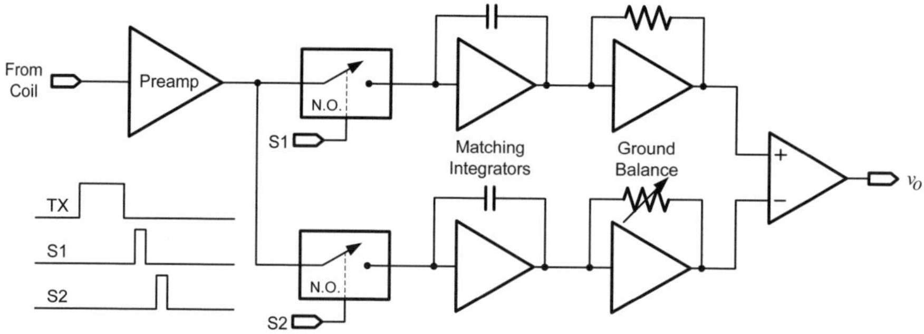

Fig. 12-19: **Ground Balance**

integrator. Metal targets with short decay time constants will not be affected. Figure 12-18 shows a differential sampling integrator.

Ground Balance

This concept can be taken a bit further to provide a method of ground balance. Most ground responses to pulsed signals are due to magnetic viscosity effects and have a fairly predictable 1/t response. By taking two samples fairly close to each other and subtracting a properly amplified second sample from the first sample (Figure 12-19), the ground response can be nullified while still passing signal responses. Because the samples are fairly close together (generally 10-20μs apart) the subtraction technique will reduce target sensitivity somewhat. However, some soils are nearly impossible to hunt without the ground balance feature, so a little target loss is better than no detecting.

A critical requirement of the method in Figure 12-19 is that the integrator time constants match well, usually to about 1%. Another approach avoids this by using only one integrator. If the S2 switch is fed with an inverted version of the preamp output, then it can directly feed the same integrator as S1. The gain of this signal (and thereby the ground balance) can be controlled either by varying the gain of the preamp inversion stage, or by varying the pulse width of S2. The latter has the advantage of integrating more noise and usually improves noise performance.

Coils

So far, all of our PI circuits have used a simple mono coil, which serves for both transmit and receive. In terms of do-it-yourself construction, this is much easier than trying to build carefully balanced coils for induction balance. However, the mono coil approach has a disadvantage: the optimal design of a transmit coil might not be optimal for the receive coil. It turns out that PI can also use separate transmit and receive coils and, unlike VLF detectors, they do not have to be inductively balanced.

All of the designs in this chapter have used a basic MOSFET coil switching circuitry, and very few changes need to be made to accommodate separate coils. Figure 12-20 shows that just splitting the circuit at the coils and adding a second damping resistor is sufficient. We can continue to use a fast (low inductance, low capacitance) transmit coil, but the receive coil need not be as fast, especially if it is inductively bal-

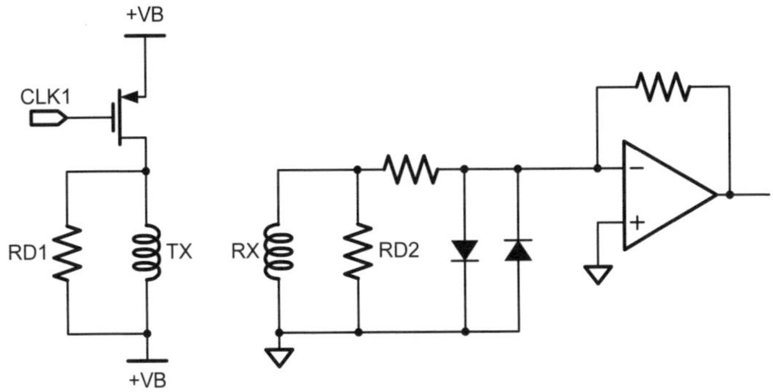

Fig. 12-20: **Separate TX & RX Coils**

anced. This gives us the freedom to increase the windings which improves sensitivity[13]. Another advantage is the RX coil can be referenced to a different point than the TX coil, making the preamp's power supply design easier.

VLF coils are almost always wound from enameled magnet wire. PI coils can also be wound using magnet wire but the thin insulation results in a large self-capacitance which slows the flyback response and limits how early the decay can be sampled. A faster sample delay is often desirable (especially for hunting small gold nuggets) so many PI designs focus on getting a fast flyback response. One way is to decrease the coil self-capacitance by using wire that has a thicker insulation or insulation with a higher dielectric material. Teflon-insulated wire is often chosen for this. Furthermore, stranded wire has shown to have a slight improvement in self-eddy generation, with Litz wire being the best.

Really Advanced Topics

We have not discussed discrimination because traditional PI designs have not had that ability, at least not like VLF. The subtractive ground balance method can be expanded to provide some level of iron discrimination. A number of experimenters have proposed sampling the decay curve at multiple points and identifying targets by their characteristic decay. This has so far not been realized, probably because the decay signal is primarily resistive and is missing the reactive component.

Another approach is to look at the target response during the transmit pulse, which requires the use of an induction balanced coil. During the transmit pulse both reactive and resistive signal components are available so target ID is easier, much akin to a VLF. However, di/dt is much lower so depth will suffer. A compromise is to combine decay sampling for raw depth and ground balance with transmit sampling for identifying shallower targets.

Another technique is to transmit two or more different pulse widths and process them through different channels. The channels can be compared for both ground and

13. More turns on the RX coil increases the signal strength of targets, but also increases the signal strength of the ground signal as well as external noise sources (interferers). Another adjunct to non-existent complementary banquets.

target information. The idea is that ground viscosity effects are generally independent of transmit pulse width, whereas target responses are not.

Finally, pulse induction is only one type of time-domain detection. There are many other transmit waveforms that can be used for time-domain, including square wave, triangle, sawtooth, half-sine, and multiple slope. Furthermore, any style waveform can be generated as a complementary bipolar signal which inherently cancels Earth field signals (and some noise) and can allow the use of higher pulse rates.

Even though pulse induction has been around for 60 years or so, it has been relegated to specialty detectors and has not been explored as deeply as induction balance. There is tremendous room for technological advances, especially in the realm of target identification and discrimination. PI has the ability to detect deeper than IB, so with better ground balance techniques and the addition of discrimination it could easily displace IB as the mainstream technology.

Construction Details — PI-1

Because this chapter has 4 projects, construction details will be limited to information needed to build the PC board and get it running. Specific packaging and interface design, plus coil construction, are left to the reader.

The component placement and PCB layout are for illustrative purposes only. You may need to adjust the layout to accommodate components available in your area. In particular, please note that transistor pinouts can vary depending on the country of manufacture, even for what appears to be an identical part number. See Chapter 15 for more details.

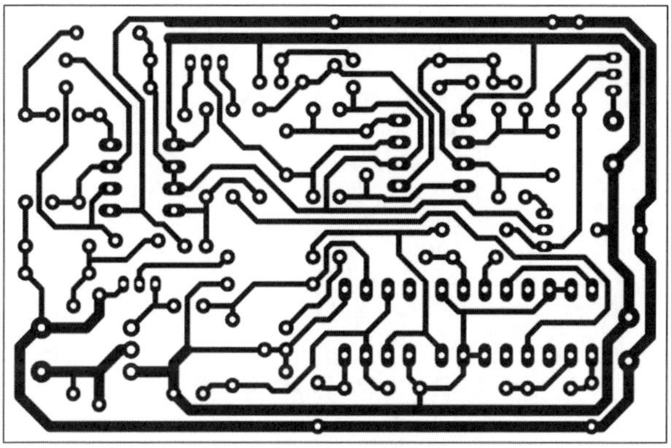

Fig. 12-21: **PI-1 PCB Layout**

The PCB layout view (Figure 12-21) is from the topside of the board[14]. The actual size of the PCB is 3.1" x 2.0" (7.9cm x 5.1cm). Note that there are 3 jumpers required in this layout. The parts placement (Figure 12-22) is also shown from the top-side.

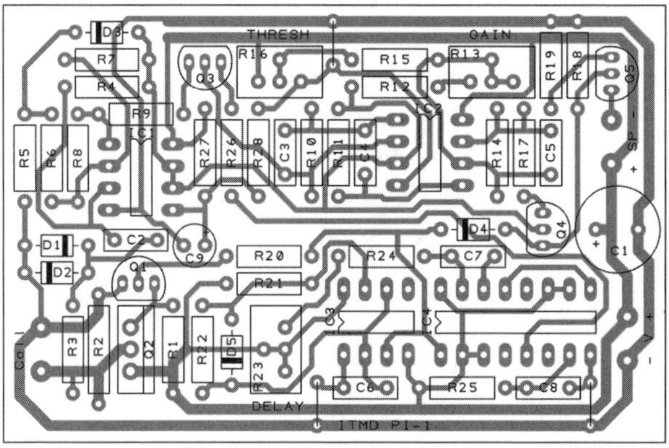

Fig. 12-22: **PI-1 PCB Parts Placement**

14. Artwork should be at 1x scale; see Chapter 15 for additional information.

Parts List — PI-1

Resistors: (5% 1/4W)

R1, R3, R7, R9, R24, R26	1k
R2	470 (1/2W)
R4, R5, R17, R18, R20	10k
R6	6k8
R8, R10, R14, R27, R28	100k
R11, R15	470k
R12	470
R19	10
R21	12k
R22	220k
R25	680

Potentiometers:

R13	10k
R16	100k
R23	2k

Capacitors:

C1	1000u elect., 16v
C2	1u
C3, C4, C5	470n
C6, C7, C8	10n
C9	10u elect., 10v

Inductors:

L1	Search coil (see main text for details)

Transistors:

Q1, Q5	2N3904 NPN
Q2	IRF9630 PMOS
Q3	2N3819 NJFET
Q4	2N3906 PNP

Diodes:

D1-D5	1N4148

ICs:

IC1	NE5532
IC2	TL072
IC3	LM555
IC4	74HC221

Switches:

S1	SPST, combined with R13

Misc:

Speaker	8-64 ohm
IC sockets (opt)	(3) 8-pin, (1) 16-pin
Control box	
Control knobs	(3)
Battery	9V battery, or (6) AA alkaline batteries & holder
Coil cable	Preferably coax, RG58 or equivalent

Building PI-1

IC sockets are recommended but not necessary. It is good practice to build the circuit in stages and test as you go. Start with the rail splitter circuit, then the 555 oscillator, additional timing, preamp, and finally add the integrator, gain stage, and audio. An oscilloscope is highly recommended for testing and troubleshooting.

Set the gain to minimum, set the pulse delay to minimum, and adjust the threshold to get a low-level audio hum. Increase the gain until the audio starts to "chatter." PI detectors are notoriously difficult to use indoors due to high levels of EMI, so don't be surprised if indoor testing is noisy. Test using targets such as foil, a thin gold ring, a silver coin, and a nail. Increase the delay and note what happens to the foil response.

Don't be afraid to experiment with the circuit. The preamp is specified to be an NE5532, but many high-speed opamps will work just fine here; try to choose one with a gain-bandwidth of 5MHz or better. Depending on the opamp used, R6 may need to be adjusted to compensate for variations in input bias current. It may be best to use a potentiometer for R6 until the value that gives a centered output signal is determined; this is best done with an oscope.

IC2 can also be substituted for, but a JFET-input opamp is preferred to avoid an induced offset with R11 due to input bias current. If a CMOS-input opamp is used here, a SPST switch added in series with R11 can be used to switch out R11 and effect a zero-motion mode. The switch will need to be occasionally shorted to account for SAT drift. As mentioned before, the 74221 chip should be CMOS compatible, not TTL. Likewise, so should the 555 timer.

R21 and R22 can be replaced with potentiometers (use a series fixed resistor to establish minimum levels) to effect a variable pulse rate and pulse width. Replacing R25 with a pot allows you to vary the sample pulse width for the JFET. This has the effect of altering the integration time, which can trade off selectivity for sensitivity.

The speaker can be just about any impedance, or headphones. Adjust the value of R19 for a desired volume level.

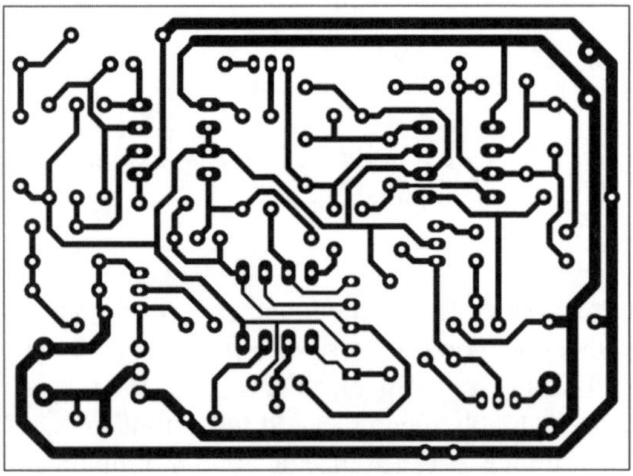

Fig. 12-23: **PI-2 PCB Layout**

The PCB layout view (Figure 12-23) is from the topside of the board. The actual size of the PCB is 2.75" x 2.0" (7cm x 5.1cm). Note that there are 2 jumpers required in this layout. The parts placement (Figure 12-24) is also shown from the top-side.

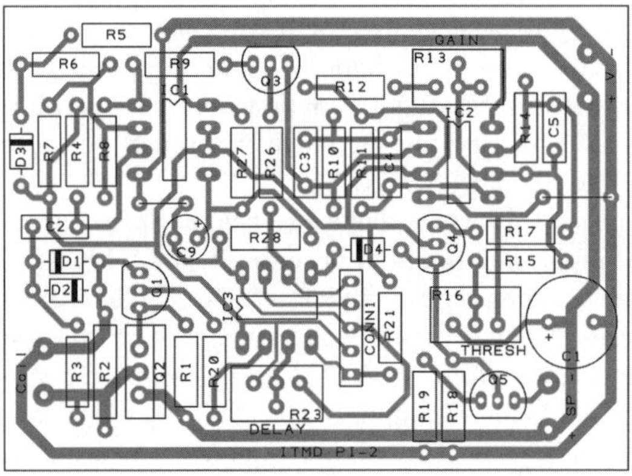

Fig. 12-24: **PI-2 PCB Parts Placement**

Parts List — PI-2

Resistors: (5% 1/4W)

R1, R3, R7, R9, R26	1k
R2	470 (1/2W)
R4, R5, R17, R18, R20	10k
R6	6k8
R8, R10, R14, R21, R27, R28	100k
R11, R15	470k
R12	470
R19	10

Potentiometers:

R13	10k
R16, R23	100k

Capacitors:

C1	1000u elect., 16v
C2	1u
C3, C4, C5	470n
C9	10u elect., 10v

Inductors:

L1	Search coil (see main text for details)

Transistors:

Q1, Q5	2N3904 NPN
Q2	IRF9630 PMOS
Q3	2N3819 NJFET
Q4	2N3906 PNP

Diodes:

D1-D4	1N4148

ICs:

IC1	NE5532
IC2	TL072
IC3	PIC 12F1840

Switches:

S1	SPST, combined with R13

Misc:

Speaker	8-64 ohm
IC sockets (opt)	(3) 8-pin
SIP header	5-pin
Control box	
Control knobs	(3)
Battery	9V battery, or (6) AA alkaline batteries & holder
Coil cable	Preferably coax, RG58 or equivalent

Building PI-2

Building PI-2 is similar to PI-1 except you will need to program the PIC. Use either the 5-pin header or a socketed PIC programmer. If you use a socketed programmer, be sure to install the PIC on the PCB with an IC socket so you can remove it and re-program if necessary. The complete source code follows this section; see Appendix B on where to download the source code for all three PIC projects.

Like PI-1, build the circuit in stages and test as you go. If the PIC is programmed correctly it should start running when power is applied to it. Keep in mind that power to the PIC is derived from half the battery voltage and should not exceed 5V, therefore the battery voltage should not exceed 10V. If you want to run a higher battery voltage then you should add a voltage regulator (such as a 78L05) for the PIC.

Use a foil target (or an oscilloscope if you have one) to test that the pulse delay is working properly. The pulse delay is under the control of the pot on pin 7. If the pulse delay is set to a fixed value in code, then the pot can be used to control some other parameter, such as the pulse frequency. Replacing the last two lines in ReadDelayPot() with the following will result in variable pulse width of 50μs to 305μs:

```
adc = ADRESH;                          // Look at the 8 MSBs; range = 0 to 255
TxPulseWidth = 65535 - 400 - adc<<3;   // 50us min, 305us max
```

The effect of increasing the pulse width is a small increase in depth with a fairly large increase in battery consumption. Also pin 3 is available to attach a second pot.

Experienced programmers will notice the use of the Watchdog Timer in the PI-2 code. In some cases micros have problems starting up properly in PI circuits, so the Watchdog Timer is used to ensure proper start-up.

C Source Code

```
//-------------------------------------------------------------------------------
// Inside the Metal Detector
// PI-2 Firmware
//-------------------------------------------------------------------------------
// This project uses the PIC 12F1840
// Clock is run off the internal 32MHz oscillator
// Each instruction takes 4 clock cycles so time resolution is 0.125us
// Timer0 controls the main loop, has 8 bits plus prescaler.
// Timer1 controls delays, has 16 bits plus prescaler.

//--- Includes ---
#include <htc.h>
__CONFIG(FOSC_INTOSC & MCLRE_ON & PLLEN_ON & WDTE_ON & PWRTE_OFF & CP_OFF \
& BOREN_OFF & CLKOUTEN_OFF & IESO_OFF & FCMEN_OFF);

//--- Defines ---
#define INPUT_PIN     1
#define OUTPUT_PIN    0

#define TxPulse       LATA5
#define Sample        LATA1
#define DelayPot      RA4                     // AN3
#define PWM           LATA2
#define ADC           RA0                     // AN0, PI-3 & PI-4

#define TrisTxPulse   TRISA5
#define TrisSample    TRISA1
#define TrisDelayPot  TRISA4
#define TrisPWM       TRISA2
#define TrisADC       TRISA0                  // PI-3 & PI-4
#define Period        47                      // 255-208 -> 1664us (~600 Hz)
#define TxPulseWidth   64735                  // 65535-800 -> 100us
#define SampleWidth    65415                  // 65535-120 -> 15us
#define Delay10us      65455                  // 65535-80 -> 10us delay
#define AudioDelay     65335                  // 65535-200 -> 25us delay
#define AudioWidth     63935                  // 65535-1600 -> 200us

// Select AN0 for signal capture (right just.)
#define SelectSample() CHS2=0; CHS1=0; CHS0=0; ADFM=1;
// Select AN3 for the pot (left just.)
#define SelectDelayPot() CHS2=0; CHS1=1; CHS0=1; ADFM=0;

//--- Variables ---
unsigned int Sample1Delay = 65535-120;  // 15us
```

```
//--- Prototypes ---
void Init(void);
void InitAdc(void);
void InitPwm(void);                         // PI-3 & PI-4
void InitTimer0(void);
void InitTimer1(void);
void Timer1Delay(unsigned int);
void ReadDelayPot(void);
void ReadSample(void);                      // PI-3 & PI-4
void ProcessSample(void);                   // PI-3 & PI-4
void CreateAudio(void);                     // PI-3 & PI-4

//--------------------------------------------------------------------------------
// Main -- enter the program here
//
void main(void)
{
  // First, set up the internal oscillator
  IRCF3=1; IRCF2=1; IRCF1=1; IRCF0=0;    // Internal osc. freq. = 32MHz
  WDTPS4=0; WDTPS3=0; WDTPS2=0;          // Set the WDT to 4ms
  WDTPS1=1; WDTPS0=0;
  CLRWDT();                             // Using the WDT prevents PI latch-up

  // Next, initialize everything
  Init();                                // below
  InitAdc();
  InitTimer1();                          // Do this BEFORE InitTimer0()
  InitTimer0();
  InitPwm();                             // PI-3 & PI-4


  // Now start the interrupt timer and idle
  TMR0IF = 0;                            // Clear Timer0 flag
  TMR0 = 200;                            // Set the timer
  while(1) {}                            // Keep the processor active
}

//--------------------------------------------------------------------------------
// Initialization routines

// Initialize the I/O pins and set some flags
//
void Init(void)
{
  TrisTxPulse  = OUTPUT_PIN;             // Set TX pulse to output
  TrisSample   = OUTPUT_PIN;             // Set main sample to output
  TrisPWM      = OUTPUT_PIN;             // Set pwm to output
  TrisDelayPot = INPUT_PIN;              // Set delaypot to input
  TrisADC      = INPUT_PIN;              // Set ADC to input (PI-3 & PI-4)

  GIE = 1;                               // Master interrupt enable
  PEIE = 0;                              // Peripheral interrupt disable
  CCP1IE = 0;                            // Capture/compare interrupt disabled
  CM1CON0 = 0x07;                        // Turn off analog comparator

  TxPulse = 0;                           // Initialize starting pulse values
  Sample = 0;
}

// Initialize the ADC.
//
void InitAdc(void)
{
  ANSELA = 0;                            // Clear the entire ADC select register
  ANSA0  = 1;                            // Set RA0 (AN0) to be an ADC input
  ANSA4  = 1;                            // Set RA4 (AN3) to be an ADC input (PI-3)
```

```
  ADCON0 = 0;                              // Clear the entire ADC control register
                                           //   This sets the active ADC channel to A0
                                           //   which is what we want to start with.

  ADPREF1=0; ADPREF0=0;                    // Use VDD for Vref

  ADCS2=0; ADCS1=1; ADCS0=0;               // ADC clock = fosc/32
  ADON = 1;                                // Turn the ADC on
}

// Defined for PI-3 & PI-4
//
void InitPwm(void)
{
  PWM = 0;
}

// Timer0 is 8 bits with Div64, so delay is (255-n)*64*4/Fosc
// For Fosc = 32MHz Timer0 has 8us of resolution
//
void InitTimer0(void)
{
  TMR0IE = 1;                              // Enable Timer0 interrupt
  TMR0CS = 0;                              // Clock off Fosc/4
  PSA = 0;                                 // Prescaler assigned to Timer0
  PS2=1; PS1=0; PS0=1;                     // Prescaler = /64 (8us resolution)
}
// Timer1 is 16 bits and clocked at 32MHz/4; use PS=1 for 0.125us resolution
//
void InitTimer1(void)
{
  TMR1CS1=0; TMR1CS0=0;                    // Use Fosc/4
  T1CKPS1=0; T1CKPS0=0;                    // Prescaler = /1 (125ns resolution)
  TMR1H=0; TMR1L=0;                        // Clear the registers
  TMR1GE = 0;                              // Gate enable is off
  TMR1IE = 0;                              // Don't allow an interrupt
  TMR1ON = 1;                              // Turn Timer1 on
}

//-----------------------------------------------------------------------------
// Processing routines

// Delay (in us) = 65535-n
// Because of the fudge factor, ~5us is the minimum delay.
//
void Timer1Delay(unsigned int n)
{
  n += 36;                                 // Adjust for fixed error (approx 4.5us)
  TMR1H = n >> 8;                          // Load the upper counter
  TMR1L = n;                               // Load the lower counter
  TMR1IF = 0;                              // Clear the flag
  while(TMR1IF == 0) {;}                   // Wait for a flag
}

// ISR() is the interrupt service routine for Timer0. The main loop is
// triggered off this interrupt and must be completed before the next
// interrupt.
//
static void interrupt ISR(void)
{
  TMR0IF = 0;                              // Clear the Timer0 Int flag
  TMR0 = Period;                           // Reset the timer (+2 fudge factor)
  CLRWDT();                                // Using the WDT prevents PI latch-up

  TxPulse = 1;
  Timer1Delay(TxPulseWidth);               // Transmit pulse
  TxPulse = 0;
```

```
  Timer1Delay(Sample1Delay);                 // TX-to-sample pulse delay
  Sample = 1;
  Timer1Delay(SampleWidth);                  // Sample pulse
  Sample = 0;

  SelectDelayPot();                          // Set the ADC to read the pot

  Timer1Delay(AudioDelay);                   // Delay for audio
  PWM = 1;
  Timer1Delay(AudioWidth);                   // Speaker pulse
  PWM = 0;

  // The remaining routines are where extra processing gets done. It is critical
  // that all processing is complete before the next TMR0 interrupt occurs, which
  // depends on the TX pulse rate, TX pulse width, and sampling time. If the pulse
  // rate = 600Hz, pulse width = 100us, and max sampling time = 35us then the
  // processing time available is 1667us - 100us - 35us = 1532us. With a 32MHz
  // clock we have an instruction cycle of 0.125us, so there is time for 12256 code
  // instructions, including calls and returns.

  ReadDelayPot();
}
// Read the pulse delay pot
//
void ReadDelayPot(void)
{
  static char adc;

  GO = 1;                                    // Start the conversion
  while(GO == 1);                            // Wait for the ADC to finish

  adc = ADRESH >> 1;                         // Look at the 7 MSBs; range = 0 to 127
  Sample1Delay = 65535 - 120 - adc;          // 15us min, 31us max
}
```

Construction Details — PI-3

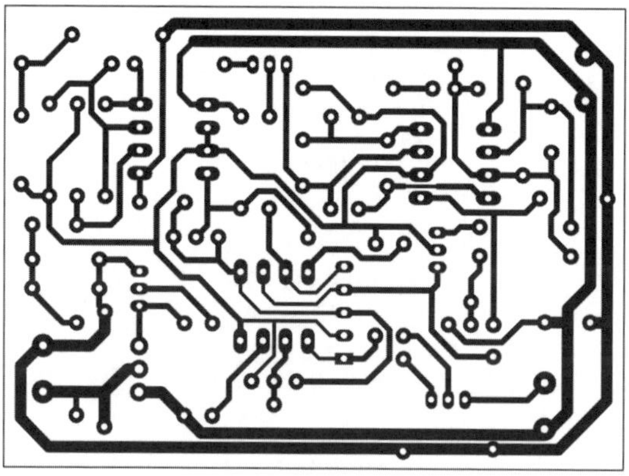

Fig. 12-25: **PI-3 PCB Layout**

The PCB layout view (Figure 12-25) is from the topside of the board. The actual size of the PCB is 2.75" x 2.0" (7cm x 5.1cm). Note that there are 2 jumpers required in this layout. The parts placement (Figure 12-26) is also shown from the top-side.

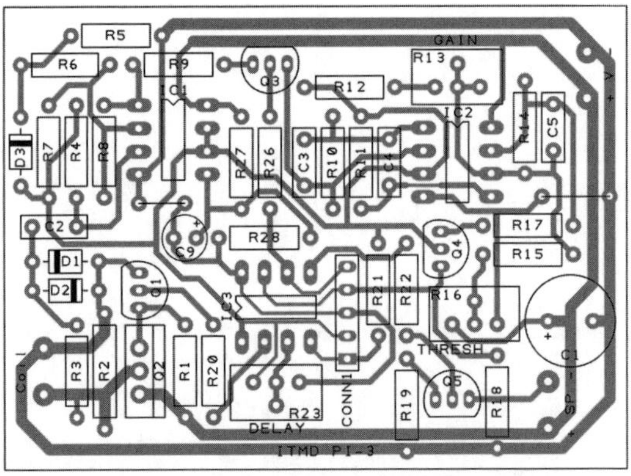

Fig. 12-26: **PI-3 PCB Parts Placement**

Parts List — PI-3

Resistors: (5% 1/4W)

R1, R3, R7, R9, R26	1k
R2	470 (1/2W)
R4, R5, R17, R18, R20, R22	10k
R6	12k
R8, R10, R14, R21, R27, R28	100k
R11, R15	470k
R12	470
R19	10

Potentiometers:

R13	10k
R16, R23	100k

Capacitors:

C1	1000u elect., 16v
C2	1u
C3, C4, C5	470n
C9	10u elect., 10v

Inductors:

L1	Search coil (see main text for details)

Transistors:

Q1, Q5	2N3904 NPN
Q2	IRF9630 PMOS
Q3	2N3819 NJFET
Q4	2N3906 PNP

Diodes:

D1-D3	1N4148

ICs:

IC1	NE5532
IC2	TL072
IC3	PIC 12F1840

Switches:

S1	SPST, combined with R13

Misc:

Speaker	8-64 ohm
IC sockets (opt)	(3) 8-pin
SIP header	5-pin
Control box	
Control knobs	(3)
Battery	9V battery, or (6) AA alkaline batteries & holder
Coil cable	Preferably coax, RG58 or equivalent

Building PI-3

PI-3 is almost identical to PI-2. Much of the PI-3 source code is re-used from PI-2, so instead of re-listing all of the code we will list only the *changes* that need to be made. Some of these lines replace existing lines in the PI-2 code, some of the code is new.

C Source Code

Add these lines to the "Variables" section

```
unsigned int CurrentSample;           // Holds the current preamp sample
unsigned int SampleAve = 0;           // Running average of 4*32 samples
unsigned int TrackAve = 0;            // Running average of 4*32*64 samples
```

Replace the InitPwm() routine with this one

```
// Initialize the PWM. Frequency is (Fosc/4)/64/255 = 490Hz;
// 64 is the TMR2 prescaler and 255 is the PR2 value.
//
void InitPwm(void)
{
  T2CON = 0;                          // Clear the Timer2 control register
  T2CKPS1=1; T2CKPS0=1;               // Set TMR2 prescaler to 64
  PR2 = 0xFF;                         // Set PR2=255
  CCP1CON = 0b00001100;               // Set the CCP register up for std PWM
  CCPR1L = 127;                       // Clear the CCPR1L register
  TMR2ON = 1;                         // Turn on TMR2
}
```

Replace the ISR() routine with this one

```
static void interrupt ISR(void)
{
  TMR0IF = 0;                         // Clear the Timer0 Int flag
  TMR0 = Period;                      // Reset the timer (+2 fudge factor)
  CLRWDT();                           // Using the WDT prevents PI latch-up

  SelectSample();                     // See #defines

  TxPulse = 1;
  Timer1Delay(TxPulseWidth);          // Transmit pulse
```

```
  TxPulse = 0;

  Timer1Delay(Sample1Delay);            // TX-to-sample pulse delay
  Sample = 1;
  Timer1Delay(SampleWidth);             // Sample pulse
  Sample = 0;

  // The remaining routines are where extra processing gets done. It is critical
  // that all processing is complete before the next TMR0 interrupt occurs, which
  // depends on the TX pulse rate, TX pulse width, and sampling time. If the pulse
  // rate = 600Hz, pulse width = 100us, and max sampling time = 35us then the
  // processing time available is 1667us - 100us - 35us = 1532us. With a 32MHz
  // clock we have an instruction cycle of 0.125us, so there is time for 12256 code
  // instructions, including calls and returns. Current processing code runs in
  // about 415us.

  ReadSample();                         // Sample the signal
  SelectDelayPot();                     // See #defines
  ProcessSample();
  CreateAudio();
  ReadDelayPot();
}
```

The remainder is all new code

```
// Read in the signal voltage
//
void ReadSample(void)
{
  GO = 1;                               // Start the conversion
  while(GO == 1);                       // Wait for the ADC to finish

  CurrentSample = ADRESH & 0x03;
  CurrentSample <<= 8;
  CurrentSample += ADRESL;
  CurrentSample &= 0x03FF;              // Mask off 6 MSBs
}

// Process the latest sample into the Sample & Tracking arrays
//
#define sampleAveShift 5
#define sampleArraySize 32
#define trackAveShift 6
#define trackArraySize 64

void ProcessSample(void)
{
  static unsigned int sampleArray[sampleArraySize];
  static unsigned int trackArray[trackArraySize];
  static char sampleIndex = 0, trackIndex = 0;
  static unsigned int tmp = 0;
  static char i, sampleCount;

  tmp += CurrentSample;                 // tmp is effectively a 12-bit value
  sampleCount++;

  if(sampleCount == 4)                  // If we've accumulated 4 samples, store
  {
    sampleArray[sampleIndex] = tmp>>2;  // Store the current sample - 10 bits
    sampleCount = 0;
    tmp = 0;

    sampleIndex++;                      // Increment the array pointer
    if(sampleIndex == sampleArraySize)  // If we're wrapping around...
      sampleIndex = 0;                  // Set the pointer back to the beginning

    SampleAve = 0;
```

```
    for(i = 0; i < sampleArraySize; i++)       // Average all the samples
      SampleAve += sampleArray[i];             // This is effectively a 15-bit value

    SampleAve >>= sampleAveShift;              // Divide back down to a 10-bit value

    trackArray[trackIndex] = SampleAve;        // Store the average

    trackIndex++;                              // Increment the array pointer
    if(trackIndex == trackArraySize)           // If we're wrapping around...
      trackIndex = 0;                          // Set the pointer back to the beginning

    TrackAve = 0;
    for(i = 0; i < trackArraySize; i++)        // Average all the averages
      TrackAve += trackArray[i];               // This is effectively a 16-bit value

    TrackAve >>= trackAveShift;                // Divide back down to a 10-bit value
  }
}

// Create a PWM based on the difference between SampleAve and TrackAve. THRESH is
// the threshold offset when SampleAve = TrackAve; when SampleAve is above TrackAve
// the PWM duty cycle is increased, which results in a louder volume.
//
#define THRESH 2
typedef union
{
  int value;
  unsigned char junk:6;
  unsigned char CCPR1L:8;
  unsigned char DC1B1:1;
  unsigned char DC1B0:1;
} DutyCycle;

void CreateAudio()
{
  static DutyCycle dc;

  dc.value = SampleAve - TrackAve + THRESH;
  if(dc.value < THRESH) dc.value = THRESH;
  if(dc.value > 127) dc.value = 127;

  PR2 = 255-dc.value;                          // VCO option

  CCPR1L = dc.CCPR1L;                          // Bits 2-9 here
  DC1B1 = dc.DC1B1;
  DC1B0 = dc.DC1B0;
}
```

Construction Details — PI-4

The PCB layout view (Figure 12-27) is from the topside of the board. The actual size of the PCB is 2.1" x 1.75" (5.3cm x 4.5cm). Note that there are 2 jumpers required in this layout. The parts placement is also shown from the top-side.

The designs in this book have all been presented with through-hole PCBs which are relatively simple for the home builder to deal with. Using surface mount technology (SMT) this design can be made exceptionally small for niche uses. The SMT layout and parts placement are shown in Figure 12-28 (all top-side views). This PCB is a tiny 1"x1.5" and, by using SMT electrolytic caps and eliminating the programming header, it could be made even smaller.

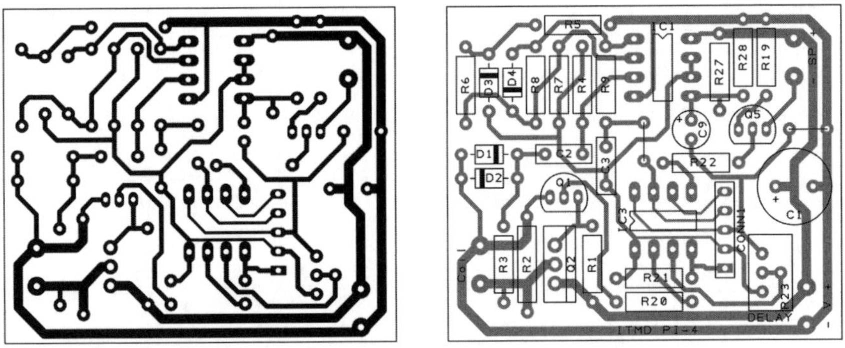

Fig. 12-27: **PI-4 PCB Layout & Parts Placement**

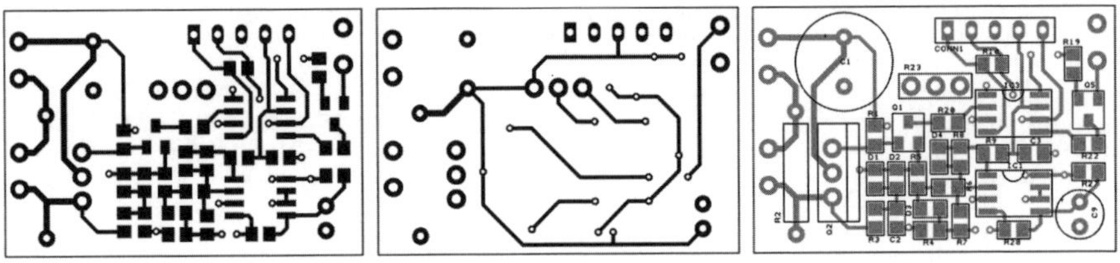

Fig. 12-28: **PI-4 SMT Layout (Top & Bottom) & Parts Placement**

Parts List — PI-4

Resistors: (5% 1/4W)

R1, R3, R7, R9	1k
R2	470 (1/2W)
R4, R5, R20, R22	10k
R6	22k
R8, R21, R27, R28	100k
R19	10

Potentiometers:

R23	100k

Capacitors:

C1	1000u elect., 16v
C2	1u
C9	10u elect., 10v

Inductors:

L1	Search coil (see main text for details)

Transistors:

Q1, Q5	2N3904 NPN
Q2	IRF9630 PMOS

Diodes:

D1-D4	1N4148

ICs:

IC1	NE5532
IC3	PIC 12F1840

Switches:

S1	SPST

Misc:

Speaker	8-64 ohm
IC sockets (opt)	(2) 8-pin
SIP header	5-pin
Control box	
Control knobs	(1)
Battery	9V battery, or (6) AA alkaline batteries & holder
Coil cable	Preferably coax, RG58 or equivalent

Building PI-4

Building PI-4 is similar to PI-1 except you will need to program the PIC. Use either the 5-pin header or a socketed PIC programmer. The source code changes from PI-3 are as follows:

C Source Code

Replace the ISR() routine with this one

```
static void interrupt ISR(void)
{
  TMR0IF = 0;                        // Clear the Timer0 Int flag
  TMR0 = Period;                     // Reset the timer (+2 fudge factor)
  CLRWDT();                          // Using the WDT prevents PI latch-up

  SelectSample();                    // See #defines

  TxPulse = 1;
  Timer1Delay(TxPulseWidth);         // Transmit pulse
  TxPulse = 0;

  Timer1Delay(Sample1Delay);         // TX-to-sample pulse delay

  // The remaining routines are where all the processing gets done. It is critical
  // that all processing is complete before the next TMR0 interrupt occurs, which
  // depends on the TX pulse rate, TX pulse width, and sampling time. If the pulse
  // rate = 600Hz, pulse width = 100us, and max sampling time = 35us then the
  // processing time available is 1667us - 100us - 35us = 1532us. With a 32MHz
  // clock we have an instruction cycle of 0.125us, so there is time for 12256 code
  // instructions, including calls and returns. Current processing code runs in
  // about 415us.

  ReadSample();                      // Sample the signal
  SelectDelayPot();                  // See #defines
  ProcessSample();
  CreateAudio();
  ReadDelayPot();
}
```

Replace the ReadSample() routine with this one

```
void ReadSample(void)
{
  GO = 1;                           // Start the conversion
  while(GO == 1);                   // Wait for the ADC to finish

  CurrentSample = ADRESH & 0x03;    // Get the high byte
  CurrentSample <<= 8;              // Move into place
  CurrentSample += ADRESL;          // Add the low byte
  CurrentSample = ~CurrentSample;   // Bit-wise invert
  CurrentSample &= 0x03FF;          // Mask off 6 MSBs
}
```

CHAPTER 13

Long Range Locators

"Our capacity for self-deception has no known limits."
— *Michael Novak*

In our overview of detector types in the *Introduction*, we briefly mentioned long range locators and flatly stated that they do not work. So why in the world do we include an entire chapter on Long Range Locators (LRLs)? Because the claims made by manufacturers are incredible and, quite frankly, *very enticing*. Just imagine[1], a device that can detect buried treasure from miles away! And can detect *only* gold while ignoring everything else! Who wouldn't be interested?

Most LRLs are also very expensive — some well over $10,000 — so buying a commercial unit and discovering that it only finds empty holes[2] is a very costly mistake. Therefore, several different "styles" are presented and thoroughly explained, so that you will know what they really are. We also take a look at some rigorous testing methods that consistently expose the failures of LRLs.

Dowsing

Because dowsing plays such a large role with many LRLs, let's take a little side trip and briefly investigate this topic. Dowsing has a rather obscure history, not helped in the least by some dowsing books which have given blatantly wrong information.

History

Dowsing most likely originated as a type of *divination*, which is really nothing more than fortune-telling. Divination goes way, *way* back in human history. It is mentioned a few times in the Bible, most notably in the use of the stones *Urim* and *Thummim*. Throughout history, prominent kings and military leaders would often seek the advice of diviners, or "seers." Divination usually involved the interpretation of something physical, such as the casting of stones (a form of *cleromancy*, such as with *Urim* and *Thummim*), or natural events that might be occurring at the time such as comets or solar eclipses.

At some point, diviners began using hand-held devices which would move, seemingly on their own, to provide some sort of indication. When this first began is

1. Imagination figures prominently with LRLs.
2. Although many LRLs are quite proficient at detecting gravity.

Fig. 13-1: **A Prehistoric Dowser?**

lost in history, but that doesn't stop dowsers from making something up[3]. For example, many dowsing books claim that an 8000-year-old cave painting in northern Algeria is the earliest known depiction of a dowser. The painting, shown in Figure 13-1[4], is known as "The Archers of Tin Aboteka" and is part of a larger group of cave art that has been dated to roughly 6000 BCE. But does this 8000-year-old painting really show someone dowsing? It certainly *appears* so at first glance. But if you look closely you will see the man is holding a bow (there is a faint bow string) and arrow (there is an arrowhead). The fact that the paintings are known as "The Archers of Tin Aboteka" is also a subtle clue. Obviously, this is no dowser.

One of the earliest indisputable publications of dowsing in the manner it is now practiced appeared in 1556 in the book De Re Metallica[5] by Georgius Agricola. In only two pages out of the whole book, he describes the use of dowsing in the search for gold in the mining regions of Germany, and includes a woodcut showing a dowser using a forked stick (Figure 13-2), just as some modern-day water-witchers still use. Interestingly, even in 1556, Agricola dismissed dowsing as folly. He recognized the "reaction" of the device as being induced by the dowser himself, and the sporadic successes of dowsers as being nothing but luck.

Georgius Agricola (1556):

> *...it must necessarily follow that the manipulation is the cause of the twig's twisting motion. It is a conspicuous fact that these cunning manipulators do not use a straight twig, but a forked one cut from a hazel bush, or from some other wood equally flexible, so that if it be held in the hands, as they are accustomed to hold it, it turns in a circle for any man wherever he stands.*

3. As we'll see, dowsers are particularly good at making things up, mostly in an attempt to explain a phenomenon that doesn't even exist.
4. Photo courtesy of David Brandt-Erichsen, Natural Arch and Bridge Society, http://www.naturalarches.org/tassili/rockart.htm
5. "Of the Metals"

Fig. 13-2: **Dowsing in Germany (Agricola)**

> *Nor is it strange that the twig does not turn when held by the inexperienced, because they either grasp the forks of the twig too tightly or hold them too loosely. Nevertheless, these things give rise to the faith among common miners that veins are discovered by the use of twigs, because whilst using these they do accidentally discover some; but it more often happens that they lose their labour, and although they might discover a vein, they become none the less exhausted in digging useless trenches than do the miners who prospect in an unfortunate locality. Therefore a miner, since we think he ought to be a good and serious man, should not make use of an enchanted twig, because if he is prudent and skilled in the natural signs, he understands that a forked stick is of no use to him, for as I have said before, there are the natural indications of the veins which he can see for himself without the help of twigs.*

One of the most popular applications of dowsing is for locating water wells[6]. It was in such widespread use in the early 1900's and generated so many questions for the U.S. Dept. of Geological Survey (USGS) that in 1917 they released a booklet entitled "The Divining Rod: A History of Water Witching" by Arthur J. Ellis. The USGS summarized the roughly 150 years of investigations that had consistently shown dowsing to be nothing but a mind trick. This booklet is not only important because of what it says about dowsing, but also because it contains an impressive 30 pages of

6. The search for groundwater is also the reason most cited by proponents for their beliefs. *"I know for a fact dowsing works, because my grandfather used to dowse for water wells, and was 100% successful."* Or something to that effect.

references dating back to the 1700's, making it an outstanding resource for anyone wanting to research the early history of dowsing.

There are other interesting elements of dowsing history, but this is not intended to be a book on dowsing, and even though dowsing books as written by dowsing proponents tend to be filled with horribly wrong information, we will have to stop here. It is sufficient to say that dowsing as we know it has been in fairly continuous use since at least the mid-16th century.

Reactions

Dowsing does not work the way dowsers believe (or wish) it works. Regardless of the dowsing device used — pendulum, L-rods, forked branch (Y-rod), bobber — its movement is entirely initiated *by the dowser*, and not by any outside physical force. There is no question that the device can *seem* to be moving on its own, and that the feeling is so overwhelmingly strong many (if not most) dowsers completely reject any suggestion they are causing the movement. But there is also no question, at least in the scientific community, that this is really the case[7].

How do we know that the dowsing response is not influenced by outside physical forces? Because we can easily test for this. Testability is the cornerstone of the scientific method, and can do a remarkably good job at exposing flaws in the human senses. For example, a good magician can easily fool our senses[8], but under scientific test controls that same "magic" would be exposed as mere illusions and trickery[9]. Likewise, dowsing can *seem* to work, and give the dowser a very eerie feeling, but in scientific testing it performs no better than chance guessing[10]. That eerie feeling is all in the mind.

To get an idea of what a dowsing reaction feels like, let's make a pair of dowsing rods and try them out.

> ***Experiment 13-1:*** Basic Dowsing Rods
>
> Make a pair of dowsing rods by bending two lengths of wire to approximately the dimensions shown below. The wire may simply be a length of clothes hanger wire, or brass stock; 5/32" (4mm) is a good diameter, and both rods can be made from a single 36" (1m) length.
>
> Hold the dowsing rods at about waist height with your elbows at your side and forearms level. The rods should be pointed straight ahead with a slight downward tilt, about 5-10°. Stretch a water hose across your back yard and turn on the water. Walk across the water hose (in a perpendicular path) and the rods should swing inward and cross when you are directly over the hose.

7. Most dowsers reading this will immediately say, "You are *wrong*!"
8. Even when we *know* we are being fooled!
9. Of course, any honest magician will readily admit that he has merely deceived your senses. A magician who claims "genuine" paranormal powers will avoid scientific testing like the plague.
10. With any statistical significance.

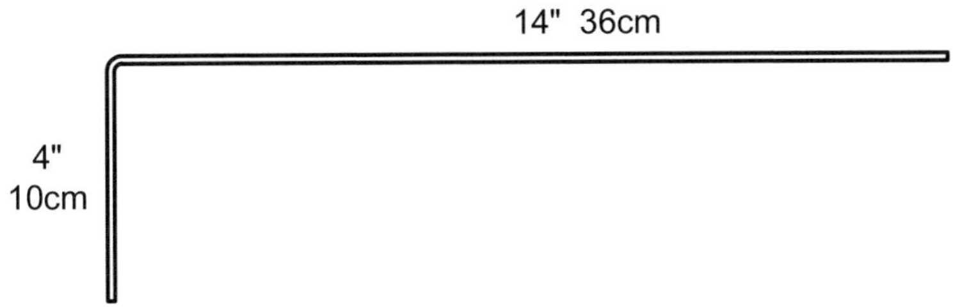

Fig. 13-3: **Basic L-Rod**

Congratulations — you are now an experienced water dowser!

Let's convert our water dowsing skills to treasure hunting. Lay a silver coin on the ground and walk over it; again, the rods should swing inwards and cross. Now lay a chunk of iron on the ground and repeat; with iron targets, the rods should swing outward away from one another.

Many people who try this exercise — even skeptics — will get a response from the rods. But if there is no physical force acting between the rods and the target then what, exactly, causes the device to react? What creates that eerie feeling, as if the rods are moving on their own? Again we look to science, in particular the science of psychology. There is an involuntary motor action known as *ideomotor* which is controlled by the subconscious and causes the dowsing device to move. Ideomotor action is well known to psychologists and manifests itself in other ways, but is beyond the scope of this book.

With dowsing, the dowser is holding a device, such as an L-rod, which is already somewhat unstable. The length and weight distribution of the L-rod makes it very sensitive to gravity and thus makes it difficult to hold steady when trying to keep it level[11]. It takes very little effort to cause such a device to begin to physically move, so the slightest wrist movement will cause the rod to swing. Once it begins to move, gravity[12] usually assists and magnifies the response.

The result is a "feeling" that the device is moving on its own, and most dowsers insist this is the case. The feeling is so compelling they are absolutely convinced that they are not causing the reaction, and that it is therefore the result of some mysterious force. But science has shown, repeatedly and indisputably, that it is ideomotor action that causes the movement of dowsing devices, and not some outside "force."

Also in the experiment, it is suggested that a silver target causes the rods to swing inward and an iron target causes the rod to swing outward. This scenario was not made up; this is what many dowsers actually do. And since dowsing is caused by a subconscious ideomotor action, it turns out that we can use suggestive statements to create particular responses. This is known as *autosuggestion*. Often if you tell a

11. "Holding it level" is required for L-rod dowsing. You could let the L-rod hang straight down, in which case it is highly stable but useless for dowsing.

12. Or applied torsion in the case of the Y-rod.

dowser to expect a particular reaction (rods swing inward) they will have that reaction. Tell a different dowser to expect the opposite reaction (rods swing outward), and that is what they will experience[13].

This is how dowsing techniques are passed along; a dowser who has developed certain responses will pass along his "methods" to a new dowser. This is also why there is little consistency in dowsing reactions; many different dowsers experience many different responses. It is all of their own fabrication, and more evidence that it is caused by psychological (imaginary), and not physical (real), forces.

Why Dowsing "Works"

If the dowsing response is purely imaginary, then why does it sometimes seem to succeed? There are several reasons. In the case of groundwater dowsing, it is well known[14] that groundwater exists in vast areas known as *aquifers*. Many dowsers mistakenly believe that water flows in narrow underground streams, and that any deviation from these streams will result in a dry well. So they dowse a location, a well is drilled, and water is found. Presto, dowsing works! But the same well could have been drilled just about anywhere in the area, with the same results. Dowsing only indicated what was *extremely* likely in the first place.

Another factor is experience. The more a dowser knows about his desired target, the more likely he is to be successful. Let's say someone who is already proficient at panning for gold tries dowsing. He walks along a stretch of stream, and the dowsing rods indicate a spot. He pans the spot and, sure enough, he finds gold. Presto, dowsing works! But it's very likely the prospector's accumulated knowledge of placer gold played a critical role in subconsciously selecting the spot. Dowsing indicated what the dowser already knew, or at least what he intuited. And gold panning also includes an element of high probability; that is, even a rank amateur prospector dowsing for gold is likely to find something — if only a few specks — just by panning anywhere in a known gold-bearing stream.

A third case is the post-hoc fallacy. Suppose someone dowses an area and gets a "hit." He pulls out a metal detector and, sure enough, within a few feet of his dowsed spot he gets a signal. He digs out a penny, and — presto, dowsing works! There are several obvious problems with this. First, what else is buried in the area that was not dowsed? Are there other items (perhaps even gold or silver) that dowsing failed to pick up? The whole area would need to be thoroughly metal detected to find out. How far from the dowsed spot was the target metal detected? One foot? Ten feet? The larger the radius, the more likely you will find *something* (not necessarily a coin), and anyone who metal detects knows very well how much metal is randomly buried in the ground. *How often* does this exercise result in success? Is it highly consistent, or sporadic? And finally, what, exactly, is "success"? If a copper penny was found while dowsing for gold, is that really a success?

All of these questions *must* be answered, honestly and continuously, in order to

13. *Bonus experiment*: teach dowsing to several friends, telling each of them different responses to expect; see if they subconsciously comply.

14. At least by hydrogeologists, though often not by dowsers.

properly evaluate the efficacy of dowsing. More than anything else, the matter of consistency should be closely looked at. Are detailed records kept of *all* dowsing attempts and the results of those attempts? People have a tendency to remember the few quasi-successes, and to conveniently forget all the spectacular failures. This is not just with dowsing, either; gamblers and lottery players are notorious for their selective memories, and their light wallets. Indeed, dowsing works every time dowsing works.

If we are to treat the apparent success of dowsing with skepticism, how in the world do we determine, *objectively*, whether it really works or not? The next section covers testing, and should be required reading of all dowsers. In fact, read it twice.

Dowsing Tests

A glaring weakness of dowsing proponents is the fact that most of them don't know how to properly design and perform a decent test. There have been countless reports of dowsers who go in their back yard, place a silver coin[15] on the ground, and walk around with dowsing rods getting a consistent response every time they pass over the target.

This is NOT a dowsing test. This is *self-deception.*

In most cases, a dowser cannot properly test his dowsing skills on his own. He needs an assistant. Since it is the dowser providing the means to make the dowsing device move, it is critical that the dowser *not know the location of the test target.* This is so important we'll repeat it: *when trying to dowse the location of a test target, the dowser must not know where it is at.* Knowing the location is all that is necessary for the subconscious to take over, and make the dowsing device respond.

So an assistant is needed to hide the target. But even this is not good enough. If the assistant is present during the dowser's attempt to locate the target, the assistant can provide cues that the dowser picks up on, aiding in the successful location of the target. This can be very subtle (such as nodding or fidgeting or making verbal noises), and can be entirely unintentional with both parties. That is, the assistant doesn't mean to give any hints, and the dowser doesn't realize he is picking up on them[16].

This is not unique to dowsing. It is not unusual for other people to look for and respond to someone else's ideomotor responses. A poker player will go to great lengths to suppress or hide their ideomotor responses, because they know all the other poker players are looking for them. Even a horse can learn to read ideomotor responses[17]. This is why proper testing is so critically important. People are easily fooled. Even a horse can fool us.

To avoid unintentional cueing, the assistant needs to place the target while the dowser is out of sight (and earshot), and then he needs to leave the area before the dowser arrives to do his thing. Unintentional cueing is magnified when there is an

15. Or a water hose, or whatever.

16. This effect was once intentionally demonstrated in a dowsing test; see Foster, Experiments on Rod Divining," *Journal of Applied Psychology*, 1923

17. Look up the fascinating case of Clever Hans.

audience, so it is equally important that anyone who gets to observe the dowsing attempts *not* be allowed to observe the target placements. In scientific testing, this is called a *double-blind* protocol.

This is still not good enough. *Where* the target is hidden is just as important as *how.* In simple tests, there are basically two ways to hide a target:

1. Place it anywhere in a large unmarked area
2. Place it in one of several individually marked locations

Each has advantages and disadvantages. Placing the target anywhere in a given area (say, a 1-acre lot) more closely replicates field conditions. However, it is critical to hide the target without leaving any clues whatsoever, and that is difficult to do. If you have to bury the target, then ground disturbance might be noticeable. Even footprints might provide clues. There is also the problem of defining success. How close does the dowser have to get? If he agrees to a 10-foot radius and manages to get within 11 feet, is that "close enough"?

With the second method, the target is hidden in one of several distinct (and clearly marked) locations. It is just like the classic "shell" game. There is no question as to where the target *might* be, and preventing clues to the actual location is much easier. Therefore, the dowser is left with dowsing only the specific marked locations, and determining which one has the target. There is no close; either he gets it right, or he gets it wrong. While this method is less like field dowsing, it is a method more typically used in scientific testing. And for the dowser, checking several known locations *should* be an easier test than trying to dowse an entire field[18], the same way that testing a metal detector on a few marked planted targets is easier than detecting a large field.

With either method there is the problem of picking a particular location to hide the target. The location chosen by the assistant needs to be completely randomized. Why? Because people have a tendency to make similar choices. *Pick a number between 1 and 10*; it has been shown that 3 and 7 are statistically the most popular initial choices. Pick a location in a 1-acre lot to hide a silver coin; whatever feature *you* zero in on for a hiding spot might be the same feature the dowser is drawn to. Complete randomization prevents the dowser from accidentally or intentionally trying to second-guess where the assistant hid the target.

How would you randomize the location selection in a field-type test? You would need to grid the field to some predetermined grid size and label the grid in some manner (this can be done on a paper map). Then randomly select a grid square, using some kind of randomizer such as a computer program. Randomization is far simpler when the test uses just a few marked locations. As an example, if six locations are used, then the simple roll of a game die would suffice. It should also be noted that, with either method, the same location can be chosen more than once in a repeated test.

The method using discrete locations is more prevalent in science because it is easier to set up, easier to randomize, and easier to statistically analyze. So let's run through an example.

18. Which might also have other unknown buried targets.

Experiment 13-2: Simple Dowsing Test

A dowser claims he can consistently detect a 1-ounce silver coin when he walks over it. So an assistant digs six small shallow holes in the ground, spaced 10 feet apart. With the dowser away, he rolls a game die which comes up, say, 4. Beginning with the first hole, the assistant uses a small trowel to disturb the dirt in every hole, placing the coin in the fourth hole under 3-4 inches of dirt.

The assistant leaves and the dowser emerges. The dowser walks over each hole as he dowses. Once he makes his choice, the assistant returns and the coin is revealed. The dowsed location, and the actual location, are recorded. The test is repeated with several more trials and the results are analyzed.

A few things to note. The procedure specifies that when the assistant hides the target he stops at *every* hole in sequence and disturbs each one. The reason for this is so, during each trial, the assistant goes through exactly the same maneuver, disturbs all the holes equally, and distributes footprints equally. That is, the procedure is identical regardless of where the target is placed. Second, it is important that there is zero communication between the dowser and the assistant until after the dowser has made his final selection. Third, as mentioned before, if there are additional witnesses to the test, it is critically important to maintain strict isolation between the "hide" and the "dowse." Those who witness the dowse cannot have *any* knowledge of the hidden location.

One of the strengths of this kind of test is that the dowsing results can be compared to non-dowsing results. Assuming that dowsing does *not* work, what results would be expected from such a test? One way to find out is to try the test on several people without the use of dowsing. This can be done even indoors, on a table using paper cups. Using good randomized blind practices, place a target under 1-of-6 inverted cups. Have someone come in and simply guess the cup they think hides the target. Repeat numerous times. A better exercise is to incorporate guessing into a real dowsing test if additional people are available. Then you have a real-time baseline.

What you should find is that, in the long run, the correct location will be selected about one-sixth of the time. This exercise is merely testing people's ability to guess, and randomized guessing games such as this form the core of an introductory probability course. These exercises and their statistical outcomes are very well-known by mathematicians, and are highly predictable. We can say, with extreme confidence, that if dowsing does *not* work, then we would expect the outcome from the described dowsing test to be equivalent to random guessing. If dowsing does work, then we should expect better-than-chance results[19].

What most people have found is that when dowsers are subjected to rigorous randomized blind tests, they never perform statistically better than chance. Skeptic organizations and even government-funded research projects[20] have tested hundreds of "experienced" dowsers, and have produced very consistent results: namely, those

19. Amazingly, some dowsers disagree that dowsing results should exceed random guessing.

20. The largest being the "Scheunen" experiments, involving over 500 dowsers.

dowsers could not do what they thought they could do, at least in scientific tests. There have been many arguments from dowsers that scientific testing is unfair or, at the very least, an unreliable way to look at dowsing. They usually argue that you can only look at the results of real-world dowsing, which seem to show success. OK, we can do that, too.

Empirical Results

Even though randomized blind testing is a very reliable way to determine the efficacy of dowsing, it is possible to study the empirical results of dowsing, providing that care is taken to avoid unsupported anecdotal accounts and that there is a valid yardstick with which to compare. One documented case involves water dowsers in Australia where, for many years, well-drilling records included information on whether each well was found by dowsing or by traditional hydrogeological methods. An analysis of those records found that dowsers actually fared worse than hydrogeologists. If the majority of those dowsers considered themselves to be "experienced" then those results are very unflattering.

Fig. 13-4: **Humongous 875-oz Australian gold nugget found with a lowly metal detector... what have all those Australian dowsers found?**

Another case to consider is treasure dowsing. Many treasure dowsers claim to be "successful," but objective results[21] seem hard to come by. How could you compare treasure dowsing to traditional treasure hunting, or even treasure found by pure luck? One way is to simply look at the reports of treasure found, and how it was found. There have been numerous specialty treasure magazines published since the 1960's, and practically all of them include reports of recovered treasure, sometimes impressively massive hoards. In looking through 40 years worth of these magazines for treasure dowsing recoveries, it quickly becomes obvious that they are practically non-existent. Only one report was found involving the discovery of a large silver bar, and that was by the manufacturer of dowsing equipment, hardly making it an objective case.

Another popular venue for reporting treasure recoveries is the Internet, namely the many treasure forums that exist. On these forums metal detectorists have reported and shown off many impressive recoveries, including gold nuggets, rare coins, and valuable jewelry. What is glaringly missing are all the dowsers showing off their

21. Meaning real, hold-in-your-hand, *recovered* treasure.

recoveries. There have been a few sporadic reports, but even assuming these are legitimately objective recoveries, and even assuming there might be a few more dowsing recoveries that did not make the forums, these results absolutely pale in comparison to people who treasure hunt using traditional equipment.

A final example concerns British treasure troves. Treasure laws in the UK require that all treasure trove recoveries be reported, and the government has the right to purchase the recovery at fair market value. The British Museum has many of these recoveries on display, and the descriptions often include the statement, "found by detectorist." Metal detectorists are, in fact, responsible for finding the vast majority of British treasure troves. What about dowsers? What have they recovered in the way of treasure troves? Sadly, there is no record of British dowsers recovering any substantial treasure trove.

Whether land caches, shipwreck treasure, monster gold nuggets, or individual rare coins, dowsing has produced pathetic results. Even accidental treasure recoveries — often the result of backhoe excavations for construction — have a far more impressive record than does dowsing. Many cases of accidental treasure recoveries have been reported in the general news, treasure magazines, and forums. So even random digging has likely produced better results than dowsing.

Many dowsers will object to these comparisons, and claim that dowsers are very secretive about their finds. Why would dowsers be any more or less secretive than metal detectorists? Or cache hunters? Or shipwreck hunters? Probably they are not, and in the case of British treasure troves secrecy is not an option. In general, treasure hunters are not very good about keeping their finds secret, even when they insist discretion is of monumental importance. In one case that occurred in the late 1990's, a long-time treasure dowser who had never had any recoveries to speak of was insistent that successful dowsers *never* reveal their results out of fear of government confiscation. He finally recovered a cache of silver U.S. half-dollars and, even though he was assisted by a written account of the location of the cache, he was understandably proud to finally have a real "dowsing" recovery, to the point that he posted his story on forums all over the Internet. So dowsers would probably be no different than any other treasure hunters should they finally make a decent recovery.

Summary

Despite the miserable results of treasure dowsing, dowsers still insist that they can dowse buried treasure. They believe in what they do because dowsing goes back thousands of years; water dowsers are successful; they get an eerie feeling and the device appears to be moving on its own; when dowsing is backed up with a metal detector, it produces actual recoveries; there are legendary stories of massive recoveries by dowsers, such as the Spanish[22].

But regardless of the dowsing device used — pendulum, L-rods, forked branch, bobber — its movement is entirely initiated by the dowser, and not by any outside

22. It is true that the Spanish were highly successful in their quest for treasure, but the reality is they plundered it. Dowsing played no role, except in legend and in poorly researched (or outright fabricated) books on dowsing.

physical force. You can videotape a dowser, and easily see that hand movement *precedes* device movement. There is no question this is happening. Furthermore, double-blind testing has repeatedly shown that dowsing is no better than guessing. If an outside physical force were initiating the movement then it would still occur in double-blind tests; physics does not suddenly stop working under such conditions. But this is not the case; there is no outside force. Dowsing is widely recognized by science as wishful thinking.

LRLs

Long Range Locators (LRLs) are also sometimes called Long Distance Locators (LDLs). There are broadly two types of LRLs: non-electronic (passive) and electronic (active). As we will see, electronic LRLs are largely dowsing-based, but lately there have been several non-dowsing models

"Passive" LRLs

The history of treasure-specific dowsing devices is obscure at best. In the early twentieth century (and possibly back into the late nineteenth century) a device called the "Spanish Dip Needle[23]" (Figure 13-5) was sold to prospectors as a gold locator. It is nothing but a Y-rod style dowsing device, and had nothing to do with the Spanish. The glass tube that supposedly holds gold nuggets actually contains pyrite,[24] plus black sand. It was a classic "snake oil" product similar to the cure-all medicinal remedies of the time.

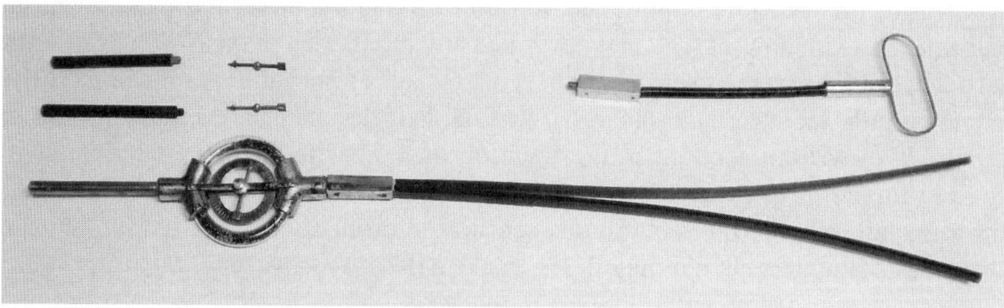

Fig. 13-5: **Antique Spanish Dip Needle**

The dowsing rod pictured in Figure 13-5 was likely made in the 1920's or 30's. It has a chamber that holds two "D" cell batteries used for illuminating a small bulb, so technically this may be one of the first "electronic" LRLs ever made, though in reality it's just a dowsing rod with a flashlight. A close variant (made with identical components) called the "Electric Flashlight Treas-o-meter" appears in the 1938 book, "Doodlebug Edition of Treasure Trails" by R. J. Santschi. Santschi also mentions in his 1927 book, "Modern Divining Rods," what may be another variant, described as a

23. Not to be confused with the "Magnetic Dip Needle," which is a legitimate scientific device. Also not to be confused with "Spanish Needles" which are another type of dowsing device.

24. "Fool's gold," aptly named in this case.

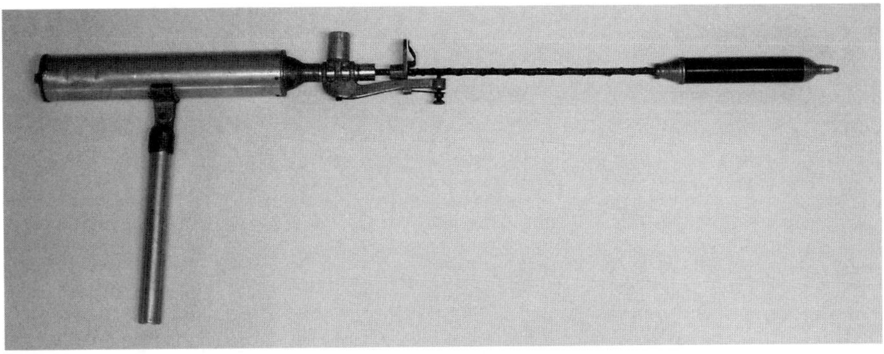

Fig. 13-6: **Treas-O-Meter Dowsing Rod**

magnetic dip needle with a flashlight indicator, and having a patent issued in 1910.

It is difficult to find any other dowsing devices marketed specifically to treasure hunters until the early 1970's, when ads began showing up in various "treasure" and "western" magazines for a device called the "Anderson Rod," shown in Figure 13-7. Passive LRLs like the Anderson Rod are fundamentally nothing but a dowsing rod, but might include some additional feature that is claimed to enhance the response. For example, the Anderson Rod also includes a so-called *witness chamber* where a sample is place (e.g., gold) that supposedly causes the rod to react only to that material. It also has a sealed portion of the chamber that supposedly contains a mixture of secret ingredients that intensifies the response. Anderson Rods are still made, and currently priced from $745 to over $2000.

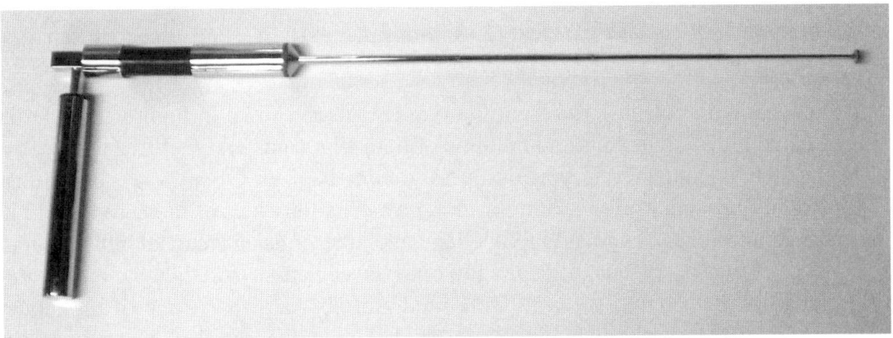

Fig. 13-7: **Anderson Dowsing Rod**

A more recent version of the Anderson Rod is called the "PPL" (Figure 13-8) priced at $1195. The manufacturer claims it "senses the gases or electron field put-off by long time buried gold." The front plastic chamber contains sand, plus several small magnets and an approximately 1-gram sample of gold. The rear chamber contains a mixture of sand, table salt, and small pellets of sulfur. Some passive LRLs are nothing but a dowsing rod with no enhancements whatsoever. The Merlin Locator rod (also Figure 13-8) has no witness chamber or any other feature that sets it apart from a simple dowsing rod, other than the fact it is machined out of plastic.

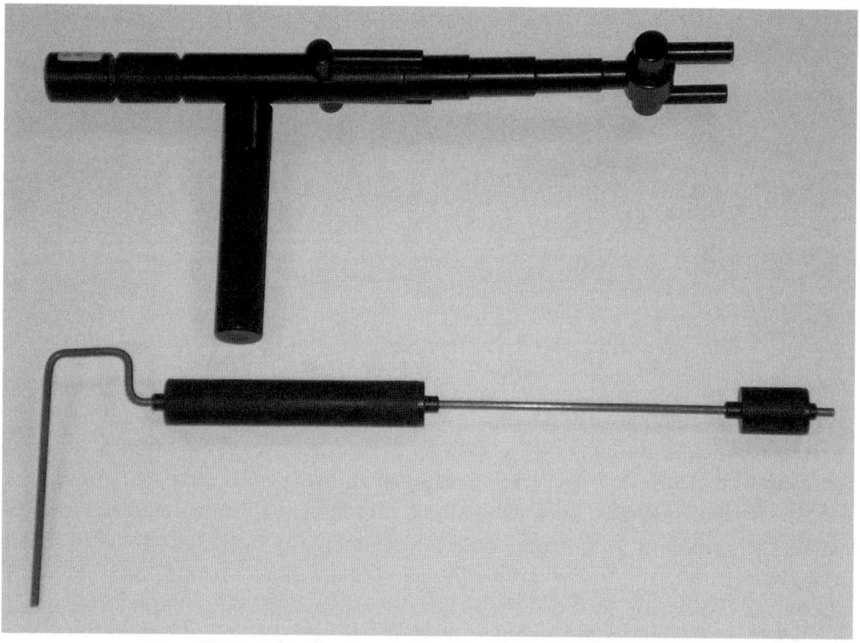

Fig. 13-8: **Merlin Locator Rod (top) & PPL rod**

Experiment 13-3: Chambered L-Rod

Required: 3/16" brass rod stock, (2) 35mm film canisters, sand, charcoal, salt, (2) silver coins.

Make a single dowsing rod from the brass stock as shown in Figure 13-9. Drill centered holes in the tops and bottoms of both film canisters (feel free to use any other small container if you don't have film canisters, which are becoming scarce). Slide one canister, bottom first, down the long part of the L-rod. Hold it upright and fill with a blend of 90% sand and 10% crushed charcoal, plus several pieces of quartz crystals. Slide the top onto the canister. Now slide the other canister onto the rod and fill with 95% sand and 5% salt. Insert one of the silver coins into the sand/salt blend. Slide the top onto the canister. Your final rod should look like Figure 13-9:

The use of two dowsing rods usually involves walking around with them waiting for them to cross. The single L-rod in this experiment may be used to "sweep" an area while standing still. Place the second silver coin on the ground and move about 30 feet (10 meters) away. Hold the L-rod in one hand, forearm level, with the rod pointed ahead and at a slight downward angle (5-10°). Slowly pivot your body and sweep the rod across, scanning the area in front of you. As the rod sweeps past the coin on the ground, you should experience a "hesitation" feeling in the rod. It should feel as if the rod is "pausing" as it passes the direction of the

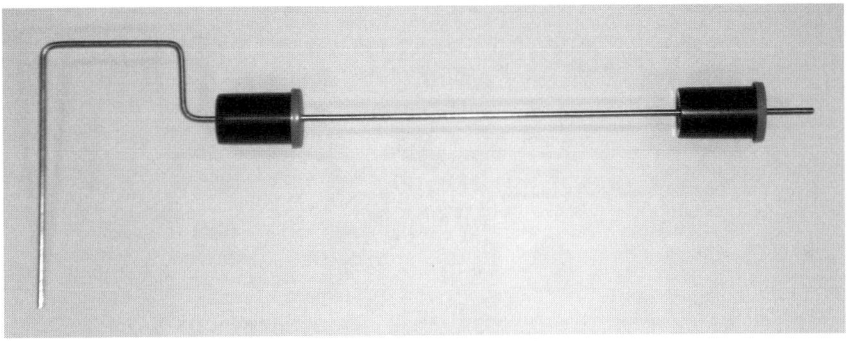

Fig. 13-9: **Chambered L-Rod**

target.

The suggested contents of the chambers is completely made up. It really doesn't matter what you put in them, nor does it matter whether you include a sample of the desired target, the so-called "bait." One common practice of dowsers is to simply write the name of the desired target on a piece of paper, and put the paper in the chamber. Want to find diamond rings? Write "diamond ring" on a piece of paper, stick it in the chamber, and go hunting![25]

Also note that the rod in Figure 13-9 has a "gooseneck" bend, similar to the PPL device shown in Figure 13-8. This doesn't change how the rod operates, it just looks cool. Feel free to bend your L-rod into even stranger shapes.

"Active" LRLs

Even with a witness chamber and secret ingredients, it's difficult for a manufacturer of a passive LRL to convince buyers that their wonder device is worth thousands of dollars. Electronic LRLs came along to entice treasure hunters to buy into another round of deception, and at a much higher price. When space-age electronics are added, buyers are suddenly impressed and feel justified in spending *many* thousands on what is still the same old dowsing nonsense.

A common form of electronic LRL is known as the *Molecular Frequency Discriminator* (MFD). In one form it consists of a pair of ordinary L-rods plus a signal generator. The signal generator typically includes two *ground probes* for "injecting" the signal into the soil. The prevalent claim is that when the generator is tuned to the "frequency" of a target (gold, for example), a *line of energy* is formed between the probes and any such buried targets (gold) in the area. The user then walks around with the dowsing rods and attempts to track the so-called "signal line" to the buried treasure.

An example of this type of MFD, the Dell Systems GS-pro ($1195) shown in Figure 13-10, has two settings, one for gold and one for silver. The signal generator (a simple 555-based oscillator in this case) produces different frequencies for each setting, supposedly the resonant frequency for that element or compound. Other LRLs

25. This isn't a joke; dowsers actually do this.

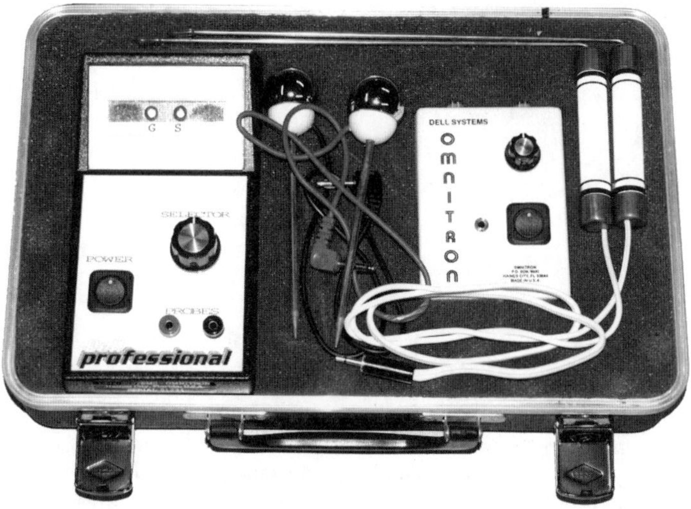

Fig. 13-10: **Dell Systems GS-pro**

may include settings for copper, tin, lead, diamond, and even paper money. *Everything* has its own frequency[26]; like a chambered L-rod, there is no limit to what you can search for with an MFD!

Experiment 13-4: MFD-LRL

Use the L-rods from Experiment 13-1 and build the circuit in Figure 13-11. This simple 555-based oscillator is actually used in many commercial LRLs (such as the GS-pro in Figure 13-10), so this project is representative of an MFD unit costing $1000 or more (no kidding). The frequency of the oscillator is 5kHz which is the frequency for gold[27], therefore a signal line will only be established

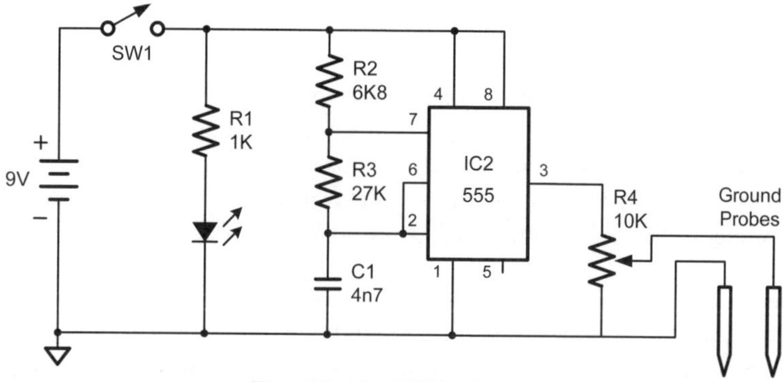

Fig. 13-11: **MFD circuit**

26. Or so purveyors of LRLs would like us to believe.

with buried gold[26]. Place the circuit in a small experimenter's box as shown below. This is a very simple project so methods and assembly are left to the reader[28]. In our version the probes are plugged in to banana jacks, though they can also be hard-wired. The potentiometer (R4) sets the output "power level" and has an integrated on/off switch (SW1).

In addition, you will need to make a pair of ground probes. These can be as simple as short lengths of 3/16" brass rods with hook-up wires attached with brass screw clamps from the hardware store. Feel free to upgrade the probe design by adding wooden or plastic T-handles, or make them fancy with the addition of PVC modules. Lay the signal generator on the ground and connect the hook-up wires for the ground probes to the signal outputs. Push the probes into the soil, keeping them at least a couple of feet apart. Turn the output power to maximum

Hold the dowsing rods and walk a circle around the signal generator and try to find a signal line emanating from the probes. If you do, track the signal line outward and mark it with flags. If you find yourself tracking a signal line a mile or

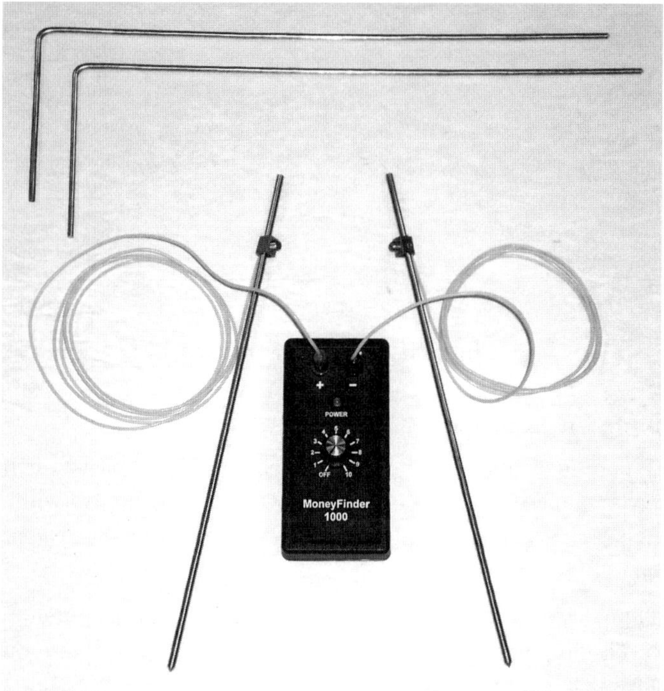

Fig. 13-12: **Completed MFD, with requisite dowsing rods**

27. Not really, but MFD proponents often say it is. We will discuss MFD frequency later in this chapter.

28. See the quotation at the head of Chapter 2. Have a child of five assemble this project to make it more closely resemble a typical manufactured MFD.

more from the signal generator, then the output power may be too high; turn it down to limit the signal line distance to nearby targets. Proponents claim that the signal line may not stop at buried gold, but may extend beyond it. Therefore, you may need to move the signal generator to a different spot and mark out a second signal line. Where it crosses the first signal line, you should dig and find buried gold![29]

Another form of MFD is one in which the LRL is fully contained in an integrated hand-held device. The most common form consists of a module (usually a plastic box) which contains some manner of electronics, and one or more antennae protruding from the front. The module is mounted on a swivel handle, so that it freely pivots. Despite liberal applications of hi-tech-looking knobs, lights, and meters, these devices are still nothing but dowsing rods. Figure 13-13 shows a well-known example of such a device, the Electroscope model 301 ($2495), which had a large presence in treasure hunting in the 1990's.

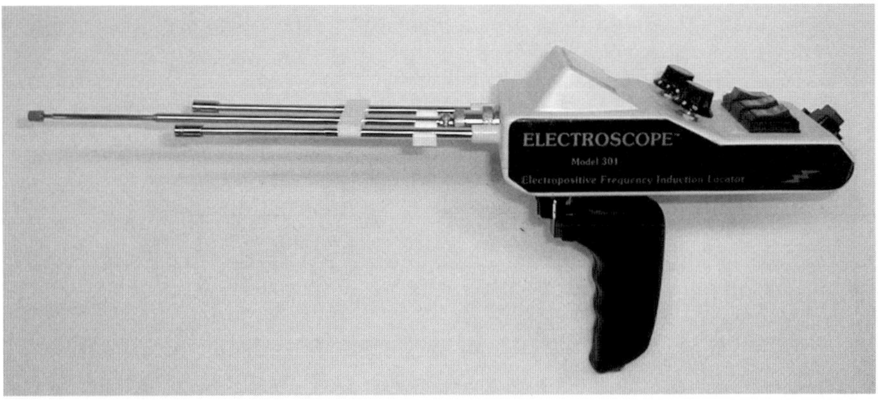

Fig. 13-13: **Electroscope 301**

Experiment 13-5: Integrated LRL

Using the same MFD circuit from Experiment 13-4, replace the probes with an antenna mounted to the box and add a swiveling handle. We will also add a toggle switch to select either gold (5kHz) or silver (8.7kHz[30]). The new circuitry is shown in Figure 13-14.

As with the previous project, methods and details are largely left to the reader. However, the pivoting handle can be a little difficult to figure out, so we have included details of this, along with the completed unit, in Figure 13-15. Besides copper tubing and copper fittings, PVC could also be used, with an adjustment in

29. Please *please* understand that this is written tongue-in-cheek. Many people will read this and believe it is supposed to Really Work this way. It does not work! It's all been fabricated by LRL proponents over the years, for the purpose of selling bogus devices.

30. Again, this number was made up by LRL proponents.

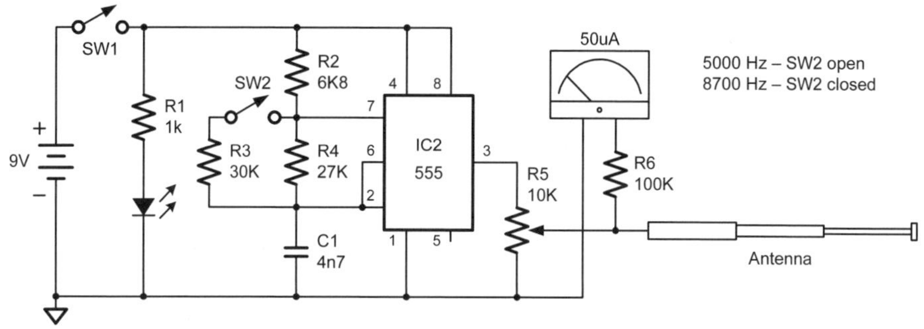

Fig. 13-14: **Integrated MFD Circuitry**

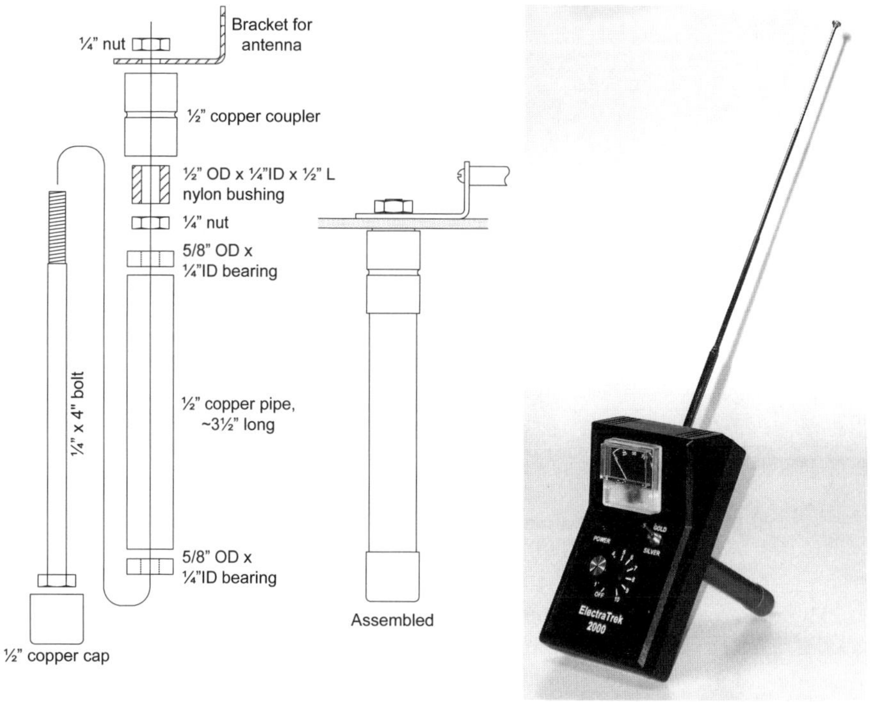

Fig. 13-15: **Integrated MFD Details**

the sizes of other components. For this project, the case used was from Bud Industries which has a strong resemblance to the Star Trek-looking cases used for Electroscope and Lectra Search LRLs. It can be found at several on-line sources.

While some LRLs of both styles contain a real circuit that really generates a signal, many of them do not. The Electroscope, for example, has a DC circuit that applies a portion of the battery voltage across the antennae, plus a resistor to drain the battery, plus a mish-mash of assorted pieces of junk[31]; but no useful circuit. A device called the Treasure Scope Raven ($1595) contains the guts from a $7 Radio Shack

AM radio and... and... *nothing else*. Lectra Search LRLs ($595 - $2495) contain a *scrap surplus circuit board*[32] with a few wires soldered to it and, again, nothing else; the antenna *isn't even wired to anything*. The Ranger-Tell Examiner (approx. $750) has a calculator taped to the top of the box, and a few nonsensically connected random components inside. That is to say, for many of these devices, all you get is an ordinary dowsing rod with some high-tech-looking window dressing; there are no signal generators, and sometimes no operational circuitry at all! If any of this sounds incredible, welcome to the Wonderful World of LRLs. It is a dark world indeed.

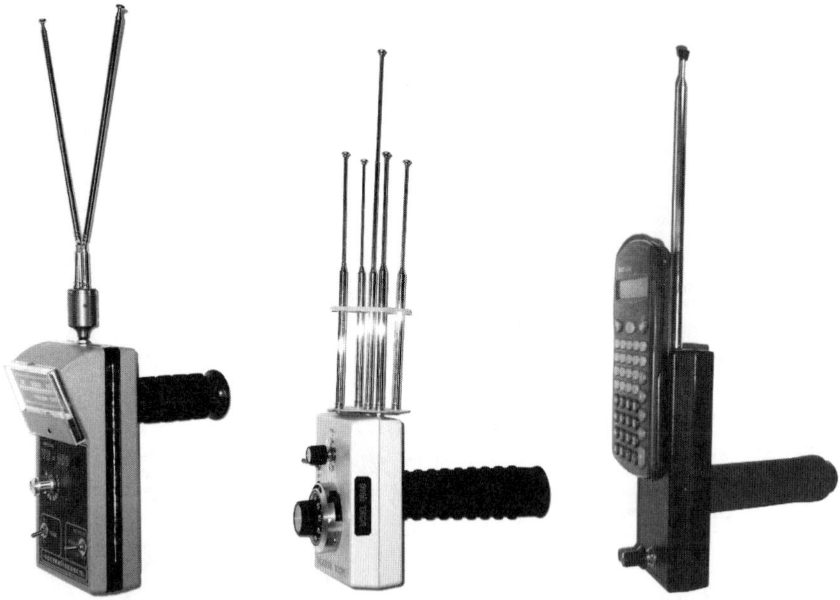

Fig. 13-16: **Lectra Search X100, Treasure Scope Raven, Ranger-Tell Examiner**

The most recent development in the area of LRLs has been to eliminate the dowsing devices (L-rods), and produce an LRL that is purely electronic. Since the use of L-rods and ideomotor action are key to the self-deception a user experiences, it's difficult at first to imagine a way to electronically induce the same level of self-deception. As it turns out, the atmosphere around us is bombarded with all sorts of man-made radio waves, and these create the perfect opportunity for another bogus product.

31. Sometimes, literally, pieces of trash.

32. One of which was identified as a board from a "Rat" guitar pedal.

A company named Mineoro in Brazil was one of the first to produce such a device. Figure 13-17 shows a Mineoro model FG80 (approx. $8,000), where the "FG" stands for "Fresh Gold." That is, this LRL is claimed to be able to detect fresh gold[33] and even comes with a gold (colored) test sample. Of course, it does not detect fresh gold, but it does have a tendency to beep when pointed in certain directions, and can give the illusion of "working." After some testing, it was determined that this particular model was tuned to around 433MHz and was simply picking up stray RF. This happens to be a popular frequency for garage door openers, so it's pretty easy to "demonstrate" this LRL successfully "detecting" gold if someone is judiciously keying a garage door opener during the test. In the next chapter we will see how to build such a device.

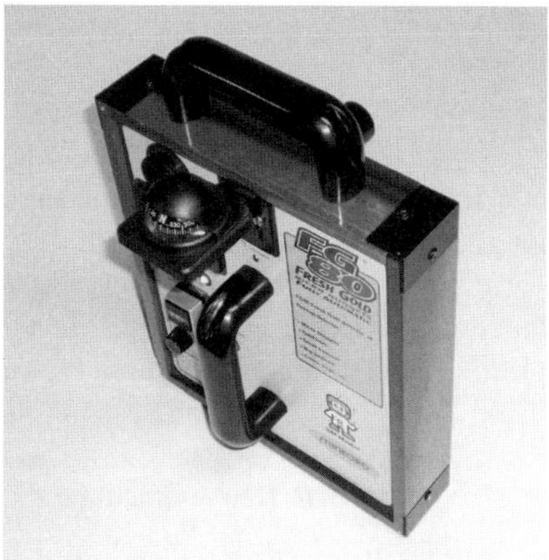

Fig. 13-17: **Mineoro FG80**

MFD Theory

What is meant by tuning to the "frequency" of a target? LRL proponents commonly claim that every element has its own vibrational frequency, and if you match that frequency with a signal generator, a "sympathetic" resonance occurs between the generator and the element. Some people even believe that like items (such as two silver coins) will establish — on their own, with no external influence — a sympathetic resonance between them. This is why some dowsing rods have a sample chamber. The idea of elemental resonance is loosely based on *RealScience*, where elements do have a characteristic *nuclear magnetic resonance* (NMR) frequency. But the NMR frequency is not even a constant; it is proportional to the ambient static magnetic field.

Experiment 13-6: Calculate the NMR frequency of hydrogen.

From Internet NMR tables (or the *CRC Handbook of Chemistry and Physics*), hydrogen has a NMR frequency of 100MHz relative to, uhh, *hydrogen*. What does this mean? It so happens that hydrogen, at a precession frequency of 100MHz, is the standard against which all other NMR frequencies are based. Now we need to find out under what conditions hydrogen has a precession frequency of 100MHz. It turns out that this occurs when the ambient magnetic field is 2.35T (Teslas). This field strength is much, much stronger than any typical Earth field, so we need to scale the NMR frequency for the local Earth field. In most parts of the United States and Europe, this is typically 50μT (±10%). The

33. One of the common excuses for LRL failures is that gold must be buried many years for an "aura" to develop.

resulting NMR frequency for a 50μT field strength is

$$f = 100\text{MHz} \cdot \frac{50\mu\text{T}}{2.35\text{T}} = 2127.66\text{Hz} \qquad \text{Eq 13-1}$$

It so happens that this is the precession frequency found in proton-precession magnetometers, where a sample of hydrogen-rich liquid (like water or alcohol) is periodically "pinged" and allowed to precess. Buried iron targets which distort the local magnetic field are detectable as variations in the NMR frequency. Note that a proton magnetometer measures the precession frequency of hydrogen, not the iron itself.

Experiment 13-7: Calculate the NMR frequency of gold.

Again from NMR tables, gold has a relative NMR frequency of 1.754MHz, which is based on the 100MHz NMR frequency of hydrogen. This means that for whatever conditions cause hydrogen to have a frequency of 100MHz, the same conditions will cause gold to have a frequency of 1.754MHz. Again, that condition is when the ambient magnetic field is 2.35T (Teslas) and, again, we need to scale the NMR frequency for the local Earth field. For our typical 50μT field strength, the resulting NMR frequency is

$$f = 1.754\text{MHz} \cdot \frac{50\mu\text{T}}{2.35\text{T}} = 37.32\text{Hz} \qquad \text{Eq 13-2}$$

Unlike iron, gold has no effect on the local magnetic field, so a proton magnetometer cannot detect buried gold. Could we instead directly detect the precession frequency of the gold? This is what MFD proponents claim, so it stands to reason that their LRLs should operate reasonably close to actual NMR frequencies. Of course, an NMR frequency will vary widely based on location as the geomagnetic field varies from about 25μT (South America) to 60μT (Canada and Siberia). For our example of gold, this results in an NMR frequency that is anywhere between 18.6Hz and 44.8Hz. So do MFDs actually use a frequency for gold in this range? It turns out that different LRLs use radically different frequencies; there is no consistency at all. Some frequencies for gold have included

• 530-817Hz[34]	Dell Systems VR800
• 222Hz	Dell Systems GS pro
• 1.518kHz	Dell Omnitron System II
• 9.194kHz	Treasure Scope Quad
• 10.54k-17.67kHz[33]	Vector Trek Mini Eliminator II
• 32.8Hz	Vector Trek SiGo

As we can see, only the "SiGo" is somewhat close to a believable NMR frequency.

34. The VR800 and Mini Eliminator have widely variable frequencies, which makes no sense at all if frequency is important.

But we also see that even LRLs from the same manufacturer have no consistency in their frequency setting. It's as if they just make up whatever they want to use, which happens to be the case.

So we now see that if an MFD were to use an actual NMR frequency it would need to adjust the frequency to account for the local ambient magnetic field. There is an LRL that actually does this. The H3Tec LRL (Figure 13-18, $5,000 - $15,000+) has a miniature magnetometer which detects the local field strength and alters the frequency of an internal signal generator to maintain a roughly believable NMR frequency.

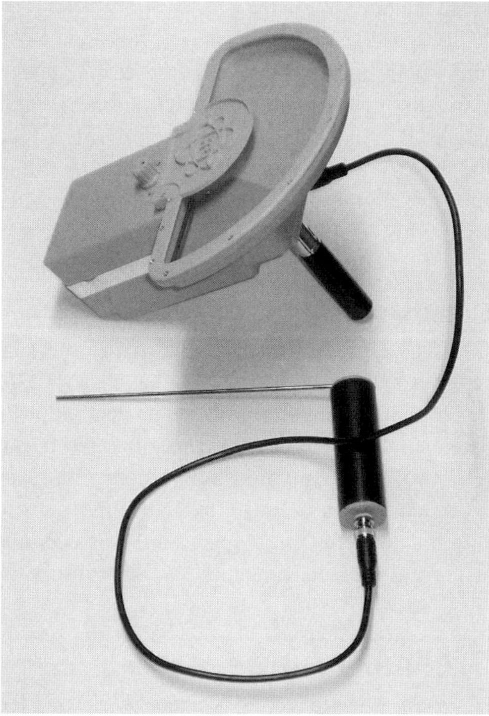

Fig. 13-18: **H3Tec**

In measurements made in Oregon where the local field strength is a little higher than 50μT (perhaps 54-55μT), the H3Tec frequency for gold was 54.5Hz, which should correspond to a field strength of 73μT. This is even higher than the strongest Earth magnetic fields occurring in the Arctic region, so it's not very accurate.

Despite having a wrong initial frequency, a simple test of bringing a magnet close to the H3Tec showed that the output frequency increased as it should. However, no measurements were made to determine whether the frequency was reasonably proportional, as a dissection of the device showed it, like most other LRLs, to be nothing but a dowsing rod. Even the most basic tests with gold and silver targets showed that the H3Tec did not work and the signal generator, correct frequency or not, would have no impact on its functionality. As with other LRL manufacturers, H3 Tec refused to demonstrate the device using a scientific test protocol.

While NMR seems like a plausible way to detect elements, the scientific reality is that NMR precession effects are vanishingly weak and detectable only for incredibly short distances. With the proton magnetometer, the hydrogen sample (e.g. water) is placed *inside* a large coil, pinged with a very strong transient magnetic field (using several amps of current), and the weak precession frequency detected with the same coil feeding very sensitive electronics. In another NMR application — medical MRI — the patient is placed *inside* a large coil in order to detect the weak precession signals (also from hydrogen atoms). NMR tables include a receptivity number which describes the relative precession sensitivity (again, relative to hydrogen) of the element. Gold has a receptivity of 0.0000277, meaning its precession reaction is 36,000 times weaker than hydrogen. So if the precession of hydrogen is difficult to measure, we can be assured that the precession of gold will be 36,000 times more difficult. The idea that these instruments can detect the precession frequency of gold from a mile away is nothing less than ludicrous.

Regardless of the proposed theories, MFD proponents are claiming to resonate and detect buried treasure, often at distances of a mile or more, and usually with a small battery-powered oscillator that produces no more than a few milliwatts of power. It is like trying to boil water with a microwave oven, by placing the water 100 meters from the oven, and then running the oven on a 9-volt battery. Furthermore, they are claiming to detect the "sympathetic resonance" with no more than dowsing rods. This is *WishScience*. Sadly, there are a lot of people who not only fall for this nonsense, but accept it, believe it, and defend it passionately.

Summary

We've presented a number of commercial LRLs and explained the theories and realities behind them. You may have noticed that the prices are often quite high, especially considering the fact that the vast majority amount to nothing but dowsing rods. This is Good News for the experimenter, because it means you can easily build these LRLs for practically nothing. It's only Bad News that they don't work. But the Good News continues because, by building your own for a few dollars, you can discover that the concepts are bogus and the effects imaginary, then toss them in the garbage with minimal loss, financially and emotionally.

Treasure Auras

Before we leave the whacky world of LRLs, let's take an opportunity to bash on a closely related topic, the so-called "treasure aura[35]." This was mentioned in a footnote a few pages back. The general concept is that buried metals [supposedly] give off an aura, which might be a gas, some kind of electrical or magnetic field, or an ion-something. It's never really clear, just that buried treasure produces a mystical aura. And the longer it's buried, the stronger the aura.

Many LRL manufacturers latch onto this myth not only to explain what their magical device is detecting, but also as an excuse as to why their LRL cannot be tested with a freshly buried target[36]: it has no aura! But LRLs have taken enough of a beating at this point, so let's look at another way folks are "detecting" auras: photography.

35. Not to be confused with a "halo," discussed in Chapter 15.

36. Yet most of them will gladly demonstrate their LRL with a fresh target.

Photographing Treasure Auras

Where the idea of capturing auras on film originated is not clear, but it appears that it began with users of the Polaroid SX-70 camera. This camera had two features which made it conducive to this effort: it could produce a photographic print in minutes rather than days, and it had a tendency to create spectacular photographic flaws. The latter was crucial to the resulting self-deception. Polaroid film is ejected from the camera through two rollers which rupture a built-in packet of developer and squeegee the developer across the print. If the roller pressure is uneven, rollers dirty, or the film is old or defective, the picture will end up with streaks, flares, and discolorations. This is not uncommon, and, with a little imagination, one could interpret these flaws as treasure auras.

Fig. 13-19: **Polaroid photo: *Amazing aura* or mundane flaw?**

In 1996 self-proclaimed dowsing expert Louis Matacia published a book called "*Finding Treasure Auras*." In it he describes a wild and vividly imaginative theory of an "evolutionary spiral" (the aura) that rises from buried treasure, complete with ghost signals which deceive treasure hunters into digging in the wrong location, explaining why this method never seems to yield a recovered treasure. He states that this evolutionary spiral "was given to us by scientists from the other side," and that the mere presence of ghosts usually indicates a nearby treasure. Matacia's book is full of bizarre nonsense and would make for wonderful fiction if he weren't serious about it all.

His book also contains a number of Polaroid shots, all of which exhibit ordinary flaws that anyone accustomed to these cameras would recognize. Polaroid cameras are no longer manufactured, and the film is obsolete. Old film, now way past its sell-by date[37], brings high prices on eBay from Polaroid purists. In order to perpetuate the myth of photographable auras, it's necessary to develop a method which does not require a Polaroid camera. David Villaneuva has done just that. In his book "*The Successful Treasure Hunter's Secret Manual (Discovering Treasure Auras in the Digital Age)*" Villaneuva demonstrates how a modern digital camera with an infrared filter can be used as a substitute. According to the book, the optimum time for taking pictures is early morning or late evening sunlight. You must position yourself to the

37. Aura purists love old, out-of-date Polaroid film, because it offers a better chance of producing a photographic flaw... err... that is, an *aura.*

north or south of the target area so that you are at right-angles to the sun. Apparently this method does not work 100% of the time (sound familiar?) and you may need to shift positions back and forth, or to the opposite side of the target area, taking many photographs before you eventually see an aura.

Villaneuva started treasure hunting using conventional metal detectors, but became interested in dowsing, LRLs and using cameras to detect treasure auras. He claims that digital cameras can be used to photograph auras on gold down to a quarter of an ounce or less (a single coin or coin nugget). Now … these are wild claims. But so are the claims made for dowsing and LRLs. Both of which are a big part of Villaneuva's repertoire.

Are these treasure auras real?

It is stated that "Treasure auras are the appliance of science", and "... you have the finest opportunity to find millions in buried and hidden wealth!". Hmm … firstly, this definitely sounds more like pseudo-science, rather than real science, and its potential to find millions in buried treasure … well, we'll come to that.

In the previously mentioned book, there are 4 photographs on pages 14 and 15, which purport to show auras emanating from buried metal. Let's take a look at each photograph in turn:

1. This shows a "probable" treasure wreck. [Note: the world "probable" is in quotes in the original text.] Therefore, we must conclude that this is just fantasy, since no wreck has ever been recovered from this "probable" site. Like so many other "probable" treasures found by dowsing and LRLs.

2. There is a buried utility pipe at this location, and we know from the text that a pipe is indeed buried there. In this case the most likely explanation of the so-called aura [read: lighter area in the picture] is sunlight reflection from an anomaly in the overlying vegetation due to different soil conditions immediately above the pipeline.

3. This is supposed to show an aura from an ounce of gold and two ounces of silver. Now (with all due respect) the photograph is so blurred that the location of these items would be impossible to pinpoint, unless you already knew where they were buried in the first place. The locations of the items are not marked, and could be almost anywhere.

4. This seems to show the rear garden of a bungalow, where lead, iron, copper-alloy and aluminum targets are buried. The picture is so blurred and over-exposed that (to be frank) it is again impossible to make out any details of the supposed auras. Even though, in this particular case, the locations are actually marked on the photograph. It's as if we're being asked to see something that is not there.

Another experiment is described later where 4 British gold sovereigns were buried, and the results compared with an Electroscope LRL. Since the Electroscope is nothing more than a plastic box full of do-nothing electronics attached to a swivel handle (aka a dowsing rod) we can safely ignore the results of this experiment as well. There are a number of other photographs scattered throughout the book, and all of them have the same problem of being blurred, over-exposed, and impossible to interpret.

Back to the actual method of using a digital camera to detect treasure auras, we should point out that digital image processing is required to reveal the "magic". With an infrared filter attached to the camera lens, the final picture is very dark. We are told that Arcsoft Photo Studio's auto-enhance feature is a recommended choice to improve the image, or alternatively a gamma filter can be used if auto-enhance is not available.

Studying this subject carefully, you cannot help but get the feeling that the basic concept is just an illusion in the same way as dowsing. There are also striking similarities with the Electronic Voice Phenomenon (EVP) which fundamentally equates to hearing something that's not really there. EVP refers to electronically generated noises, which resemble speech, but are not the result of intentional voice recordings. There are many possible causes for EVP, such as static, stray radio transmissions, or simply background noise. This phenomenon has a name — "apophenia" — which means finding significance in insignificant phenomena. This is what is most likely happening in the case of treasure auras. Although these treasure auras should more correctly be called "visual pareidolia", which is a specific form of apophenia in which images can be found in random stimuli. Humans have a tendency to seek out patterns in nature, and the human brain is exceptionally good at this task, seeing identifiable patterns in almost anything. Even in the grain of a plank of wood, or the tiles on a bathroom wall.

But before we leave this subject entirely, there is a proven scientific method known as Forward Looking Infrared (FLIR) that is used by the military to guide pilots and drivers at night, or in the fog, by detecting the heat of surrounding objects. This is different to night vision, which operates in the visible and near infrared region of the spectrum. FLIR could conceivably detect the heat signature from large buried metal targets not far below the surface, but would not be able to detect single gold nuggets or coins. The conclusion is that, if FLIR cannot do it, then a digital camera with an infrared filter attached, and some image processing software, doesn't have a chance. Whether you agree with these findings or not, it can still be an entertaining exercise to try and photograph these so-called treasure auras, even though ultimately you will be wasting your time. After a while you might even start to convince yourself there's really something there.

Now, where's my metal detector?

CHAPTER 14

Pistol Detector

The TOTeM Project

"Here there be dragons!"
— Classical phrase used by ancient
Roman and medieval cartographers.

The quotation in the title section of this chapter is particularly relevant to this project. It is a phrase used to denote an unexplored territory, and refers to the Roman medieval practice of putting sea serpents and other mythological creatures in blank areas of maps.

We have already explored several types of long range locator (LRL) in the previous chapter, but the pistol detector (PD) is a curious beast that warrants a chapter all to itself. Firstly, the term "pistol detector" does not refer to an ability to detect pistols, but instead describes the mechanical construction and the way it is held. Traditional metal detectors have a search coil attached to a long stem that is swept back and forth across the surface of the ground. In contrast the PD is held at chest height, some considerable distance above the ground, in the manner of a pistol; see Figure 14-1 for a typical example. The search coil on the device points forwards in a horizontal fashion, and not downwards. The general idea behind this device is to detect any anomalies, which may be present in the background electromagnetic noise, which might be associated with longtime buried treasure.

It is easy to dismiss LRLs as working by self-deception, because the final recov-

Fig. 14-1: **Typical example of a Pistol Detector**

ery is nearly always performed using a conventional metal detector. However, in the case of the PD, the device incorporates a TR-type detector that can be used to pinpoint the target. Therefore it is difficult to use the same argument as for the other LRLs, because there is no doubt, in the case of a recovered target, that the PD was involved. The important question is whether this device has any credibility as an LRL, or is this just another case of wishful thinking? Note that many LRL proponents claim the PD is really a medium range locator (MRL) with detecting distances measured in meters rather than kilometers.

As usual, many theories abound purporting to explain the abilities of the PD, including the usual ionic nonsense. In general though, most proponents agree that the PD is detecting some "phenomenon" surrounding longtime buried objects (in particular, gold and silver) but they cannot explain what it is. The idea of this "phenomenon" is a useful argument in the PD supporter's toolbox, when experimenters fail to obtain any long (or medium) range detection in the laboratory. It appears the extra range can only be obtained in the field, and in the presence of objects buried for several years, during which time the "phenomenon" is said to build up in strength. Another excuse, often presented, is that conventional metal detectors destroy the "phenomenon," and it takes a number of days before it can return to full strength. The general idea behind the PD is to search for an anomaly in the background noise, and to follow this "signal line" to the target. Despite the claim that a PD can only be set up correctly in the field, there are a number of tests that can apparently be performed in the laboratory. One such test involves shorting out a 1.5V battery with a length of wire from several meters away, and checking that it elicits an audio response from the PD. Likewise, the detection of a TV (with a CRT) from 6 meters is touted as a good indication. There is one other lesser known test available that we will describe later in this chapter.

The purpose of the project presented here is to design and build an experimental PD based on information available from the public domain. There are very few detectors of this type for sale in the open market, as the method of construction is relatively complex and fairly difficult to calibrate, given the general lack of information, and there is no accepted scientific theory to back up the claims made for this device.

We are aware of one particular prototype that has been back-engineered and cloned by several amateur experimenters, and there are even videos on the internet of this device in operation. Throughout this chapter we will refer to this prototype as the *original PD*. As far as we can ascertain, the clones that have been constructed have either failed to function properly, or have fallen short of the claims for the original device. Even the videos of the *original PD*, showing it in operation, are not totally convincing, and could easily be explained away as wishful thinking. The *original PD* consists of a TR detector circuit, which is used for pinpointing and final recovery, and a separate receiver circuit that uses a coil wound on a ferrite rod. Investigations have shown that the TR circuit is an exact copy of a Heathkit GD348, which is a design originating from the 1970s. At this time it is unclear as to whether the ferrite receiver is intended to be used in conjunction with or separate from the TR circuit, due to some confusion concerning the wiring of the selector switch. Of course, anyone with an oscilloscope could find the answer in about 5 minutes, but the *original PD* is considered so precious that this may remain one of its secrets. Even PD proponents are in disagreement over this aspect of the design. In our own PD the intention is to provide both options for maximum flexibility, as the main purpose of the project is to act as an experimental platform in this unexplored territory.

One important point to note is the size of the search coil mounted on the front of the PD. This has a very small diameter of less than 4", and the *original PD* used a search coil housing from a Garrett Groundhog, which again was a detector from the 1970s. This probably puts a date on the construction of this device, and begs the question as to why these PDs are virtually unknown in the metal detecting world today? Perhaps it is such a closely guarded secret that other detector manufacturers are kept in the dark, or simply they do not work as claimed.

The attraction of such a device is clear. It is small in size, conveniently carried out of sight in a rucksack, and can be used to discretely detect treasure from a distance. There is a strong incentive for certain people to believe the claims made for the PD, and a willingness to pay a premium for this capability.

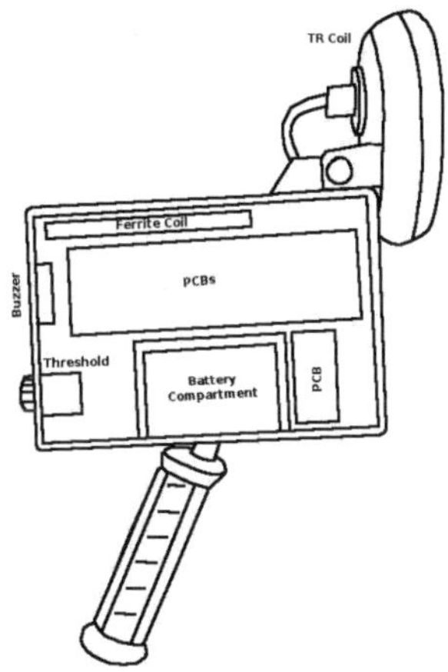

Fig. 14-2: **Inside the Pistol Detector**

It has been speculated that the PD is simply a miniature version of a 2-box detector, as the TX coil points forward. However, neither the RX coil in the search head nor the ferrite coil point downwards. In this case the 2-box idea must certainly be incorrect.

The only way to settle the controversy surrounding the PD is to build one. Although there is little point in creating yet another clone of the *original PD*, as this has already been done by others, and the results were at best inconclusive. Since the TR section of the design is fairly straightforward, let's start with the ferrite receiver.

Ferrite Receiver

From the information available we know that the coil on the ferrite is tuned to the same frequency as the transmitter, which implies that these two circuits are intended to work together. But, as stated earlier, due to some confusion over the internal wiring of the *original PD*, the intention is to provide both passive and active modes in our design. It is claimed by certain LRL proponents that a frequency of between 60-70 kHz is ideal for long range gold detection. Consequently we have designed our PD to operate at 65 kHz. Note the Heathkit GD348 operated at 100 kHz, but it is known that the *original PD* did not use this frequency. The ferrite rod used in this project was 8 cm long and 1 cm in diameter, with a yellow marking on one end. A small plastic former from a ferrite core was used to support the coil, which consisted of 100 turns of 0.56 mm thick enamelled wire, resulting in an inductance of 870 μH. The coil was positioned in the center of the ferrite rod, but could be adjusted either way, if desired, to alter the inductance.

Fig. 14-3: **Ferrite Coil**

The receiver circuit consists of a 2-stage pre-amp operating from a single-ended power supply. The reason for this is simply to remove the complication of a dual supply, but to operate in this mode it is necessary to provide a phantom ground. Initially this was created using a simple potential divider, but later in the project it was discovered that unwanted oscillations were occurring due to positive feedback through the power supply. Consequently a third dedicated opamp was used to generate the phantom ground. Stage one of the pre-amp uses a differential configuration with a gain of 50. The output is AC coupled to the second stage, again with a gain of 50. C3 and R3 provide a break frequency of 241 Hz, and the upper cutoff frequency is limited by the gain bandwidth of the opamp to about 80 kHz. C1 and C2 are used to tune the ferrite coil to approximately 65 kHz.

Initial tests appeared to indicate the receiver was directional, tending to indicate a strong null when pointing north. After further investigation, with an oscilloscope connected to the pre-amp output, it was discovered that a 20 kHz signal was being detected to the east. This was eventually traced to an inductive electric toothbrush charger which had been left connected in an adjacent room. After disconnecting the charger the noise level dropped considerably, and the north direction null was no longer detected.

TX Circuit

A separate forced oscillator transmit circuit, running at 65 kHz, was built to allow some preliminary testing, and used to drive a coil of 80 cm diameter (75 turns of 0.56 mm thick enamelled wire) with no electrostatic shielding. Even at a distance of 3 meters the pre-amp output was 400 mV peak-to-peak. The TX signal is not a perfect sine wave, but this is unimportant. In fact, the Heathkit GD348 produces a series of underdamped pulses, and it is known that many different types of TX circuits are used by LRL experimenters. To increase flexibility during experimentation, R21 (10K) could be replaced with a 10K preset in series with a 4K7 resistor, to allow the

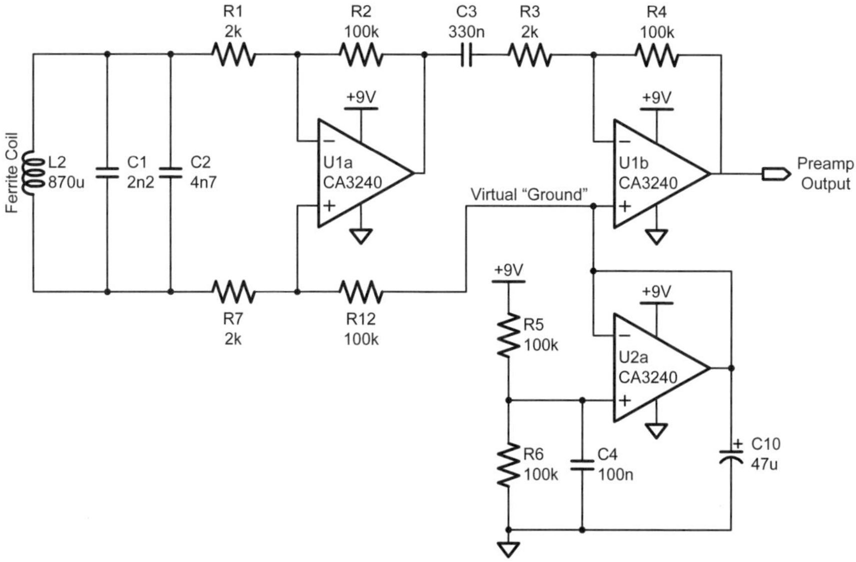

Fig. 14-4: **Pre-amp with single-ended supply**

TX circuit to be tuned into resonance with the RX coil. The TX frequency can be calculated as follows:

$$f = \frac{1.46}{(RA + 2RB)C} = 69.5\text{kHz} \qquad \text{Eq 14-1}$$

The actual frequency was measured in practice to be 65 kHz. No attempt was made to achieve resonance, so this may be an area worthy of further investigation.

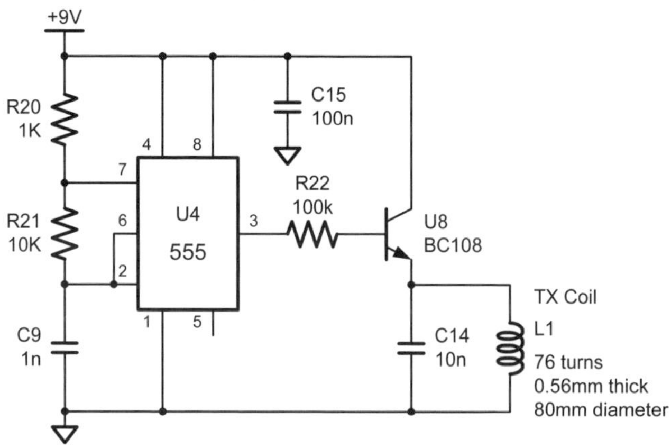

Fig. 14-5: **Transmit Oscillator**

The internal layout of the *original PD* implies the existence of a null point at the position of the ferrite, otherwise the pre-amp would be overloaded by the TX signal, but experiments have shown no such null point exists at that location. Interestingly not many people are aware that there is a "null line" which projects away from the edge of the coil at a 45 degree angle. Placing a ferrite coil anywhere along this line will produce a null. In fact, the line projects away from the TX coil all the way round the circumference, both behind and in front, forming two "null cones." The position of the null is quite easy to find, and with careful adjustment can provide some good sensitivity to non-ferrous targets while at the same time rejecting ferrous items. So the important question here is why is the ferrite coil positioned away from the "null line" in the *original PD*?

There are several possible answers:

1. The ferrite coil is only used in passive mode, with the transmitter disabled. In which case a null point is not required.
2. The omega coil configuration, used by the Heathkit GD348 circuit, has a different electromagnetic field pattern that bends the "null line" away from its usual position.
3. A different method is being used to null the ferrite coil.

Let's look at each possibility in turn:

1. Our own experimental PD will have active and passive modes selectable from the front panel, allowing both modes of operation to be tested.
2. The omega coil does appear to distort the electromagnetic field, but the distortion is insufficient to bend the "null line" into the required position.
3. Close examination of the ferrite coil of the *original PD* reveals there are in fact two ferrite rods (or one rod cut into two parts) connected together, but leaving a small gap. There are also two coils wound onto the ferrites in anti-phase. The conclusion is that the coil closest to the TX coil is used for nulling. This coil is not driven directly by the TX circuit, but relies on the current induced in the first ferrite by the TX coil. Nulling is achieved by adjusting the gap between the two ferrites. This indicates there is a strong possibility that the *original PD* must operate in active mode, otherwise there would be no requirement for the second nulling coil.

The discovery of a nulling coil shows the constructors of the *original PD* were either unaware of the already existing "null line," or they deliberately placed the fer-

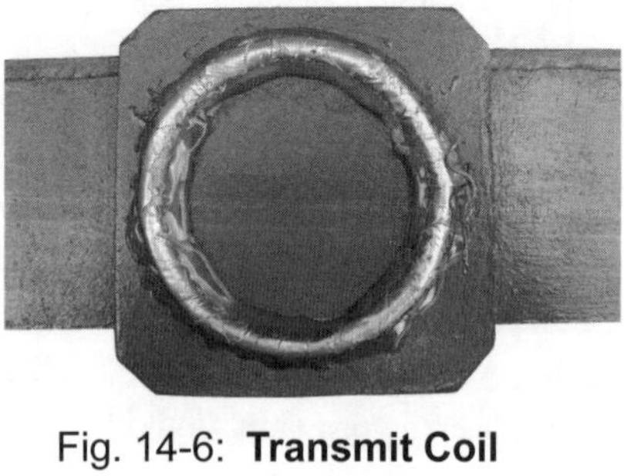

Fig. 14-6: **Transmit Coil**

rite coil higher up to reduce the size of the PD enclosure.

Experiments have shown it is possible to place the ferrite coil on the "null line" using an enclosure similar to the original, but its consequent proximity to the PCB and associated wiring causes tremendous instability. Therefore an alternative solution was found by placing the ferrite coil on the "null line" above the TX coil. This provides the advantages of ease of nulling (the two ferrite solution is known to be difficult to implement) and improved stability, but with the slight disadvantage of increased enclosure size.

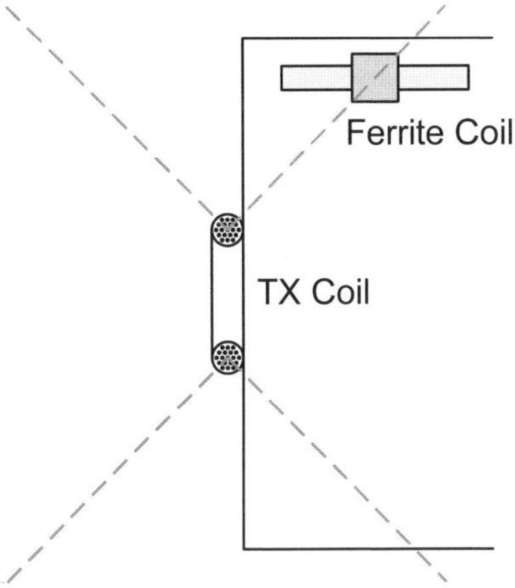

Fig. 14-7: **"Null lines" around TX coil**

It would also be reasonable to question why a ferrite coil is used for the receiver when an air coil could possibly perform the same function? It is highly likely that the answer has its roots in the old miner's tale of a man who was able to detect gold using only an AM radio tuned off-station. Whenever he detected a null in the audio hiss it was an indication that he was close to an ore deposit. This is the basic idea behind the passive LRLs which contain ferrite rods. So somewhere along the way, the use of a ferrite coil has become synonymous with long range detection. You may also question why the ferrite rod is not positioned broadside, rather than pointing end-on, to the front of the unit? Surely AM radios are always positioned so the ferrite antenna is broadside to the radio transmitter?

First let's look at the basics of a ferrite rod antenna. Ferrite is an iron-based magnetic material with a high permeability. This causes the magnetic component of the transmitted signal to be concentrated in the rod, which has the advantage of making it directive. This means that reception is highest when the magnetic lines of force are in line with the long axis of the antenna. Conversely the antenna receives minimum signal when it is end on to the radio transmitter. So doesn't this mean the ferrite rod in the PD is positioned incorrectly, and really needs to be rotated by 90 degrees?

Not at all!

Remember that a radio transmitter uses a long vertical pole as an antenna, and this produces a magnetic field pattern of concentric rings that radiate outwards horizontally in all directions. For the ferrite antenna to concentrate the maximum lines of force it must be positioned horizontally and broadside to the transmitter. In the case of the PD the magnetic field pattern is very different. For the ferrite antenna to receive maximum signal it must be positioned end on to the TX coil. The simplest way to imagine what's happening is to picture the receive coil on its own without the ferrite rod. This is exactly the same situation that occurs in a concentric coil. In fact, it then becomes obvious the two ferrite solution in the *original PD* is a direct extrapolation of the concentric coil concept with its TX, RX and nulling coil.

The Sky and Compass Effects

Several PD builders have discovered their clones of the *original PD* will beep when pointed to the sky. That is, anywhere above the horizon. There is also a compass effect whereby the PD is aligned to the north / south direction, which also causes a beep. Some experimenters (but not all) believe the *original PD* designer discovered that an unshielded coil is able to detect the "phenomenon," but both the sky and compass effects tend to mask the signal. The conclusion is that the purpose of the ferrite coil is to eliminate both of these effects, thus allowing the signal to be detected more easily. To achieve this situation the ferrite and TR coils must be set to a critical balance.

Interestingly you can experience the sky effect with both a standard Heathkit GD348 detector and a Micronta 4001. Both of these detectors are of the TR variety that use an omega configuration with unshielded coils. However, a C.SCOPE 220 (which is also a TR) does not exhibit the sky effect at all. This could be due to the C.SCOPE search coil having a DD configuration. But being a more modern design, the coils are also likely to be shielded. Unfortunately we were unable to dismantle the search head to confirm this assumption.

According to other LRL experimenters, the ferrite coil is used to cancel the sky effect produced by the omega coil. But for this to be true, the ferrite coil must also be detecting this effect, otherwise it would not be possible to produce a cancellation signal. Here is the paradox — from our own experiments it is clear that the ferrite coil does not exhibit either the sky or compass effects. In other words, how can any cancellation occur when neither of these effects are detected by the ferrite coil? The obvious conclusion is that the sky detection problem, in particular, is an unwanted side effect of the unshielded RX coil, and the even more obvious solution is to use only the TX and ferrite coil without the RX. You might then ask, why does the *original PD* retain the RX coil? It appears the LRL proponents are searching for a complicated solution to explain their inability to clone the *original PD*, when in fact the answer is quite simple. The TR detector is for short range pinpointing, and is only there to reinforce the user's belief that the PD is capable of medium and long range detection, because the final recovery is made using the same device. In addition, the internet videos of the *original PD* show the compass effect to still be present. So what happened to the so-called cancellation? The intended role and interactions between the TR circuit and the ferrite coil are extremely unclear, even amongst those with first-hand experience of the *original PD*. Discussions involving this topic will no doubt rattle on for several years to come, until the prototype is accidently lost or destroyed, thereby creating yet another LRL myth.

Detecting the Anomaly

One common feature of PD-type LRLs is the use of a pulse extender. Its purpose is to allow any short duration signal, that appears above the background noise, to become audible to the operator.

First it is necessary to compare the output from the pre-amp with a fixed threshold. In this project we have used an identical circuit to that used to generate the phantom ground. The duplication is required to prevent positive feedback occurring that would result in unwanted oscillation. Two separate comparators are used. One is used to drive the yellow LED, which indicates the strength of the input signal. This facility is somewhat limited in practice, but it was retained in order to make use of the second comparator in the LM393. The second comparator drives the pulse extender that consists of a 555 timer configured as a monostable which provides a pulse of 80 ms. The pulse extender in the *original PD* provided a pulse width of 363 ms, but the shorter 80 ms pulse was found to be more effective at determining the direction of the received signal. The output of the pulse extender drives both a red LED and a piezoelectric buzzer.

After much effort was expended in trying to eliminate the buzzer, and replace it with a more standard audio oscillator and speaker combination, it was found that the speaker caused a lot of interference, making this approach useless as an alternative solution. Clearly this is the reason why the buzzer is the preferred approach by LRL experimenters. Because the buzzer can be somewhat annoying, and may attract undesired attention while using the device, an on-off switch was incorporated to disable the audio output, For the occasions where an audio output is not desired, and the ambient light is too bright to see the LEDs, a meter was added to provide an extra visual indication. The most effective result was obtained by putting the meter on the

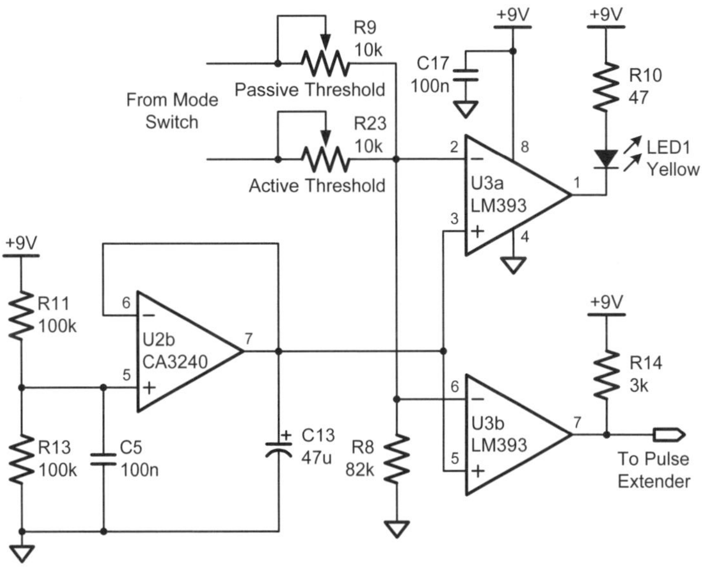

Fig. 14-8: **Comparator circuit**

same output as the buzzer. Originally the meter was connected to the same output as the yellow LED, but the response was very poor. In the final implementation the diode D1 limits the voltage across the meter and R24, which in turn limits the current through the meter to the 250μA required for full scale deflection (FSD) and prevents the meter needle from hitting the end stop. For added flexibility C18 can be switched in or out of circuit to provide either a slow or fast meter response.

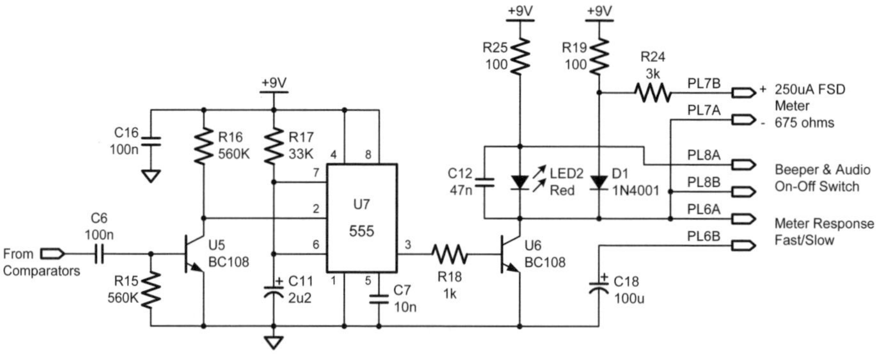

Fig. 14-9: **Pulse extender and audio output**

The Front Panel Controls

- On-off switch
  Simply turns the PD on or off.
- Active and passive thresholds
  Since the PD is capable of being used as a passive receiver or an active TR, there are two separate 10-turn pots for the two modes.
- Active / passive switch
  Allows operation in passive mode (receive only, no transmitter) or active mode (transmit and receive).
- Audio on-off
  Turns the buzzer on or off.
- Meter response
  Controls how the meter responds to the input signal.

TR Mode of Operation

Although the *original PD* could also be used as a mini version of the Heathkit GD348 for pinpointing purposes, it was found that our own PD could be used in the same way without the complication of adding another receive coil and associated circuitry. Having tested a GD348 firsthand, it was easy to confirm the target response of the Heathkit is relatively poor. Considering the *original PD* uses exactly the same circuit, but with a much smaller search coil (less than 4" as opposed to the original 10" of the Heathkit) it was clear that the PD's short range detecting ability would be even less effective. Even with the built-in TR circuit the *original PD* often required a con-

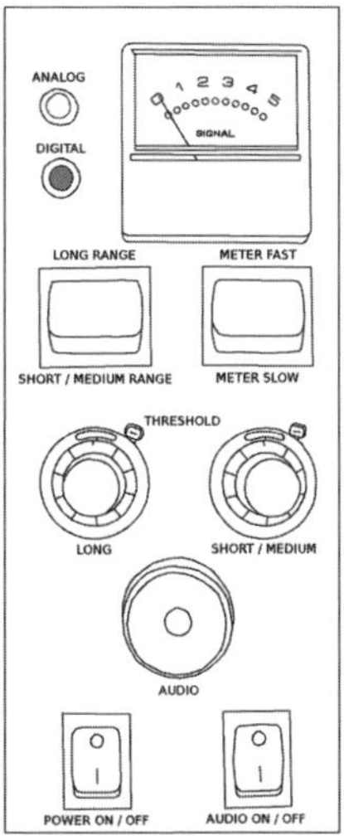

Fig. 14-10: **Control panel layout**

ventional metal detector to pinpoint and recover the target. The response at short range of our own PD was found to be at least the same, if not better, than the original, but using the ferrite receiver for short range as well as medium and long range detection. Also, as described earlier, an unshielded RX coil produces the unwanted sky effect, and in some cases a compass effect as well.

Nulling the Ferrite Coil

As mentioned earlier it is critical to correctly position the ferrite coil to achieve maximum sensitivity and to provide ferrous rejection. It is important that the enclosure is constructed of wood. Definitely do not use metal, and the use of a plastic box is not recommended by the LRL experimenters who believe the "phenomenon" to be ionic-based. If you construct the enclosure according to the measurements shown in the diagram, the nulling of the ferrite coil should be reasonably straightforward.

The prototype was constructed using 5 mm thick medium density fibreboard (MDF) but plywood would also be suitable. First attach the ferrite coil to a small piece of MDF using hot glue, and place it in the middle of the upper compartment as shown in the photos. Make sure the passive/active switch is set to "active", and adjust the active threshold pot until an audible sound is heard from the buzzer. Carefully adjust the position of the ferrite coil until the sound stops. Try to locate the coil in the center of the null. Then readjust the position of the active threshold, and repeat the procedure until no more improvement can be obtained. At this point you should be able to detect non-ferrous targets placed near the TX coil, while rejecting ferrous targets. Previously we discussed shorting a 1.5V battery using a length of wire in order to test the PD's ability to detect a spark. This should be detectable from at least 3 meters. An additional test, particularly for the passive mode, is to move a magnet near the ferrite, which should produce a clear audio response. If you do not have a magnet

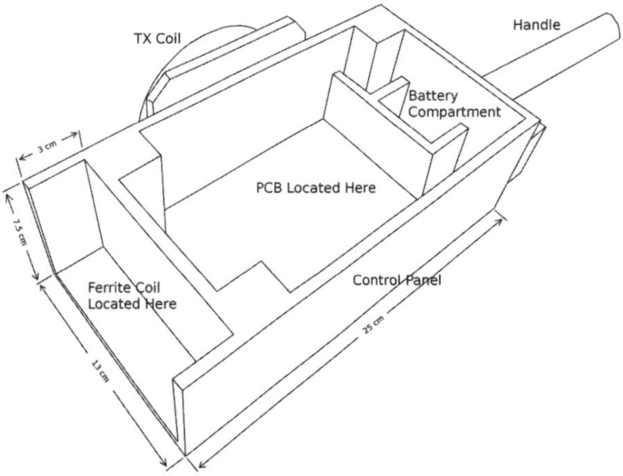

Fig. 14-11: **Inside the TOTeM PD**

available, then a magnetized screwdriver can be used as a suitable substitute. The audio response is caused by discontinuous "jumps" due to domain wall movement in the ferrite material as the domain walls become pinned and released from microstructural obstacles. Each abrupt "jump" produces a brief burst of magnetic noise, known as Magnetic Barkhausen Noise (MBN). Once you are satisfied that the ferrite coil is correctly positioned, secure it firmly using hot glue. Small adjustments can be made afterwards by moving the wires that connect the ferrite coil to the PCB, before also securing these with more hot glue.

Fig. 14-12: **Ferrite coil in enclosure**

But Does it Actually Work?

This of course depends on your definition of "work." It is clear that the electronics in our PD actually do something, unlike many so-called LRLs which are filled with nonsense do-nothing electronics to fool the technically challenged. These scam devices often have no power supply, and rely on some highly dubious pseudo-scientific theory to explain their method of operation. In most cases, these LRLs are nothing more than dowsing rods dressed up in fancy clothing.

Does it work as a metal detector? At least in the active mode at short range, the answer is yes. When used in the active medium range mode there are definitely signals being detected, and you can even occasionally appear to be following a "signal line." In the passive long range mode the same can be said to be true, and it certainly acts as a very sensitive electromagnetic field detector.

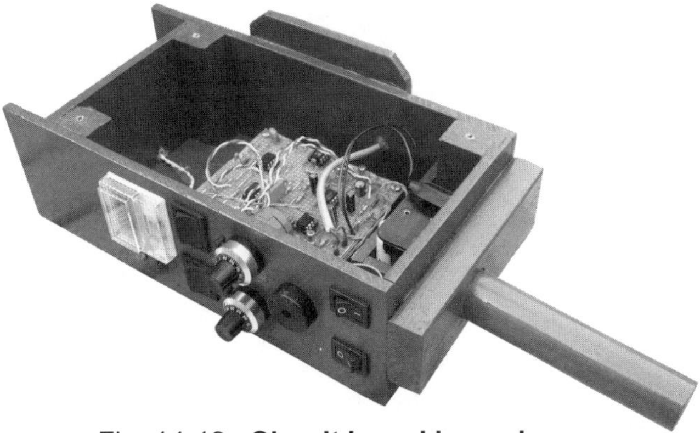

Fig. 14-13: **Circuit board in enclosure**

But does it detect treasure (in particular, gold) at long or even medium distances? Well, this is the big question. From the skeptical point of view the answer is almost certainly no, but as we stated at the start of this chapter, this project is highly experimental, and the intention was to keep an open mind during the development of this device.

According to LRL proponents, longtime buried targets are surrounded by a "phenomenon" that can be stimulated by a suitable transmitter operating in the frequency range between 60-70 kHz. How this is actually supposed to work is somewhat vague, and such a "phenomenon" is completely unknown according to accepted science.

From a skeptical point of view the most likely explanation is that LRLs function by a combination of wishful thinking, self-deception and selective memory. When using the passive long range mode it is possible to imagine you are following a "signal line." This could be due to many things, including multipath interference. Eventually you reach an area where the "signal line" appears to vanish or become erratic. LRL proponents attribute this to being close to the target, and at this point a conventional metal detector is employed to pinpoint the treasure. Usually something is located and recovered in the area, but the glory goes to the LRL for having detected it from several meters away. The very existence of the conventional metal detector being conveniently forgotten. Sometimes nothing is discovered, and this is where selective memory plays a role. Alternatively, you can always conclude that the target must be too deep to find with the metal detector, or there is some elusive micro-gold that is causing an erroneous signal.

The purpose of this project was to develop a "working" experimental platform for anyone wishing to investigate this grey area of metal detecting. This is a highly speculative area of research, and there is absolutely no guarantee of success. Remember... here there be dragons!

Fig. 14-14: **The completed TOTeM PD**

Meaning of TOTeM

At the very start of this chapter it states "The TOTeM Project", but what is the meaning?

Is it?

1. Because you can see a long way from the top of a totem pole? (ie. A tenuous reference to long range locating.)
2. An abbreviation of Totally Electromagnetic?
3. Or simply a "Trick Of The Mind"?

Whatever you believe, have fun building and experimenting with this device. Full construction details can be found in the remainder of this chapter, including a stripboard layout.

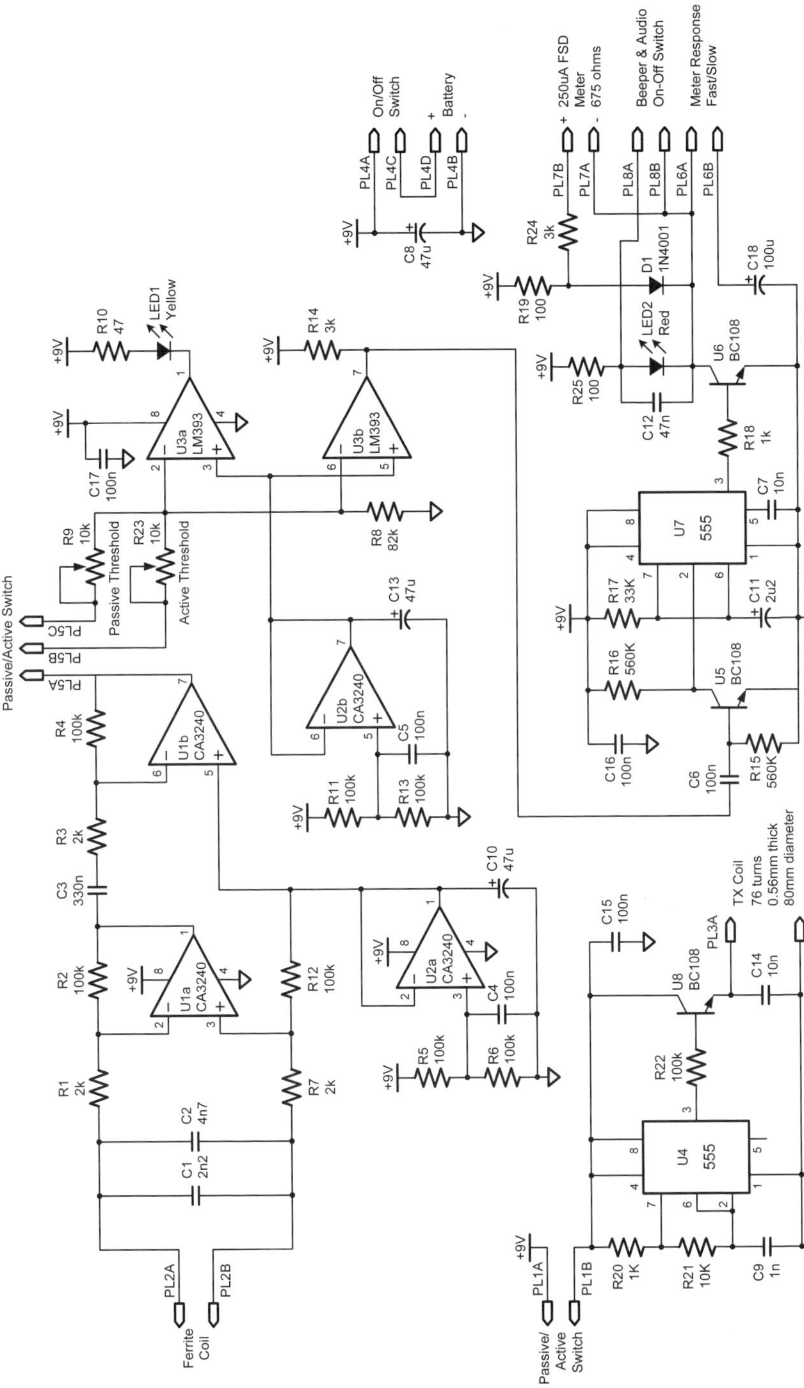

Fig. 14-15: **Final TOTeM schematic**

Construction Details

The component placement and layout are for illustrative purposes only. You may need to adjust the layout to accommodate components available in your area. In particular, please note that transistor pinouts can vary depending on the country of manufacture, even for what appears to be an identical part number. See Chapter 15 for more details.

Due to the experimental nature of this project, a decision was made to use stripboard for the layout, rather than a PCB. There is unfortunately a lot of track cutting and jumpers required, but at least you will not need to resort to using nasty chemicals.

Note the connectors are designated as follows:

PL1 = Passive / Active switch (TX power)
PL2 = Ferrite coil
PL3 = TX coil
PL4 = Battery and on-off switch
PL5 = Passive / Active switch (Threshold controls)
PL6 = Meter response
PL7 = Meter
PL8 = Buzzer and audio on-off switch

However, in the prototype there were no connectors used. The wiring was soldered directly to the PCB.

Parts List

Resistors: (5% 1/4W)

R1, R3, R7	2k
R2, R4, R5, R6, R11, R12 R13, R22	100k
R8	82k
R9, R23	10k pot (10-turn)
R10	47
R14, R24	3k
R15, R16	560k
R17	33k
R18, R20	1k
R19, R25	100
R21	10k

Capacitors:

C1	2n2
C2	4n7
C3	330n
C4, C5, C6. C15, C16, C17	100n
C7, C14	10n
C8, C10, C13	47u elect., 6v3
C9	1n
C11	2u2 elect., 10v

C12	47n
C18	100u elect., 10v

Inductors:

L1	TX coil
L2	Ferrite coil (see chapter 14 for details)

Diodes:

D1	1N4148
LED1	Yellow LED (with holder)
LED2	Red LED (with holder)

Transistors:

U5, U6, U8	BC108

ICs:

U1, U2	CA3240
U3	LM393
U4, U7	LM555

Switches:

S1	SPST (Power On-Off)
S2	SPST (Audio On-Off)
S3	DPDT (Active / Passive)
S4	SPST (Meter Response)

Misc:

(5) 8-pin IC sockets
Piezo-electric buzzer
Control knobs 10-turn precision
Battery holder to hold 8x AA alkaline batteries
Meter 250uA (full scale) 675 ohms

Component Placement

The actual size of the stripboard is 4.0" x 3.0" (10.2cm x 7.6cm). PCB software was used to design the layout, with components placed on a 0.1" grid to suit the stripboard used in this project. The jumpers on the top side were drawn as tracks. Some of the components (in particular, the transistors and certain electrolytic capacitors) did not have a footprint that would align correctly to the grid. In these cases the components may appear to be slightly off grid, requiring a small copper stub in the layout. These should be relatively easy to spot, but please be aware when placing components. U6 is a typical example.

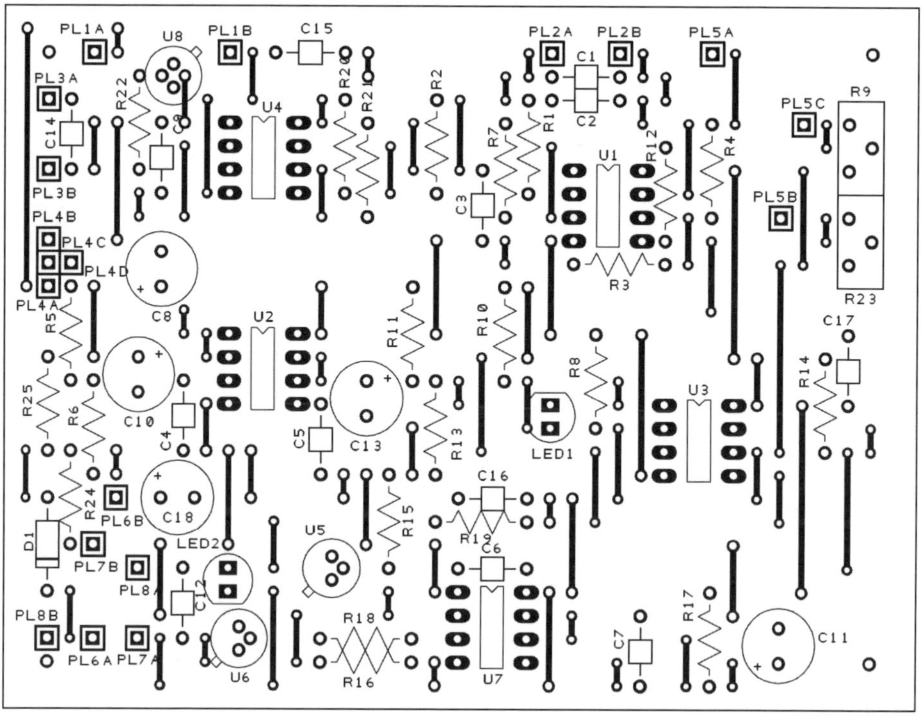

Cutting the Tracks

There are a number of breaks required in the stripboard tracks, as shown in the diagram below. You can either use a special tool designed for the job, known as a spot face cutter, or a hand drill with a suitable sized bit. Be very careful to cut the tracks in the correct place without leaving any bridges across the break. It would be advisable to double-check with an eye glass. Likewise with the jumpers. If you get any of these in the wrong place it could prove very difficult to find the cause of the problem.

The copper tracks run horizontally across the board, and the triangles indicate the positions where breaks are required. The view is from the underside of the board for convenience.

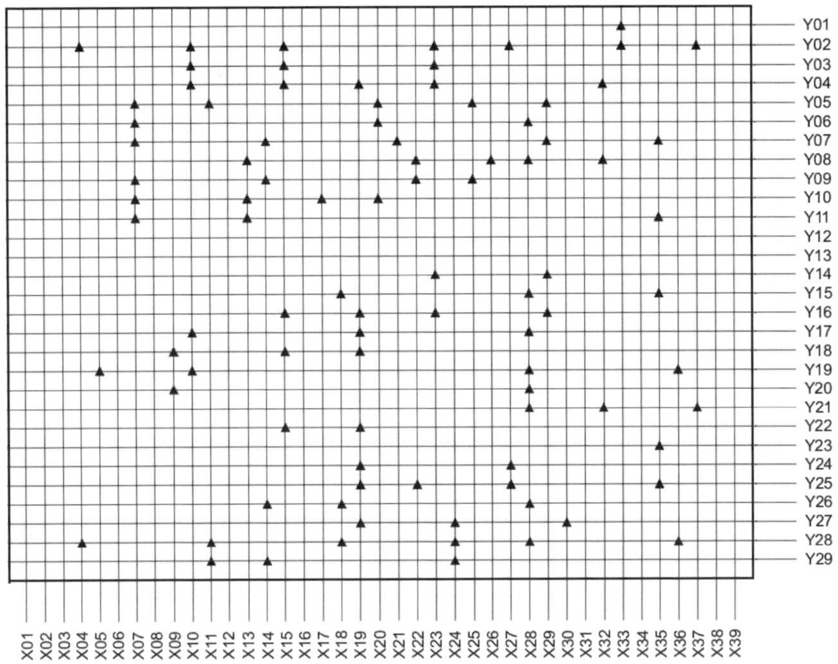

The 3-dimensional view provides an idea of what the final product will look like in real life, except the two presets in the top right, which are replaced by multi-turn pots on the control panel.

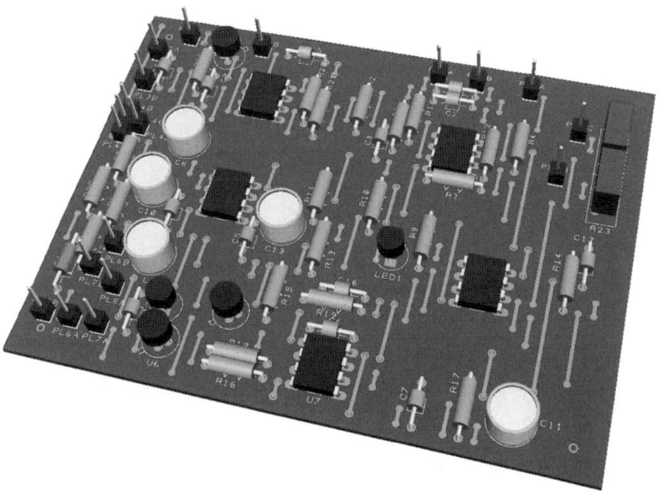

The diagram below shows the copper side of the board overlaid with the breaks shown previously. This will assist with correct positioning of the components. Again, the board is viewed from the underside.

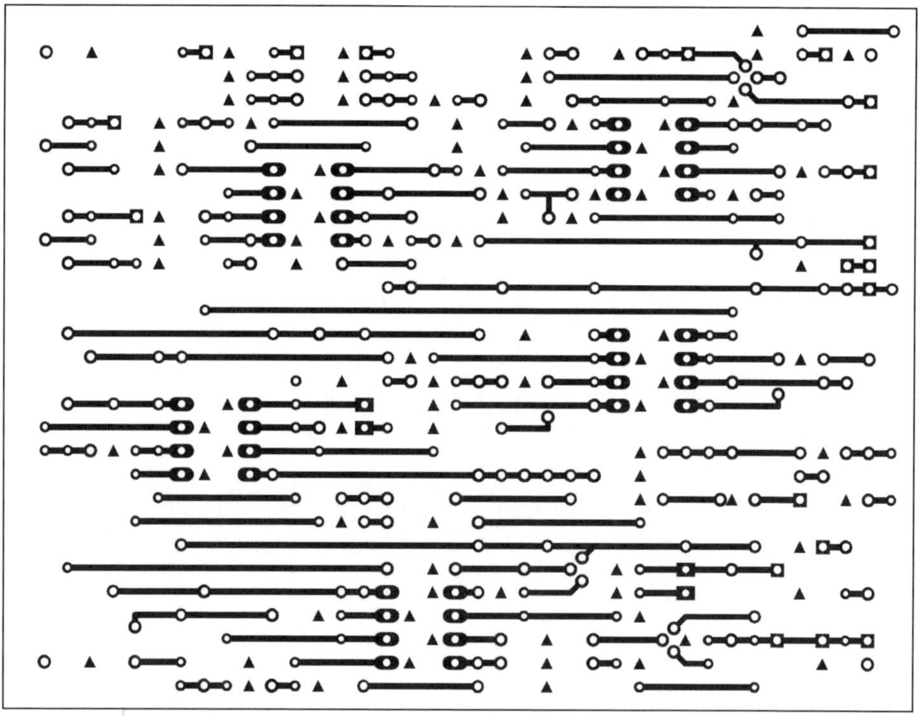

Populating the Board

In order to minimise errors and mistakes during construction, please follow these instructions:

1. Make the breaks in the copper tracks on the bottom of the board.
2. Fit the IC sockets, which will act as a visual guide for the other components.
3. Add the wire links on the top of the board.
4. Use a continuity tester to check there are no unintentional shorts.
5. Build the transmitter and test this is working as expected when connected to the TX coil.
6. Insert the components in the pre-amp stage and the phantom ground circuit, and test for correct operation using an oscilloscope.
7. Fit the rest of the components.
8. Connect the board to the control panel, and test the comparators and audio stage.
9. Install and null the ferrite coil as described in previously in this chapter.
10. Perform the laboratory-based tests (spark detection, TV emission and MBN).

Control Panel Wiring Details

Passive / Active switch wiring

The switch view is from the rear. Switch down = active, and switch up = passive.

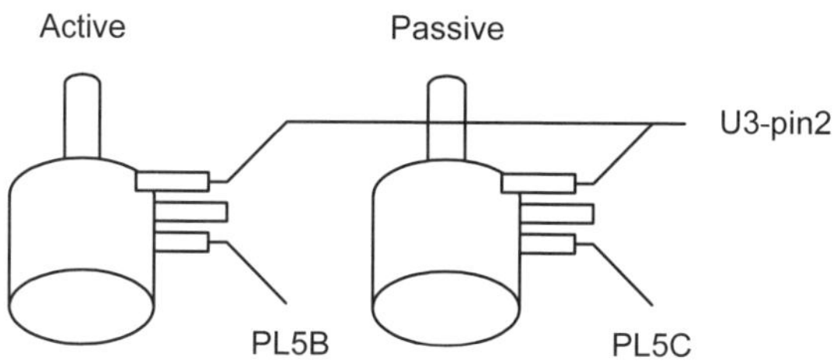

Threshold pot wiring

The external Active and Passive threshold pots are wired in place of the presets shown on the circuit board layout. One important point to note is that the multi-turn pots have the center pin located at one end, and not in the middle like a standard pot.

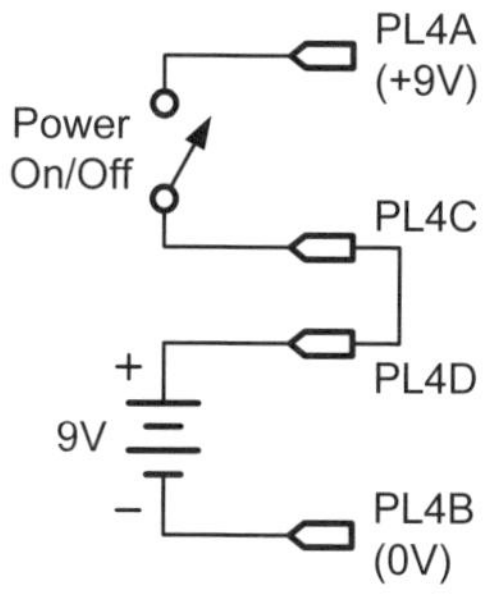

Power switch & battery wiring

Conclusion

Unlike the original PD, the TOTeM project is easily replicated with a little care and attention. It easily passes all the laboratory-based tests used by LRL experimenters, and certainly appears to react in the same way as the device shown in the internet videos. Whether it will lead you to treasure or not is maybe another story, but at least you will have the opportunity to explore the pseudo-scientific world of long range locators for yourself, and make up your own mind on the matter.

CHAPTER 15 Loose Ends

"Engineering is the art of modelling material we do not wholly understand, into shapes we cannot precisely analyse, so as to withstand forces we cannot properly assess, in such a way that the public has no reason to suspect the extent of our ignorance."

— British Institute Of Structural Engineers, 1976

This is one of those books where you could keep on writing and writing, and never get finished[1]. At some point, you have to draw a line in the sand and say, "This is where we stop." In doing so, there will be a lot of information that never gets discussed, or topics that only receive elementary coverage.

So far, we have presented quite a few metal detector designs which represent most of the techniques that have been used in industry, and have done a fairly decent job in working through the fundamentals of those designs. But if you compare these designs to state-of-the-art detectors being sold, they obviously don't have the complexity or features you can currently buy. There is only so much that a book like this can hope to present, and a good foundation in the basics of detector design is our foremost goal. Advanced designs are built on these same basics, so this book puts an electronics enthusiast pointed in the Right Direction.

Nevertheless, there are a few other things we need to mention. The purpose of this chapter is to wrap up a few topics, and to present some ancillary material that just doesn't fit anywhere else.

Multi-frequency Design

After reading up to this point, you will notice a glaring omission: there has been no mention of multi-frequency designs. Isn't this one of the Hot Technologies in metal detecting? Yes, it is, but designing and building a multi-frequency detector is challenging, and (currently) beyond the scope of this book. But we will attempt to explain the fundamentals behind multi-frequency (MF) techniques.

An example of a simple MF detector is shown in block diagram form in Figure 15-1. The transmitter applies a 5kHz square wave to the TX coil. The square wave is rich in harmonics so it contains a strong 5kHz fundamental frequency, plus odd harmonics; therefore, it will have a reasonably useful harmonic at 15kHz. The received signal is split into two channels, one for the 5kHz fundamental and one for the 15kHz harmonic. The signals are I/Q demodulated and multiplexed to an ADC. A microprocessor does the remaining work

1. And we almost did that!

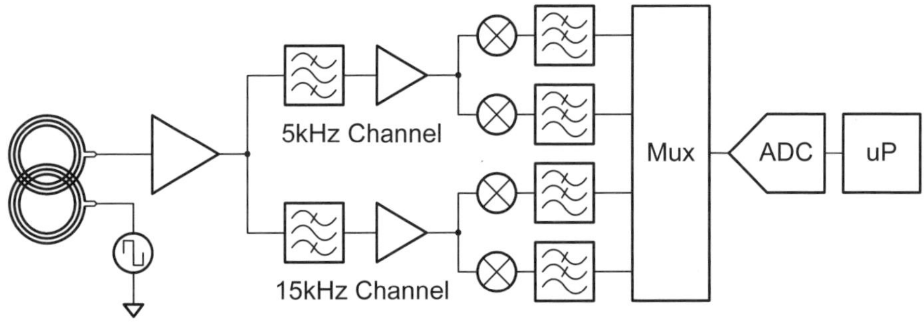

Fig. 15-1: **Multi-frequency block diagram**

Because there are now two channels, both channels will need to be calibrated for ground (ferrite). Additionally, specific targets give different phase (and amplitude) responses at different frequencies, so both channels will need to be calibrated so that their responses can be combined (or correlated) in some useful way. Finally, using two frequencies (with enough spread) allows you to use a subtraction method to notch out salt responses. That is, you can simultaneously ground balance to ferrite *and* eliminate salt, which is why MF detectors are preferred when salt water discrimination is needed.

All this calibration and correlation and subtraction can be accomplished in an all-analog design, but it is a rather tedious exercise that a good microprocessor renders unnecessary. Even so, the code development is still challenging, which is why MF is beyond the scope of this book. By the way, the example described above is essentially identical to the Fisher CZ series of MF detectors.

Myths & Fallacies

Is it an antenna?

Many detectorists refer to the coil as an antenna. It is not, at least in the classical way that an antenna is used. Like an antenna, it converts electrical current into electromagnetic (EM) energy, and vice versa. But the EM energy produced by an antenna behaves differently depending on how far away from the antenna it is. There are two regions: the near-field region and the far-field region. With antennae, effective radio-wave propagation occurs in the far-field region, while the near-field region is strongly reactive.

The transition between the two regions is gradual, but basically occurs between one and two wavelength away. For a typical VLF operating at 10kHz the wavelength is 30,000 meters (roughly 18-1/2 miles); obviously, a metal detector does not operate in the far-field region. In either region the EM energy gets weaker as it travels farther away, because the energy is spreading out into a greater and greater expanse. In the far-field region the EM energy falls off as a function of $1/d^2$ (where d is the distance away), just like you would expect from basic algebra, where the surface area of a sphere is proportional to r^2 (radius). But reactive effects cause the near field energy to fall off as a function of $1/d^3$.

From classical antenna theory, elements like dipoles and circular loops are most efficient at transferring energy when they have dimensions (length or circumference) equal to a quarter wavelength. For example, consider a classic Yagi-Uda TV aerial, the kind stuck on many rooftops. The low channel VHF band is around 75MHz, so the wavelength is about 4 meters, and the quarter-wavelength is 1 meter. This makes for a reasonably sized antenna and, indeed, this is about the size of a basic TV aerial. For our 10kHz metal detector, we would need a loop with a diameter of about 1-1/2 miles. Obviously our coils are nowhere near the optimal size for classical radiowave propagation and that is not what we are using them for.

Instead of an antenna, a metal detector coil behaves exactly like a transformer; it transfers energy via induction. You can think of the system as being a double transformer; the TX coil and the target form one transformer, and the target and the RX coil form a second transformer. In fact, for the purposes of analysis and simulation, metal detectors are often modeled exactly this way.

The Halo Effect

Most detectorists have experienced the situation where they've dug a much deeper hole than usual and found a coin at the bottom, but a later air test shows that the coin is not detectable from the same distance. Even placing the coin back in the bottom of the hole does not allow the detector to see the coin. This is the origin of the Halo Effect. Some detector manufacturers even put information in their user manuals, stating that detection ranges can vary depending on how long the object has been buried. The claim is that the coin interacts with salts in the ground, thereby making it appear larger to the detector. But is this effect real, or just another detecting myth?

It is true that some coins will interact with ground salts, and this is certainly true for iron objects, but for gold and silver coins the amount of corrosion is either minimal or non-existent. The Halo Effect may actually have many different sources. The most obvious would be that the coin was never detected at the depth claimed, but it was instead located in a side pocket of the hole, and as the hole is gradually enlarged in the search for the elusive target, it eventually falls into the bottom of the hole, and the detectorist experiences the so-called Halo Effect. This is most likely to occur when using a DD coil, as pinpointing tends to be less accurate than with a concentric.

Since iron targets are most susceptible to corrosion, and large iron targets can often defeat the discrimination circuitry of a detector, it is possible that it is not the coin that is being detected, but a nearby large iron object and its associated halo. As digging starts, the ground matrix is disturbed, thus destroying the halo around the deep iron object, and the coin, which happens to be lying above it, is discovered. Hence an erroneous conclusion is reached.

The Halo Effect is one of those subjective experiences that is difficult to quantify, and it is highly possible that the capabilities attributed to this effect are over-rated. A skeptical mind might be tempted to believe that it is simply a convenient ploy by some detector manufacturers to explain why their detectors do not perform well in an air test. However, there is no need to worry, because it will go much deeper when you use it in the field.

In general, most people believe in the existence of the Halo Effect, at least as far as iron is concerned. But gold and silver are a different story.

Metal or Plastic Stem Bolt?

Some people have suggested that using a metal bolt to fix the stem to the search coil can result in phantom signals when using the detector. In theory the presence of the metal bolt will upset the coil balance, unless it was originally balanced while bolted to the stem. But otherwise it should not result in false signals, because the bolt does not move relative to the coil, and is therefore not detected. However, the bolt will distort the electromagnetic field from the coil, and could possibly reduce sensitivity. Some simple tests have been made by replacing the metal bolt on a particular make of detector with a plastic one, but no significant improvement could be discerned. Since plastic bolts are readily available, there seems little point in using a metal one and then having to deal with coil balancing problems.

Electrostatic Shielding

There is often confusion about the real purpose of the electrostatic shielding used on metal detector coils. Even some of the cheap detector manufacturers appear to think that it is unnecessary, and construct their search heads without any shielding at all.

When two objects are brought into close proximity to each other there is an electrostatic field potential between them. In other words, you are creating a capacitor, with one plate being the detector coils and the second plate being the ground. When the metal detector is being used in the field, the search coil can brush against wet grass. or other objects, and the electrostatic field produced can cause unwanted signals in the detector. That is why a Faraday shield is used in the search head to shield the coils from nearby objects. For obvious reasons the Faraday shield is also known as an electrostatic shield.

The problem is not so much that an electrostatic field potential exists across the coil-ground capacitance, but the fact that it changes when the coil is in motion. These changes can be sensed by the detector circuitry and give a false indication. Even moving your hand across the face of an unshielded coil and can create an audio response.

Although this has been mentioned previously, it is worth repeating. There must be a break in the electrostatic shield that surrounds the coil, otherwise it will act as a shorted turn, and your detector will be next to useless. If you are building a DD coil, then be careful to insulate the points where the coils overlap, as this can also create a loop in the shield.

Power Output

If we just want more depth, then why not simply increase the transmitted power? Is there a limitation on the transmitter power in metal detectors?

Many people point to FCC regulations (in the US) and claim that detector output power is limited to a certain level. A number often given is 100mW of power. Exactly where this number comes from is uncertain, but it is erroneous. Yes, there are legal limitations to transmitted field strength in metal detectors, but the reality is that detectors are not in any danger of exceeding these limitations.

While we are on this topic, let's also dispatch with the term "output power" when it comes to metal detectors. In electronics, an ideal inductor does not dissipate power, so output "power" has no real meaning. Detectors only develop a local magnetic field and, unlike radios, don't propagate electromagnetic energy. Magnetic field strength is determined by the size of the coil and the "ampere-turns" — that is, the number of turns in the coil, and the peak amperes flowing through those turns. You might think that a certain amount of "power" from the transmit circuitry is needed to achieve a desired magnetic field strength, but that will depend on the design of the circuitry. A transmit coil in a resonant circuit will effectively recycle much of the coil current, making it far more fuel efficient than a brute-force non-resonant "driven" coil. In fact, an ideal LC tank circuit is purely reactive so even the power consumption will approach zero.

The practical limit on detector field strength has more to do with signal-to-ground ratio and practical battery life than FCC regulations. Signal-to-ground ratio is the amount of target signal compared to the amount of ground signal. As you crank up the transmitted field strength both go up proportionally. Because deeper targets have more ground above them, this is a losing situation for achieving greater depths. The only case where it can help is in very mild ground mineralization.

Even in zero mineralization (air) the depth gained from a stronger TX field is downright depressing. When we discussed coil-vs-antenna earlier, we stated that the near-field strength falls off as a function of $1/d^3$. The same is true for the target's returned signal strength, so the round-trip falls off as a function of $1/d^6$. What this means is that in order to double our detection depth we would need to increase the transmitted field strength by a factor of 64. Put another way, doubling the field strength will increase detection depth by a mere 8%, so if you could barely detect a target at 6" you would barely detect it at 6.5" if the coil drive is doubled. That's a pretty minimal depth increase for a lot of additional signal, and mineralization will only make it worse.

As you can see, the FCC doesn't really need to limit detector "power," physics does it for us!

Coil Sensitivity

For the case of a simple circular coil, the detection depth for a small coin-shaped target is roughly equal to the diameter of the search coil. Whereas the sensitivity is approximately equal to the cube of the target diameter. The sensitivity is also dependent on the distance between the coil and the target, and varies according to the sixth power.

In practical terms, doubling the depth of the target reduces the sensitivity to one sixty-fourth, and reducing the target size by a half reduces the sensitivity to one eighth. Although using a larger diameter coil may seem an obvious way of increasing depth, the sensitivity is also reduced to one eighth if the coil size is doubled. The result is that small targets become increasingly difficult to detect with a large coil. On the other hand, a smaller diameter search coil will increase sensitivity to small targets, but will lack depth. This is the reason why general purpose coils are in the range of 7 to 10 inches.

The Concentric Ring Experiment

In Chapter 5 it was briefly mentioned that there is a difference in response between a solid round target like a coin and an open round target like a ring. But what is the real cause of this difference, and what happens if you cut the ring?

If you take a coin and a ring of the same diameter and material as the coin, a metal detector will be able to detect the ring at a greater distance despite the ring having less mass. It has been proposed that this effect is due to the ring's ability to sustain a greater eddy current flow, and hence provide a stronger response.

Eddy currents are produced in a metal target when they intersect the changing magnetic flux from the transmit coil. These currents consist of circulating eddies that produce their own magnetic fields which oppose the applied field (Lenz's Law). The term eddy current is analogous to the circulating currents seen in flowing water. This initial picture of the circulating eddy currents would seem to indicate that a coin should produce a stronger target response due to the larger number of vortices, but a simple experiment can be carried out whereby a second identical coin has its center removed. Comparison tests clearly show that the drilled-out coin has the better response. Therefore it is likely that superposition of the large number of intersecting vortices result in some cancellation, such that larger eddy current rings are formed in the target. The result is a set of concentric rings that follow the circumference of the coin, much like the cross-section of an onion, as shown in Figure 15-2.

To test this theory an experiment was performed using a set of concentric rings to

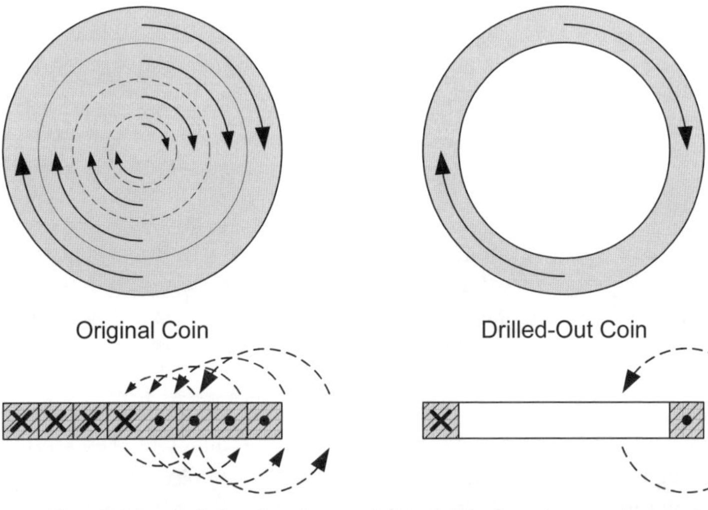

Fig. 15-2: **Original coin and its drilled-out counterpart**

see what would happen as sets of rings were added or removed. The rings were constructed using 22 swg (0.71 mm) tinned copper wire with diameters of 37, 30, 25 and 15 mm. They were arranged as shown in Figure 15-3. Using the Raptor IB detector from Chapter 9, the RX amplitude was recorded at the output of the synchronous demodulator in the DISC channel (TP4) for different combinations of the rings. It was a simple matter to show the best response was obtained by using all 4 rings, whereas any other combination gave a lower response. The same measurements were then made using two 2 coins - one unmodified, and the other with its center removed (in other words, a ring).

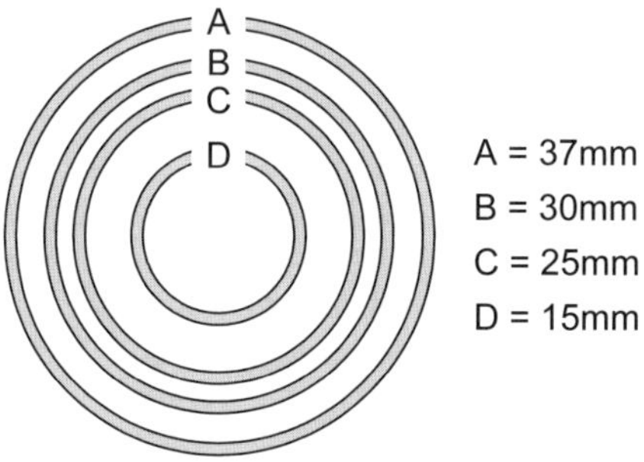

Fig. 15-3: **Concentric rings**

Now this is where it gets interesting...

It was immediately obvious from the measurements that the ring had a lower amplitude response than the original coin. This was contrary to expectation, particularly since the Raptor detector was able to detect the ring at a greater distance.

The answer was found by repeating the experiment, but this time monitoring the output of the synchronous demodulator in the GEB channel (TP6). In this case it was discovered that the ring produced a greater phase-shift than the original coin, and consequently an improved response in the GEB channel. In the repeated concentric ring experiment the combination of rings A+B gave the best response when compared to A+B+C and A+B+C+D, or even A on its own. Which implies there is an optimum ring thickness at which the phase-shift will be a maximum. Trying other ring combinations showed the A+C combination to provide the most phase-shift overall, which could be explained by considering the individual rings to be tuned circuits.

The concentric rings were found to essentially behave in the same manner as the solid target (coin) and its drilled-out counterpart, indicating that eddy currents are indeed established as a set of concentric rings. It was also shown that the enhanced response of the drilled-out coin was due to it producing a greater phase-shift in the received signal.

So what happens if we cut the drilled-out coin to break the ring (Figure 15-4c)?

According to the concentric ring theory the target should now give a very poor response, as only the original small eddy current vortices can be established in the target, which should dramatically kill the target response. Indeed this is exactly what happens.

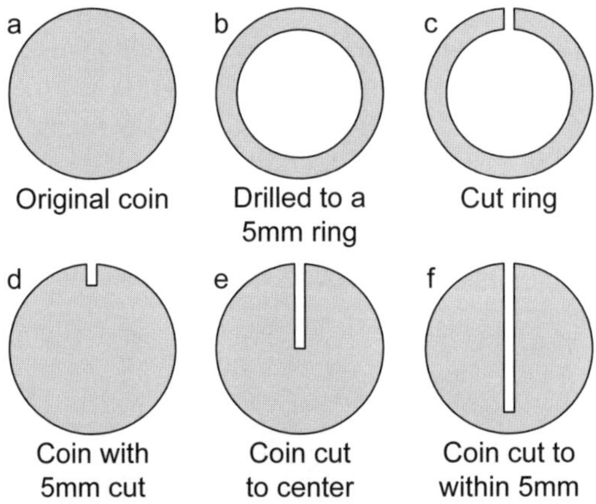

Fig. 15-4: **Final coin experiments**

In one final experiment, a 5 mm cut was made into the edge of the original coin (Figure 15-4d) which reduced the target phase-shift. This is consistent with the idea that the outer eddy current ring(s) have been cut. The coin was progressively cut through, and at each stage the response was found to be reduced. The final cut almost reduced the coin to two semi-circles (Figure 15-4f) thus forcing the eddy currents to form alternate paths within the two halves of the coin.

Of course we have only considered the coin-shaped targets as 2-dimensional objects oriented with their flat surfaces parallel to the search head. If the target has a less homogeneous composition, and is inclined at a different angle, then the situation becomes more complicated. But the general principle appears to be that the smaller eddy current vortices will tend to form larger loops within the target body, providing a greater response than would generally be expected. In cases where the formation of these larger loops is encouraged (for example, in the drilled-out coin) then an even greater response can be obtained. You may have also noticed this with large iron rings, which can be detected at great depths, completely defeating the detector's discrimination circuitry.

Nuances in International Electronics

Occasionally problems can occur when engineers from different countries collaborate during the design process. One spectacular failure (due to a mix-up between metric and imperial measurements) happened in 1999, when the Mars Climate Orbiter was destroyed as it entered the Martian atmosphere. There was a navigation error that resulted on some of the spacecraft data being reported in imperial units rather than metric. Of course, this was an extreme, expensive and highly public failure, but less critical errors frequently happen with multi-disciplinary teams spread across the globe.

Even in this book (which is a U.K./U.S.A. collaboration) we have frequently mixed metric and imperial units. Despite these differences we have attempted to follow a strict set of rules regarding symbol notation. Common errors by engineers are (for example) using an uppercase "S" for seconds, when they should use a lowercase "s". The uppercase "S" is reserved for siemens, which is the unit of conductance. When inexperienced electronics engineers use SPICE (Simulation Program with Integrated Circuit Emphasis) they forget that SPICE is not case sensitive. In this instance "M" and "m" mean the same thing, and "M" needs to substituted for "MEG" or "meg". These examples are only a few of the traps that lay in wait for the unwary, and several others are outlined below.

Measurement	Unit	Symbol
electric current	Amp	A
electric charge	Ampere hour	Ah
power or intensity	Bell	B
electric charge	Coulomb	C
capacitance	Farad	F
inductance	Henry	H
frequency	Hertz	Hz
energy	Joule	J
temperature	Kelvin	K
force	Newton	N
resistance	Ohm	Ω
angle	radian	rad
angular velocity	radian per second	rad/s
time	second	s
flux density	Tesla	T
voltage	Volt	V
power	Watt	W
magnetic flux	Weber	Wb

There are many more symbols we have not mentioned, but these relate to other disciplines than electronics. In general, the rule is that symbols named after famous people have a capital letter. To add further to the confusion, Americans refer to the measurement of 0.001 inches as "one mil" (the British call this "one thou") but outside America, 1 mm is also referred to as "one mil". No wonder we have problems!

Component Marking Standard

Now just when you thought you knew the rules, we put a spanner in the works!

In schematic diagrams there are several methods used by engineers to mark the component values, such as (for a resistor) 1.2kΩ, 1.2k or 1k2. First note that you should use a lowercase letter "k" to denote kilo (1000x) and not "K", which indicates a temperature measurement (Kelvin). It is acceptable to leave out the symbol as most people will implicitly understand that a resistor is measured in Ohms, a capacitor in Farads, and an inductor in Henries. Many companies have now adopted the de facto standard of writing 1k2. The reason is that the decimal in 1.2k can easily be misread, or lost during copying, so that it ends up being incorrectly interpreted as 12k. This standard is derived from the practice of marking physical components to eliminate the dot problem. Over time the use of 1k2, 4u7, 6m8, etc., in schematics has become commonplace, even though it is not an official standard. Finally, when a resistor does not require a multiplier, such as 100 ohm, then it may be written as 100R instead of 100. This also allows precision resistor values to be specified using the same format, such as 10R5.

When building any of the designs presented in this book, it would be advisable to test the transistors first before soldering them to the PCB. Even though you might think a particular model of transistor would have the same pinout from all manufacturers, you would be wrong. Even from the same component supplier we obtained transistors, during development of the Raptor project, that had completely different pinouts. Transistors that are constructed in TO92 format are particularly bad in this respect. The problem is related to the country of manufacture, and Figure 15-5 shows some examples of the pinouts you might encounter. The only way to be certain is to test each one separately. Most digital meters nowadays can do more than simply measure voltage, current and resistance. Even the cheaper models now have frequency measurement, logic, diode and transistors testers built-in. A transistor that was inserted according to the silkscreen layout could well turn out to be inserted incorrectly. This can be a nasty fault to find, as we can vouch from firsthand experience.

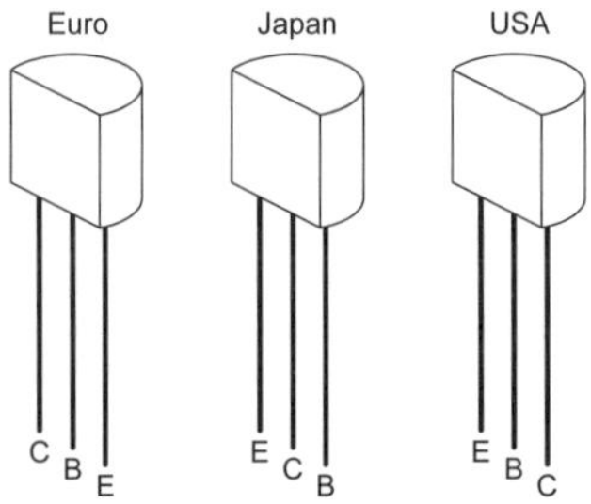

Fig. 15-5: **TO-92 transistor pinouts**

Electromagnetic Terminology

There are many terms used in the field of electromagnetism, and these can often get confusing. Here is a simple explanatory guide through the electromagnetic smog.

Permittivity

As the name suggests, permittivity is a measure of a material's ability to *permit* entry of an electric field. The symbol for permittivity is the Greek letter epsilon (ε) and is measured in farads per meter (F/m). So permittivity is the capacitance per unit length.

There are two types of permittivity, ε_r (relative permittivity) and ε_0 (permittivity of free space). The value of ε_0 is 8.85×10^{-12} F/m. Relative permittivity relates to a particular medium (usually a dielectric) and is often referred to as the dielectric constant.

Permeability

Whereas the previous term relates to the electric field, permeability relates to the ability of a material to support a magnetic field. In other words, it indicates how *permeable* the material is to the field. It is represented by the Greek letter mu (μ) and is measured in henries per meter (H/m), which is the inductance per meter.

As you might expect there are again two types of permeability, μ_r (relative permeability) and μ_0 (permeability of free space). The value of μ_0 is $4\pi \times 10^{-7}$ H/m and is referred to as the magnetic constant.

However, it should be noted that neither the dielectric constant nor the magnetic constant are in fact constants. Both ε_0 and μ_0 can vary with position within the medium, and are dependent on frequency, humidity, temperature, etc., etc.

Diamagnetism

Materials that create a magnetic field in opposition to an external applied field are said to be diamagnetic. This effect can be demonstrated by dropping a magnet down the inside of a copper pipe. Although copper is not attracted to a magnet, it will repel the magnet as it moves down the pipe, and will slow the descent. There is a significant time difference between dropping the same magnet down either a copper or a plastic pipe. Superconductors are extremely diamagnetic and can be used to levitate a magnet. Diamagnetic materials have a relative permeability of less than one.

Paramagnetism

Paramagnetic materials are attracted to a magnet but cannot be used to create a permanent magnet. They have a relative permeability greater than unity. A ferrite rod (as used in the Pistol Detector, Chapter 14) is a good example of a paramagnetic material.

Ferromagnetism

Iron is an example of a ferromagnetic material. You will note that we have referred to iron targets as *ferrous* throughout this book (the Latin for iron is *ferrum*). This material is attracted to a magnet, and can be turned into a permanent magnet. The strict definition of a ferromagnetic material is one where all the magnetic ions contribute to the overall magnetism.

Ferrimagnetism

Ferrimagnetic materials have similar properties to ferromagnets, but in this case some of the magnetic ions act in opposition so that the overall magnetization is reduced.

Antiferrimagnetism

Antiferrimagnetic materials are those where aligned and opposing magnetic ions exactly cancel so that the overall magnetization is zero. You may be wondering why these materials are not simply referred to as non-magnetic. The simple answer is that this cancellation effect only occurs below a certain temperature.

Magnetic Viscosity

This refers to the delay between applying or removing a magnetic field to a ferromagnetic medium, and a change taking place in the material. This delay is sometimes called the magnetic lag. The underlying mechanism that causes the delay is the time required for magnetic domain movement to occur in the material. Magnetic viscosity is important in archaeology, humanitarian demining, and for researchers interested in soil and rock magnetics. It has no direct application for the average treasure hunter.

Remanence

Quite simply, this is the magnetism left behind in a material after the external magnetic field is removed. Only ferromagnetic and ferrimagnetic materials exhibit remanence.

Coercivity

This is a measure of the resistance exhibited by a ferromagnetic material when an attempt is made to completely demagnetize it using an external magnetic field. Materials with a high coercivity (measured in amps per meter, A/m) can be used to make permanent magnets, whereas those with a low coercivity may be used to make transformer cores, for example.

Reluctance

Reluctance is sometimes referred to as magnetic resistance. In fact there is a magnetic equivalent of Ohm's law, in which resistance = reluctance, magnetomotive force = volts, and magnetic flux = current. Just as current tends to follow the path of least resistance in an electrical circuit, the magnetic flux tends to follow the path of least reluctance in a magnetic circuit.

PC Boards

Most projects in this book are easily built on a breadboard or veroboard, but we also include a single-sided PCB layout so you can make a more professional-looking circuit if you like. PCBs can either be ordered from a board house or made at home. Figure 15-6 shows an example of a PI detector[2] PCB made both at a board house and at home. The professionally made PCB has a few major advantages: the copper is nickel-plated, the holes are pre-drilled, and the reverse side has a silkscreen of the component placement. Of course, it costs more.

If you only need a one-off PCB, then making it at home is not difficult. An excellent product for making a single-sided PCB is *Press-n-Peel Blue*, an iron-on transfer film. Use a standard dry toner laser printer (or copier[3]) to print the PCB image on the

2. Not from a project in this book.
3. Beware of very slight size expansion/reduction in copiers, which can mess up component pad spacing.

Fig. 15-6: **Store-bought vs home-made PCB**

film, then simply[4] iron it on to the copper surface. The laser toner becomes the resist mask, and the board is ready to etch. Press-n-Peel Blue typically costs about $2 USD per sheet. The home-made PCB in Figure 15-6 was made with this product.

When you need more than 2 or 3 of the same PCB then you should consider using a board house. Some of these companies specialize in low-volume prototype PCBs, and small boards can be made for $15 - $20 USD each. Board houses will require Gerber files, and Gerbers for all the PCBs in this book are available at http://www.geotech1.com/itmd.

4. OK, "simply" is understating it a bit... it's critical to get the copper really really *really* clean, then carefully and thoroughly iron the transfer. It may take a few tries to get the hang of it.

APPENDIX A

Air-Cored Coil Calculator

"On two occasions I have been asked [by members of Parliament], 'Pray, Mr. Babbage, if you put into the machine wrong figures, will the right answers come out?' I am not able rightly to apprehend the kind of confusion of ideas that could provoke such a question."
— Charles Babbage

Probably the main difficulty encountered, when starting to build your own coils, is deciding how many turns are required to achieve a certain inductance. There are many online coil calculators to be found on the Internet, but the majority are inaccurate when applied to air- cored inductors. In general, these calculators are based on the widely-known equation by Wheeler[1]:

$$L = \frac{4r^2 \cdot N^2}{l + 0.9r} \qquad \text{Eq A-1}$$

where L = inductance in Henries
r = coil radius in meters
l = coil length in meters
N = number of turns

The above formula is accurate to within 1% for $l > 0.8r$, and provides a good estimate for the simple case of a single layer solenoid coil. However, metal detector coils are usually multi-layer, and the Wheeler equation can give incorrect results.

An interesting question that is often asked is, "How can I wind a coil that has the maximum possible inductance for a given length of wire?" Maxwell discovered that a coil with a square cross-section, and a mean diameter of 3.7 times the dimensions of the cross-section, or $2d = 3.7c$, is the optimum configuration. In 1931, Brooks wrote a paper in which he calculated the ideal value for the mean radius r as very close to $3c/2$. In practice, the inductance is rather weakly related to this ratio, and the exact geometry can deviate quite significantly before the accuracy of the calculated inductance becomes unacceptable. According to Brooks, the inductance of any air core coil can be approximated by the following equation:

$$L = 400\pi \cdot 10^{-9} \cdot r \cdot N^2\left(\left(0.5 + \frac{l^2}{48r^2}\right)\ln\left(\frac{32r^2}{l^2}\right) - 0.84834 + 0.051\frac{l^2}{r^2}\right) \qquad \text{Eq A-2}$$

where, again

1. H.A. Wheeler, "Simple Inductance Formulas for Radio Coils," Proc. I.R.E., vol 16, pp 1398-1400, Oct 1928

L = inductance in Henries
r = mean radius of the winding in meters
l = coil length (which equals thickness c) in meters
N = number of turns

In conclusion, although a Brooks coil is a special case of the circular coil inductor with a rectangular cross section, that is constructed to achieve maximum inductance with a given length of wire, the resulting inductance value is not heavily dependent on the relationship between the coil radius, length and thickness. But the equation is sufficiently complex, and a computer program would be a useful tool to assist with estimating the correct number of windings required.

Three versions of the program are provided here. The first is written in Visual Basic which, of course, assumes that you have access to this commercial program, whereas the second version is written in Tcl/TK which is freely available to download from the Internet. The third version is written using JavaScript, but you will need to add the necessary HTML code yourself. If you have some knowledge of another programming language such as C or Java, for example, then it should be a relatively simple matter to duplicate the functionality provided by the following programs.

The Visual Basic Version

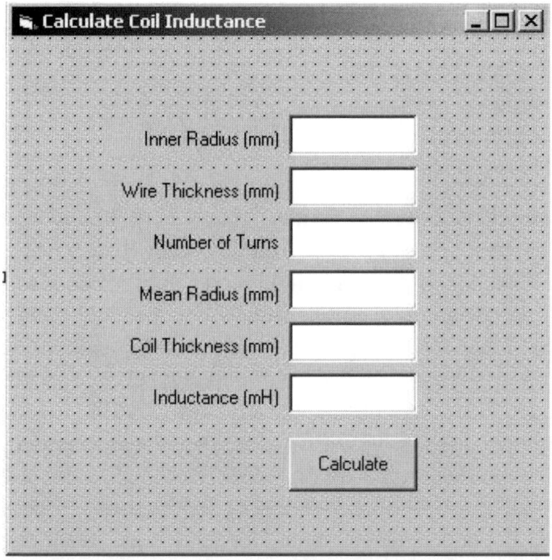

Creating forms in Visual Basic is a simple graphical procedure. The three text entry areas are called txtInnerRadius, txtWireThickness and txtNumberOfTurns respectively. The text boxes containing the calculated results have the "enable" property set to "false". Everything else should be self-evident from the source code, such as the name of the Calculate button (cmdCalculateInductance).

If you don't wish to write the program yourself using the source code provided, then an executable version can be downloaded from *http://www.geotech1.com/itmd.*

The Visual Basic version will run in Microsoft Windows, as long as you have the VB runtime dynamic linked library (DLL) installed. It will also run in Linux using Wine.

```
Option Explicit
Dim a, c, w, x, L, pi As Double
Dim N As Integer
Private Sub cmdCalculateInductance_Click()
  'Calculate inductance of coil
  pi = 3.14159
  N = txtNumberOfTurns.Text
  w = txtWireThickness.Text
  c = Sqr(N*w)
  a = txtInnerRadius.Text + c/2
  x = (c/2/a)^2
  L = 0.0000004*pi*a*N^2*((0.5+(x/12))*Log(8/x)-0.85+0.2*x)
  txtMeanRadius.Text = Format(a, "0.00")
  txtCoilThickness.Text = Format(c, "0.00")
  txtInductance.Text = Format(L, "0.00")
End Sub
```

The Tcl/TK Version

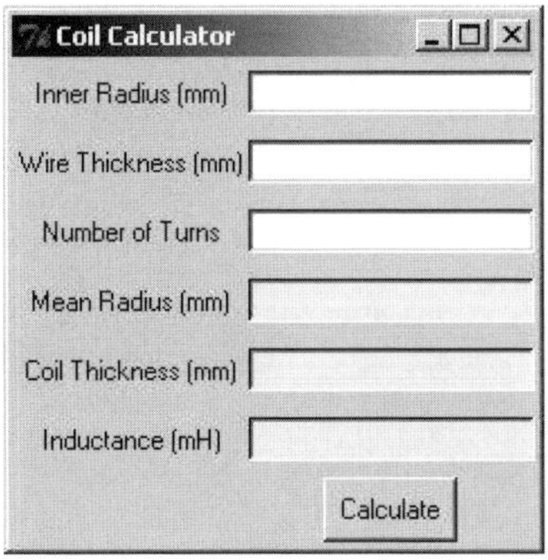

The Tcl/TK dialog boxes are not produced graphically. Even programs such as Visual Tcl and SpecTcl are not GUI builders in the same sense as Visual Basic. Although the coil calculator program shown below appears, on the face of it, to be much larger than the previous version, this is purely because the VB GUI code is hidden from the user, whereas the Tcl/TK GUI code is not. In order to test that the program is working correctly, enter the following data:

Inner Radius = 216mm
Wire Thickness = 0.56mm
Number of Turns = 45

The calculated result should be:

Mean Radius = 217.88mm
Coil Thickness = 3.76mm
Inductance = 2.741mH

Tcl/TK programs are platform independent and will run in Microsoft Windows and Linux. This program can be downloaded from http://www.geotech1.com/itmd.

```
proc CoilCalculator_ui {root args} {
  if {$root == "."} {
    set base ""
  }
  else {
    set base $root
  }
  label $base.lblInnerRadius -text {Inner Radius (mm)}
  entry $base.entInnerRadius -justify center -textvariable innerRadius
  label $base.lblWireThickness -text {Wire Thickness (mm)}
  entry $base.entwireThickness -justify center -textvariable \
        wireThickness
  catch {
    $base.entwireThickness configure \
      -font {-*-MS Sans Serif-Medium-R-Normal-*-*-80-*-*-*-*-*-*}
  }
  label $base.lblNumOfTurns -text {Number of Turns}
  entry $base.entNumOfTurns -justify center -textvariable numberOfTurns
  label $base.lblMeanRadius -text {Mean Radius (mm)}
  entry $base.entMeanRadius -background bisque -justify center \
        -state disabled -textvariable meanRadius
  catch {
    $base.entMeanRadius configure \
      -font {-*-MS Sans Serif-Medium-R-Normal-*-*-80-*-*-*-*-*-*}
  }
  label $base.lblCoilThickness -text {Coil Thickness (mm)}
  entry $base.entCoilThickness -background bisque -justify center \
    -state disabled -textvariable coilThickness
  label $base.lblInductance -text {Inductance (mH)}
  entry $base.entInductance -background bisque -justify center \
    -state disabled -textvariable inductance
  button $base.cmdCalculate -command cmdCalculate -text Calculate
# Geometry management
  grid $base.lblInnerRadius -in $root -row 1 -column 1
  grid $base.entInnerRadius -in $root -row 1 -column 2
  -row 2 -column 1
  grid $base.lblWireThickness -in $root
  grid $base.entwireThickness -in $root -row 2 -column 2
  -row 3 -column 1
  grid $base.lblNumOfTurns -in $root
  grid $base.entNumOfTurns -in $root -row 3 -column 2
  grid $base.lblMeanRadius -in $root-row 4 -column 1
  grid $base.entMeanRadius -in $root -row 4 -column 2
  -row 5 -column 1
  grid $base.lblCoilThickness -in $root
  grid $base.entCoilThickness -in $root -row 5 -column 2
  grid $base.lblInductance -in $root -row 6 -column 1
  grid $base.entInductance -in $root -row 6 -column 2
  grid $base.cmdCalculate -in $root -row 7 -column 2
# Resize behaviour management
  grid rowconfigure $root 1 -weight 0 -minsize 30
```

```
  grid rowconfigure $root 2 -weight 0 -minsize 30
  grid rowconfigure $root 3 -weight 0 -minsize 30
  grid rowconfigure $root 4 -weight 0 -minsize 30
  grid rowconfigure $root 5 -weight 0 -minsize 30
  grid rowconfigure $root 6 -weight 0 -minsize 30
  grid rowconfigure $root 7 -weight 0 -minsize 30
  grid columnconfigure $root 1 -weight 0 -minsize 30
  grid columnconfigure $root 2 -weight 0 -minsize 30
# additional interface code
  set .title "Coil Calculator"
  proc cmdCalculate {} {
    global innerRadius wireThickness numberOfTurns
    global meanRadius coilThickness inductance
    set pi 3.14159
    set N $numberOfTurns
    set w $wireThickness
    set c [expr sqrt([expr $N * $w])]
    set a [expr $innerRadius + $c/2]
    set x [expr pow([expr $c/2/$a], 2)]
    set s1 [expr 0.0000004 * $pi * $a]
    set s2 [expr pow($N,2)]
    set s3 [expr $s1 * $s2]
    set s4 [expr 0.5 + [expr $x/12]]
    set s5 [expr log(8/$x)]
    set s6 [expr 0.85 + 0.2 * $x]
    set s7 [expr $s4 * $s5 - $s6]
    set L [expr $s3 * $s7]
    set meanRadius $a
    set coilThickness $c
    set inductance $L
  }
# end additional interface code
}
CoilCalculator_ui
# end of program
```

Note: Save the program as "CoilCalculator.tcl". Then double-clicking on the file name will automatically invoke the Wish interpreter, assuming that you have the Tcl/TK application software installed on your system.

The JavaScript Version

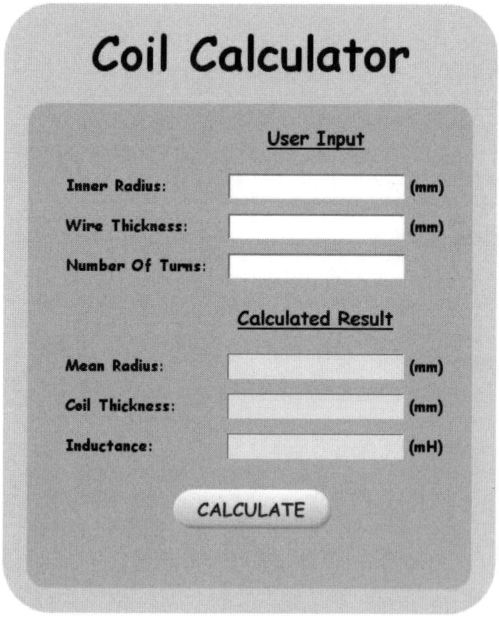

The main advantage of programs written using JavaScript is their ability to run in any Internet browser that supports the JavaScript language. Like the Tcl/TK version, this means they are also platform independent, but in order to duplicate the user interface you will need to write some additional HTML. The version shown in the above figure makes use of cascading style sheets (CSS) and some JavaScript. How this is achieved is beyond the scope of this book. However, the source code (as with the previous examples) is freely downloadable from http://www.geotech1.com/itmd.

```
function calculate(data) {
  var innerRadius = data[0];
  var wireThickness = data[1];
  var numberOfTurns = data[2];
  var c,a,x,s1,s2,s3,s4,s5,s6,s7;
  c = Math.sqrt(numberOfTurns * wireThickness);
  a = innerRadius + c/2;
  x = Math.pow(c/2/a, 2);
  s1 = 0.0000004*Math.PI*a;
  s2 = Math.pow(numberOfTurns, 2);
  s3 = s1 * s2;
  s4 = 0.5 + x/12;
  s5 = Math.log(8/x)/Math.LN10;
  s6 = 0.85 + 0.2*x;
  s7 = s4 + s5 - s6;
  data[3] = Math.round(a*100)/100;
  data[4] = Math.round(c*100)/100;
  data[5] = Math.round(s3*s7*1000)/1000;
}
```

Accuracy

All available coil calculator programs will only provide an estimate for the inductance. A lot depends on how carefully you construct the coil, and how tightly you tie the bundle. The main purpose of these programs is to assist you in approximately determining the required parameters. Often you will need to either add or remove a few turns to achieve the exact result. In general, all these programs will underestimate the value of inductance, and the Brooks equation is no exception, but in practice it has been found to return more accurate results than Wheeler's equation.

APPENDIX B

Resources

In the Introduction we mentioned that information on metal detector technology is remarkably difficult to find, especially compared to other electronic devices. Metal detectors represent a very small market with only a handful of companies who actively design them, so there is a natural tendency to closely guard intellectual property.

However, over the years there have been a few books that cover detector technology, plus a number of detector designs presented in electronic hobby magazines. For the most part these have dealt with only simplistic designs, often basic BFO and TR circuits. A far better resource for advanced developments is the patent office. While patents can offer insight to the latest developments, they are sometimes written in an (often intentionally) obscure manner, on top of the legalese.

This chapter presents a number of resources and references for the curious who want to dig deeper into detector technology.

Web Sites

Geotech http://www.geotech1.com — This web site is run by the authors of this book and contains the single largest collection of metal detector information anywhere. Gerber files for the various PCBs and software for the PI projects can be downloaded here (http://www.geotech1.com/itmd), and errata and updates to the book are also posted. Geotech also hosts a large discussion forum on a variety of technical topics and projects at http://www.geotech1.com/forums.

Findmall PI Tech forum http://www.findmall.com/list.php?34 — This is a tech forum for discussion of PI detectors, established by PI guru Eric Foster.

THunting.com Tech forums http://www.thunting.com/smf/index.php?c=15 — These are the tech-related forums of the larger THunting.com web site.

PulsDetektor.de http://www.pulsdetektor.de/apboard/main.php — These forums are in German, appears to deal mostly with PI.

MD For You http://www.md4u.ru/forum — These forums are in Russian.

Books

Modern Divining Rods — R.J. Santschi, 1927

How to Build Proximity Detectors and Metal Locators — John Potter Shields, 1965

The Electronic Metal Detector Handbook — E.S. "Rocky" LeGaye, 1969

Official Handbook of Metal Detectors — Dr. Arnold Kortejarvi, 1969

How to Build Your Own Metal & Treasure Locators — F.G. Rayer, 1976

How to Build Metal/Treasure Locators — Robert Traister & John Traister, 1977

The Complete VLF-TR Metal Detector Handbook — Roy Lagal & Charles Garrett, 1979

How to Build Gold & Treasure Detectors — ETI, 1981 (Collection of former magazine projects.)

The Advanced Handbook of Metal Detectors — Charles Garrett, 1985

Building Metal Locators: A Treasure Hunter's Project Book — Charles D. Rakes, 1986

Treasure: The Business and Technology — Phillip S. Olin, 1991 — Chapter 3: Tools for Archaeology covers some advanced methods.

Inside Treasure Hunting — Ty Brook, 1999 — Chapters 1-5 cover history & technology.

Ortungstechnik für Profis — Wolfgang Schüler, 1999 — Written in German.

Magazine Articles

Searching for the Bullet - Bell & Garfield — Harper's Weekly, Aug 13 1881

Bell Induction Balance — Lancet, Jan 20 1883

Rucker's Magnetometer / "Divining Rod" — Scientific American, Jul 5 1890

Electrical Ore Finder — Scientific American, Aug 1892

Unscientific & Scientific Divining Rods — Scientific American, Aug 20 1892

Unscientific & Scientific Divining Rods — Scientific American, Jan 20 1906

Buried Shells Found By Induction Balance — Popular Mechanics, Feb 1916

The Divining Rod I — Scientific American, Feb 23 1918

The Divining Rod II — Scientific American, Mar 2 1918

The Divining Rod Made Respectable — Scientific American, Sep 24 1921

Detection of Minerals by Electric Methods — Science, Jun 6 1924

Buried Treasure - Apparatus for Locating Minerals — Scientific American, Dec 1925

Strange Radio Devices Locate Buried Treasure — Popular Science Monthly, Apr 1928

Build a Radio Treasure Hunter — Radio News, Sept 1928
Henry's Experiments on Electromagnetic Induction — Science, Jan 22 1932
How to Build the Radio "Treasure" Finder — Clyde J Fitch — Radio Craft, June 1932
The Divining Rod and Fakers — Science, Jul 10 1932
Science Aids Quest for Gold — Scientific American, Oct 1932
Finding Buried Gold with the Aid of Magnets — Popular Mechanics, Apr 1933
How to Build the New Treasure Finder — E.F. Sarver — Radio Craft, July 1933
Another "Treasure Finder" — Scientific American, Sept 1933
How to Build the New Treasure Finder — E.F. Sarver — Radio Craft, Nov 1933
How to Build the New Treasure Finder — E.F. Sarver — Radio Craft, Apr 1934
An Improved Treasure Locator — C.W. Palmer — Radio Craft, Aug 1934
How to Build an Inexpensive Metal Locator — Popular Mechanics, Sept 1935
Newest in Treasure Locators — Gerhard Fisher — Radio Craft, Dec 1936
Prison Gun Detector — Radio World, Dec 1936
Treasure Hunting by Radio — Science & Mechanics, Feb 1938
A Practical Metal Detector — WC Broekhuysen — Electronics, Apr 1938
Finding Hidden Treasure — Radio News, May 1938
Build Your Own Treasure Hunter — Radio News, Sept 1938
Plans for a Radio Treasure Finder — Gerhard Fisher — Science & Mechanics, Feb 1939
How to Make a Modern Radio Treasure Locator — Allan Stuart — Radio Craft, Sept 1939
How to Build an Inexpensive Metal Finder — Popular Mechanics, Jan 1940
Building a Modern Miniature-Tube Metal-Treasure Locator — G.M. Bettis — Radio Craft, Dec 1940
FM Metal Treasure Locator — G.M. Bettis — Radio News, Oct 1942
Metal Locator — G.M. Bettis — Radio Craft, Apr 1943
Metal Detectors — W.H. Blankmeyer — Electronics, Dec 1943
Mine Locators — Connery Chappell — Radio News, Mar 1944
Treasure Locators — Radio Craft, July 1944
Non-metallic Mine Locator — T.E. Stewart — Electronics, Nov 1945
Vehicular Mounted Mine Detectors — H.G. Doll et al — Electronics, Jan 1946
How Mine Detectors Work — Eric Leslie — Radio Craft, July 1946
Electron-coupled Zero-beat Metal Detector — Harry A Fore — Popular Mechanics, Aug 1946

Treasure Finding Modernized —W.E. Osborne — Radio & Television News, Sept 1946

Try This 1946 Treasure Finder — W.E. Osborne — Radio News, Nov 1946

The Modern Divining Rod — Radio News, Apr 1947

The Modern Divining Rod — Alvin B Kaufman — Radio & Television News, Apr 1947

Small-space Metal Locator — John Haynes — Radio Craft, Sept 1948

Underground Prospecting Tests with Radio Waves — Electronics, July 1949

How Mine Detectors Work — Eric Leslie — Radio Craft, July 1949

Industrial Metal Detector Design — Electronics, Nov 1951

A Modern Metal Locator — Rufus P Turner — Radio & Television News, Sept 1954

Hidden Metal Detector — Thomas A. Blanchard — Radio-TV Experimenter, 1955

Two Transistorized Metal Locators — Edwin Bohr — Radio Electronics, Mar 1955

Portable Metal Locator — Harvey Pollack — Popular Electronics, June 1955

How a Metal Detector Works — Popular Science, Aug 1955

Metal Detector Finds Ducts & Pipes — Carl David Todd — Electronics, Jan 1957

Detectascope — Popular Electronics, Mar 1957

Two box TR Detector — Popular Science, July 1960

Underwater metal detector — C.L. Henry — Science & Mechanics, Aug 1960

Transitone Locates Hidden Wiring — Harry Parker — Radio Electronics, Dec 1960

Simple Metal Locator — Frederick H. Calvert — Electronics World, July 1961

Transistorized Metal Detector — W.E. Osborne — Electronics World, Mar 1962

Lodestar Metal Detector — Popular Electronics, Sept 1962

Simplified Metal Locator — B.F. Miessner — Radio Electronics, Sept 1962

Finding Buried Stuff — C. Beeler — Radio Electronics, Apr 1966

Underwater Metal Hunting for Fun or Profit — Olle Klippberg — Radio Electronics, June 1966

A Treasure Finder You can Build — Ronald Benry — Popular Science, July 1966

Summer Fun with a Sensitive Metal Finder — GH Gill — Radio Electronics, July 1966

Electronic Metal Locators: Basic Types & Design Factors — Don Lancaster — Electronics World, Dec 1966

Underwater Metal Detector — James E. Pugh — Radio-TV Experimenter, 1967

Build the IC-67 Metal Locator — Don Lancaster — Popular Electronics, Jan 1967

Underwater Metal Detector — Skin Diver, May 1967

Build the Beachcomber — D Meyer — Popular Electronics, July 1967

Build a Treasure Finder — Charles Rakes — Radio Electronics, Nov 1967

Gold Grabber — Radio-TV Experimenter, Sept 1968

Metal Detector — D. Bollen — Practical Electronics, Jan 1969

Build a "Different" Metal Locator — Leslie Huggard — Popular Electronics, Feb 1969

Metal Detector — F.G. Rayer — Practical Electronics, Jan 1970

Treasure Tracer — Practical Wireless, Aug 1971

Metal Detector — D. Bollen — Everyday Electronics, May 1972

Treasure Detectors for Land Use — L George Lawrence — Popular Electronics, Sept 1972

PLL Metal Detector — Practical Electronics, Dec 1973

Metal Pipe or Wiring Locator — C. Whitehead — Practical Electronics, Dec 1976

BFO Metal Detector — D Waddington — Wireless World, Apr 1977

Induction Balance Metal Detector (Project 549) — ETI, May 1977

Seekit — W. Opel — Practical Wireless, May & June 1977

Treasure Locator — Everyday Electronics, Oct 1977

CMOS Twin Oscillator Detector — Mark Anglin — Electronics, Dec 22, 1977

Metal Locators Are Fun — W. Neville — Electronics Australia, Dec 1977

Discriminitive Metal Detector — RCV Macario — Wireless World, July 1978

Detecteur de Metaux Sensible — Elektor, Jan 1979

Sandbanks Metal Detector — PJ Wales — Practical Wireless, Jan 1979

Low Cost Metal Locator — Robert Penfold — Everyday Electronics, June 1979

Treasure Hunter Sound Adapter — Everyday Electronics, Feb 1979

Prospector Metal Locator — Electronics Australia, Nov 1979

Build a Metal Locator — Robert Krieger — Popular Electronics, Jan 1980

Pulse Induction Metal Detector — JA Corbyn — Wireless World, Mar, Apr, & May 1980

Pipe & Metal Locator (Project 566) — Phil Wait — ETI, Apr & Oct 1980

Metal Detector — P.R. Williams — Practical Electronics, Apr 1980

Off-Resonance Metal Detector — G Wareham — Wireless World, June 1980

Magnum Metal Locator — Andy Flind — Practical Electronics, Aug & Sept 1980

Houndog IB Metal Locator — Elementary Electronics, Sept & Oct 1980

All About Metal Detectors — R Gallagher — Radio Electronics, Nov 1980

Metal Detector — Jennings Kimberley — Elektor, July & Aug 1981

Coinshooter — William Lahr — Popular Electronics, Aug 1981

HE "Diana" Metal Detector — Hobby Electronics, Sept 1981

Metal Detector — Elektor, Nov 1981

Beachcomber Metal Detector — Colin Dawson — Electronics Australia, Dec 1984

Metal Detector — Elektor, May 1987

Construction Details of an IB Detector — Everyday Electronics, May 1987

EE Buccaneer IB Metal Detector — Andy Flind — Everyday Electronics, July 1987

Pipe & Cable Locator — Everyday Electronics, Apr 1988

Phaser TR Metal Detector — Practical Electronics, Oct 1988

Find Some Treasure... — Keith Bindley — ETI, Jan 1989

Metal Detector — Robert Penfold — Everyday Electronics, May 1989

EE Treasure Hunter — Mark Stuart — Everyday Electronics, Aug 1989

Twin Loop Treasure Seeker — Robert and David Crone — ETI, Sept 1989

Probe Pocket Treasure Finder — Andy Flind — Everyday Electronics, Sept 1989

Circuit Circus — Charles Rakes — Popular Electronics, June & Aug 1991

Metal Detector — Kamil Kraus — Electronics World, Sept 1992

Metal Detector — Kamil Kraus — Wireless World, Sept 1992

David Hughes: Electromagnetic Pioneer — James P Rybak — Popular Electronics, Nov 1992

Induction Balance Metal Locator — Silicon Chip, May & June1994

Microcontroller PI Treasure Hunter — Mark Stuart — Everyday with Practical Electronics, June 1994

Mini Metal Detector — Elektor, Oct 1996

Simple Metal Detector — Robert Penfold — Everyday with Practical Electronics, Apr 1998

A Metal Detector for Metal Objects — John Clarke — Silicon Chip, May 1998

BFO Metal Detector — Rachel & Steve Hageman — EDN, Sept 1998

Metal Detectors — Gavin Cheeseman — Electronics & Beyond, May & June 1999

Fortune Finder — J. Clarke — Silicon Chip, Dec 1999

Frequency Meter Metal Detector — Andrei Chtchedrine & Yuri Kolokolov — Circuit Cellar, May 2001

Metal Detection — Charles Rakes — Poptronics, Aug 2001

Metal Detecting Circuits II — Charles Rakes — Poptronics, Sept 2001

Matchless Metal Locator — Thomas Scarborough — Silicon Chip, June 2002

Bounty Treasure Hunter — Thomas Scarborough — Everyday with Practical Electronics, Oct 2002

Back-to-Basics Metal Detector — Bart Trepak — Everyday with Practical Electronics, Mar 2003

EPE Mini Metal Detector — Thomas Scarborough — Everyday with Practical Electronics, July 2003

Beat Balance Metal Detector — Thomas Scarborough — Everyday with Practical Electronics, May 2004

Poor Man's Metal Locator — Thomas Scarborough — Everyday with Practical Electronics, June 2006

Minimalist Induction-Balance Metal Detector — Elektor, Oct 2007

Metal Locator — John Clarke — Everyday with Practical Electronics, July 2011

LRC Beat Balance Metal Detector — Thomas Scarborough — Everyday with Practical Electronics, Feb 2012

Patents

Frequency Shift

US3355658 — *Differentiating Metal Detector for Detecting Metal Objects and Distinguishing Between Detected Diamagnetic and Non-Diamagnetic Objects* — Gardiner, VLF BFO

US3467855 — *Object Detector and Method for Distinguishing Between Objects Detected Including a Pair of Radio Oscillators* — Basic BFO

US3519919 — *Frequency Stabilizing Element for Metal Detectors*

US3546628 — *Oscillating Metal Object Detector* — Simplest detector on earth

US3601691 — *Metal Detector Responsive to Small Metallic Objects for Differentiating Between Ferrous and Non-Ferrous Objects* — Gardiner, Z-response

US3626279 — *Metal Detector Utilyzing Radio Receiver and Harmonic Signal Generator* — Harmonic technique increases sensitivity; also printed spiral coil

US3662255 — *Apparatus for Locating Concealed or Buried Metal Bodies and a Stable Inductor Usable in Such Detectors* — Garrett Zero-Drift BFO

US3742341 — *Inductively Coupled Metal Detector Arrangement* — Appears to be an off-resonance technique

US3823365 — *Metal Detecting Apparatus Having Improved Ground-Effect Immunity* — Uses induction balance

US3875498 — *Metal Detector for Distinquishing Between Precious Metal Objects and Other Metals* — D-Tex, your real basic BFO

US3896371 — *Metal Detector With a Resonating Circuit Being Driven by a Frequency Higher Than Its Natural Resonance Frequency* — A.H. Electronics, off-resonance type

US3961238 — *Selective Metal Detector Circuit Having Dual Tuned Resonant Circuits* — Gardiner, very basic Z-response

US3986104 — *Dual Frequency Metal Detector System* — Gardiner

US4130792 — *Metal Detector With Feedback Tuning* — Z-response with feedback

US4196391 — *Metal Locator with Stereotonic Indication of Translateral Position Stereo*

US4204160 — *Metal Detector With Automatic Optimum Sensitivity Adjustment*

US4255710 — *Plural Search Frequency Directional Metal Detector Apparatus Having Enhanced Sensitivity* — Uses two gated oscillator frequencies

US4263553 — *Discriminating Metal Detector With Compensation for Ground Minerals* — A.H. Electronics, off-resonance detector

US4321539 — *Digital BFO Metal Detecting Device with Improved Sensitivities at Near-Zero Beat Frequencies*

US4439734 — *Metal Object Locator Including Frequency Shift Detector* — Induction balance *and* ground balance

US4678992 — *Electronic Metal Detector* — A.H. Electronics, off-resonance type

US5025227 — *Metal Detection Circuit* — Appears to be off-resonance

US7068028 — *Method and Apparatus for Metal Target Proximity Detection at Long Distances* — Appears to be a variation of energy theft, which is akin to frequency shift methods.

Induction Balance

US3405354 — *Apparatus for Limiting Phase-Angle Response Range, Particularly in Eddy Current Testing Apparatus* — Uses synchronous demodulation to determine target phase, see 3848182

US3471772 — *Instrument for Measuring the Range and Approximate Size of Buried or Hidden Metal Objects* — Synchronous detectors, lots of fundamental theory & equations

US3471773 — *Metal Detecting Device with Inductively Coupled Coaxial Transmitter and Receiver Coils* — Basic TR with 4B-style coil (NOT coaxial!)

US3686564 — *Multiple Frequency Magnetic Field Technique for Differentiating Between Classes of Metal Objects* — Multi-frequency detector from 1972

US3826973 — *Electronic Gradiometer* — Technos PRG (Phase Readout Gradiometer)

US3835371 — *Apparatus for Detecting the Presence of Electrically Conductive Material Within a Given Sensing Area* — Focuses on submersible probe & cabling method

US3848182 — *Apparatus for Limiting Phase-Angle Response Range, Particularly in Eddy Current Testing Apparatus* — See 3405354

US3872380 — *Metal Detector Distinguishing Between Different Metals by Using a Bias Circuit Actuated by the Phase Shifts Caused by the Metals* — Gardiner, basic phase response IB

US4016486 — *Land Mine Detector with Pulse Slope, Width and Amplitude Determination Channels*

US4024468 — *Induction Balance Metal Detector with Inverse Discrimination* — White's TR Discriminator (1975), includes good description with phase diagrams and circuitry with component values

US4030026 — *Sampling Metal Detector* — White's, the basis of most of their early-80's analog detectors

US4053828 — *Metal Detector with First and Second Nested Rectangular Coils* — Not too practical for a handheld

US4096432 — *Metal Detector for Discriminatory Detection of Buried Metal Objects* — Basic phase response IB

US4099116 — *Metal Detector With Phase Related Selective Discrimination Circuit* — Nautilus, feedback method for ground balance and discrimination

US4110679 — *Ferrous/Non-ferrous Metal Detector Using Sampling* — White's

US4128803 — *Metal Detector System With Ground Effect Rejection* — PNI (the old Bounty Hunter), probably the Red Baron series (RB3/5/7)

US4249128 — *Wide Pulse Gated Metal Detector With Improved Noise Rejection* — White's

US4263551 — *Method and Apparatus for Identifying Conductive Objects by Monitoring the True Resistive Component of Impedance Change in a coil System Caused by the Object* — Title says it all, looks at target response vs. frequency

US4300097 — *Induction Balance Metal Detector with Ferrous and Non-ferrous Metal Identification* — Techna, now First Texas Mfg. (Bounty Hunter)

US4303879 — *Metal Detector Circuit with Mode Selection and Automatic Tuning* — Garrett (ADS?)

US4325027 — *Metal Detector for Locating Objects with Full Sensitivity in the Presence of Distributed Mineral Material* — Compass

US4334191 — *Metal Detector Circuit Having Momentary Disabled Output* — Garrett

US4334192 — *Metal Detector Circuit Having Automatic Tuning With Multiple Rates* — Garrett, probably their Master Hunter VLF

US4344034 — *Selective Ground Neutralizing Metal Detector* — Gardiner patent for phase discriminator

US4348639 — *Transmitter-Receiver Loop Buried Metal Object Locator with Switch Controlled Reference Voltage* — Discovery Electronics' two-box detector

US4423377 — *Compact Metal Detector of the Balanced Induction Type* — Garrett, handheld with integrated double-D coil, see also 4488115

US4470015 — *Metal Detector System With Undesirable Target and Mineralized Ground Discrimination* — Teknetics, lots of diagrams and waveforms

US4486712 — *Frequency Dependent Pulsed Gain Modulated Metallic Object Detector*

US4486713 — *Metal Detector Apparatus Utilizing Controlled Phase Response to Reject Ground Effects and to Discriminate Between Different Types of Metals* — Tesoro, lots of signal & phase diagrams

US4488115 — *Low Battery Voltage Indicator Circuit for a Metal Detector* — Garrett, see also 4423377

US4507612 — *Metal Detector Systems for Identifying Targets in Mineralized Ground* — Teknetics, tons of info

US4514692 — *Metal Detector and Discriminator Using Differentiation for Background Signal Suppression* — Fisher, basis for early X models

US4563645 — *Inductively Balanced Oscillatory Coil Current for Metal Detection*

US4628265 — *Metal Detector and Classifier with Automatic Compensation for Soil Magnetic Minerals and Sensor Misalignment* — Fisher

US4659989 — *Inductively Balanced Metal Detector Circuit with Orthogonal Balancing Signals and Including Phase and Polarity Detection*

US4677384 — *Target-Identifying Metal Detector* — Teknetics

US4700139 — *Metal Detector Circuit Having Selectable Exclusion Range For Unwanted Objects* — Garrett patent with a pretty good explanation of I&Q signal processing

US4709213 — *Metal Detector Having Digital Signal Processing* — Garrett

US4783630 — *Metal Detector With Circuits for Automatically Screening Out the Effects of Offset and Mineralized Ground* — White's

US4868910 — *Metal Detector with Microprocessor Control and Analysis* — White's, ton's of info on their target ID

US4881036 — *Phase Shift Compensation for Metal Detection Apparatus*

US4894618 — *Metal Detector Using Cross-Correlation Between Components of Received Signals* — Minelab, good explanatory text

US4912414 — *Induction-Type Metal Detector with Increased Scanning Area Capability* — Describes a large array for use with a submersible, includes IB and PI methods

US4942360 — A *Method and Apparatus of Discrimination Detection Using Multiple Frequencies to Determine a Recognizable Profile of an Undesirable Substance* — Minelab, multiple frequency (Sovereign)

US4975646 — *Detector System for Recognizing a Magnetic Material Multi-frequency*

US5148151 — *Metal Detector Having Target Characterization and Search Classification* — Garrett patent on VDI, explanation on phase response and VDI flowchart

US5506506 — *Metal Detector for Detecting and Discriminating Between Ferrous and Non-ferrous Targets in Ground* — Minelab, combines synchronous demodulators with PI, calls it multi-frequency, compare with US5576624

US5523690 — *Metal Detector with Bivariate Display* — White's patent detailing VDI, includes 6805 assembly code (see also 5596277)

US5596277 — *Method and Apparatus for Displaying Signal Information from a Detector* — White's patent detailing VDI, includes 6805 assembly code (see also 5523690)

US5642050 — *Plural Frequency Method and System for Identifying Metal Objects in a Background Environment Using a Target Model* — White's multi-frequency detector (see 5654638)

US5654638 — *Plural Frequency Method and System for Identifying Metal Objects in a Background Environment* — White's multi-frequency detector (see 5642050)

US5691640 — *Forced Balance Metal Detector*

US5721489 — *Metal Detector Method for Identifying Target Size* — Garrett's target ID patent, lots of signal analysis and flow charts, see also 5786696

US5729143 — *Metal Detector With Nulling of Imbalance* — Bucking signal calibrated for amplitude & phase

US5786696 — *Metal Detector for Identifying Target Electrical Characteristics, Depth and Size* — Garrett's target ID patent, lots of signal analysis and flow charts, see also 5721489

US5969528 — *Dual Field Metal Detector* — Garrett's two-box add-on

US6172504 — *Metal Detector Target Identification Using Flash Phase Analysis* — White's, applies phase info to what's essentially a flash ADC; see 6421621

US6421621 — *Metal Detector Target Identification Using Flash Phase Analysis* — White's, applies phase info to what's essentially a flash ADC; see 6172504

US6583625 — *Metal Detector and Method in Which Mineralization Effects Are Eliminated*

US6879161 — *Method and Apparatus for Distinguishing Metal Objects and Employing Multiple Frequency Interrogation* — White's

US6911823 — *Metal Detector Employing Static Discrimination* — White's

US7078906 — *Simultaneous Time-Domain and Frequency-Domain Metal Detector* — Combo PI & phase sampling

US7088103 — *Metal Detector Having a Plurality of Phase Discrimination Regions with Corresponding Selectable Exception Spaces Therein* — White's

US7126323 — *Systems and Methods for Synchronous Detection of Signals*

US7432715 — *Method and Apparatus for Metal Detection Employing Digital Signal Processing* — Minelab

US7579839 — *Metal Detector* — Minelab, multi-frequency demod technique

EP0580396 — *Metal Detector with Display* — White's VDI

Pulse Induction

US3707672 — *Weapon Detector Utilyzing the Pulsed Field Technique to Detect Weapons on the Basis of Weapons Thickness*

US4868504 — *Apparatus and Method for Locating Metal Objects and Minerals in the Ground with Return of Energy from Transmitter Coil to Power Supply* — Fisher Impulse PI, recycles power from the coil

US4894619 — *Impulse Induced Eddy Current Type Detector Using Plural Measuring Sequences in Detecting Metal Objects*

US5047718 — *Improving the Discrimination of an Impulse Technique Metal Detector by Correlating Responses Inside and Outside of a Cut-Off Peak Area*

US5414411 — *Pulse Induction Metal Detector* — White's, block level

US5537041 — *Discriminating Time Domain Conducting Metal Detector Utilizing Multi-Period Rectangular Transmitted Pulses* — Minelab pulse

US5576624 — *Pulse Induction Time Domain Metal Detector* — Minelab, compare with US5506506

US6326790 — *Ground Piercing Metal Detector Having Range, Bearing, and Metal-Type Discrimination* — A PI probe

US6326791 — *Discrimination of Metallic Targets in Magnetically Susceptible Soil*

US6452396 — *Method for Detecting the Metal Type of a Buried Metal Target* — "Periscope" model

US6452397 — Ground Piercing Metal Detector Method for Detecting the Location of a Buried Metal Object — "Periscope" model

US6456079 — *Circuit for Detecting the Metal Type of a Metal Target Object* — "Periscope" model

US6529007 — *Temperature Compensation for Ground Piercing Metal Detector* — "Periscope" model

US6586938 — *Metal Detector Method and Apparatus*

US6636044 — *Ground Mineralization Rejecting Metal Detector (Receive Signal Weighting)* — Minelab, multi-width pulse

US6653838 — *Ground Mineralization Rejecting Metal Detector (Transmit Signal)* — Minelab, multi-width pulse

US6686742 — *Ground Mineralization Rejecting Metal Detector (Power Saving)* — Minelab, multi-width pulse

US6690169 — *Interference Cancelling Metal Detector Including Electronic Selection of Effective Sensing Coil Arrangement* — Minelab

US6724305 — *Pulse Induction Silverware Detector*

US6853194 — *Electromagnetic Target Discriminator Sensor System and Method for Detecting and Identifying Metal Targets*

US6927577 — *Digital Nulling Pulse Inductive Metal Detector*

US6967574 — *Multi-Mode Electromagnetic Target Discriminator Sensor System and Method of Operation Thereof*

US7078906 — *Simultaneous Time-Domain and Frequency-Domain Metal Detector* — Combo PI & phase sampling

US7148691 — *Step Current Inductive Antenna for Pulse Inductive Metal Detector*

US7474102 — *Rectangular-Wave Transmitting Metal Detector* — Minelab PI, see 7791345

US7652477 — *Multi-Frequency Metal Detector Having Constant Reactive Transmit Voltage Applied to a Transmit Coil* — Minelab

US7710118 — *Resonant Pulse Induction Metal Detector that Transmits Energy from High Voltage Flyback Pulses* — Discriminating PI

US7791345 — *Rectangular-Wave Transmitting Metal Detector* — Minelab PI, see 7474102

EP0654685 — *Arrangement and Method for Detecting Metal Objects* — PI with DD coil

EP0732600 — *Active Impulse Magnetometer* — Looks more like a PI method

Coils

US3549985 — *Metal Detector Device Having a Disk-Shaped Head for Housing a Coil System* — Seems to mostly cover plastics & fillers

US3753185 — *Metal Detector Search Coil* — Bill Mahan, D-Tex

US3882374 — *Transmitting-Receiving Coil Configuration* — IB

US4255711 — *Coil Arrangement for Search Head of a Metal Detector* — Compass, concentric IB

US4276484 — *Method and Apparatus for Controlling Current in Inductive Loads Such as Large Diameter Coils* — Pulse method

US4293816 — *Balanced Search Loop for Metal Detector* — White's concentric loop

US4345208 — *Anti-falsing and Zero Nulling Search Head for a Metal Detector* — Daytona search coil

US4552134 — *Equipment for Determining the Position of a Metal Body in a Medium With Low Electric Conductivity* — Particular IB coil arrangement

US4862316 — *Static Charge Dissipating Housing for Metal Detector Search Loop Assembly* — White's concentric loop

US4890064 — *Metal Detector Sensing Head with Reduced Eddy Current Coils* — Minelab

US5038106 — *Detector of Metalliferous Objects Having Two Pairs of Receiving Loops Symmetrical and Orthogonal to a Driving Loop* — see US5039946

US5039946 — *Metalliferous Objects Detector Having a Pair of Angularly Positioned Driving Loops and a Pair of Parallel, Coaxial Receiving Loops* — see US5038106

US5245307 — *Search Coil Assembly for Electrically Conductive Object Detection*

US5498959 — *Metal Detector With Multipolar Windings Shaped So As To Eliminate the Neutralizing Effects When Several Metal Masses Are Passing Through Simultaneously* — Walk-through type coil arrangement

US5859532 — *Method of and Measuring Arrangement for Metal Detection With a Coil Device Having Several Separately Controllable Regions* — Coil arrangements for walk-through type

US5863445 — *Etched Coil Unibody Digital Detector* — Handheld wand, EP0249110

US6791329 — *Portable Metal Detection and Classification System* — Coil methods for a powered rover

US6822429 — *Inductive Sensor Arrangement Comprising Three Sense Coil Cooperating with Said Three Field Coils to Perform Three Field/Sense Coil Pairs and Method for Detecting of Ferrous Metal Objects* — Say that 3 times fast

US7075304 — *Variable Damping Induction Coil Metal Detection* — Uses a MOSFET for the damping resistor

US7176691 — *Switched Coil Receiver Antenna for Metal Detector*

EP0249110 — *Sensors for Metal Detectors* — Some various coil arrangements

EP0764856 — *Sensor for a Metal Detector* — Several IB coil arrangements

Eddy techniques

US3337796 — *Eddy Current Testing Device with Means for Sampling the Output Signal to Provide a Signal Proportional to the Instantaneous Value of Said Output Signal at a Particular Phase* — Helluva title, the basis for most modern sampled discriminators

US3478263 — *Wide Frequency Range Eddy Current Testing Instrument*

US4006407 — *Non-destructive Testing Systems Having Automatic Balance and Sample and Hold Operational Modes*

US4095180 — *Method and Apparatus for Testing Conductivity using Eddy Currents*

US4188577 — *Pulse Eddy Current Testing Apparatus for Magnetic Materials, Particularly Tubes*

US4191922 — *Electromagnetic Flaw Detection System and Method Incorporating Improved Automatic Coil Error Signal Compensation*

US4230987 — *Digital Eddy Current Apparatus for Generating Metallurgical Signatures and Monitoring Metallurgical Contents of an Electrically Conductive Material*

US4303885 — *Digitally Controlled Multifrequency Eddy Current Test Apparatus and Method*

US4424486 — *Phase Rotation Circuit for an Eddy Current Tester*

US5508610 — *Electrical Conductivity Tester and Methods Thereof for Accurately Measuring Time-Varying and Steady State Conductivity Using Phase Shift Detection*

US5952879 — *Device for the Simultaneous Demodulation of a Multifrequency Signal, Particularly for an Eddy Current Measurement*

Security

US3676772 — *Metallic Intrusion Detector System* — Pass-through type

US3950696 — *Trapezoidal Coil Configuration for Metal Detector in the Shape of an Inverted U* — Walk-through type

US4012690 — *Device for Selectively Detecting Different Kinds and Sizes of Metals* — Walk-through type

US4605898 — *Pulse Field Metal Detector with Spaced, Dual Coil Transmitter and Receiver Systems* — Walkthrough PI

US4779048 — *Metal Detector for Detecting Metal Objects* — Pass-through type

US4821023 — *Walk-Through Metal Detector*

US4866424 — *Metal Detector Coil* — Walk-through type

US4906973 — *Walk-Through Metal Detector* — White's

US5121105 — *Metal Detector* — Walk-through type

US5521583 — *Metal Detection System* — Walk-through type, see US5680103

US5680103 — *Metal Detection System* — Walk-through type, see US5521583

US5726628 — *Metal Detector System* — Walk-through type

US5790685 — *Apparatus and Method for Detecting and Imaging Metal* — Pass-through type

US6133829 — *Walk-through Metal Detector System and Method* — Fisher

US6696947 — *Metal Detector* — Foldable security walk-through

US6970086 — *Wide Area Metal Detection (WAMD) System and Method for Security Screening Crowds*

EP0611970 — *Multiple Aerial for Metal Detector* — Cylindrical coil for walk-through

Mine Detection

US5307272 — *Minefield Reconnaissance and Detector System* — Pulsed radar

US5680048 — *Mine Detecting Device Having a Housing Containing Metal Detector Coils and an Antenna* — Appears to combine a metal detector and GPR in one search head

US7532127 — *Motion and Position Measuring for Buried Object Detection* — Mine detector

Misc.

US3836960 — *Sensor System* — VHF/UHF

US3976564 — *Combination Digger and Sifter for Use With Metal Detector*

US4006481 — *Underground, Time Domain, Electromagnetic Reflectometry for Digging Apparatus* — High frequency wide spectrum detector mounted to a digging tool

US4529937 — *Metal Detector With Spring Loaded Hinged Support*

US4540943 — *Belt-Supported Swingable Metal Detector*

US4560935 — *Remote Actuator for Metal Detector Discriminating Adjust Switch*

US4594559 — *Metal Detector Audio Amplifier*

US4641091 — *Device for Testing and Calibrating Treasure Hunting Metal Detectors*

US4644290 — *Metal Detector Audio Amplifier*

US4719421 — *Metal Detector for Detecting Product Impurities* — Auto phase adjustment

US4779777 — *Support Bracket for Metal Detector*

US4797618 — *Caddy for Metal Detector*

US4983281 — *Metal Detector Scoop Sifter*

US5045789 — *Detector for Detecting Foreign Matter in Object by Using Discriminant Electromagnetic Parameters*

US5138262 — *Metal Detector Having Detachable Battery and Speaker Housing* — Garrett

US5247257 — *Electronic Metal Detector Return Signal Phase Changer* — A nail and two coils, weird

US5501283 — *Hole Cutting Device for Recovering Targets Located with a Metal Detector Audio Amplifier*

US5696490 — *FM (VHF) Infrared Wireless Digital Metal Detector*

US5896031 — *Quad Coil Vibration Cancelling Metal Detector*

US5963035 — *Electromagnetic Induction Spectroscopy for Identifying Hidden Objects* — Multi-frequency method, interesting data

US5994897 — *Frequency Optimizing Metal Detector*

US6791329 — *Portable Metal Detection and Classification System* — Detector on wheels

US6838886 — *Method and Apparatus for Measuring Inductance* — Vehicle method

US6870370 — *Electromagnetic Induction Detection System* — Airborne method

US7081754 — *Metal Detection System With a Magnetometer Head Coupleable to Conventional Footwear and Method of Use* — Metal detector on your shoe

US7123016 — *Systems and Methods Useful for Detecting Presence and/or Location of Various Materials* — A mish-mash of techniques, see 6724191

US7132943 — Moving Belt Sensor

US7288927 — *Remote Substance Identification and Location Method and System* — Supposedly an infrared molecular locator (LRL?)

US7310586 — *Metal Detector with Data Transfer* — Minelab, transfer of operating parametrics

US7575065 — *Metal Detector with Excavation Tool* — Garrett pinpointer

EP0790507 — *Metal Detector with Pivoting Detector Coil...* — Lockable coil pivot

Pre-1970

US269439 — *Apparatus for Finding Torpedoes* — 1882, C.A. McEvoy (First US detector patent?)

US1126027 — *Apparatus for Detecting Pipe Leads or Other Metallic Masses Embedded in Masonry* — 1915, Dr. Max Jüllig (Figure 8 coil)

US1812392 — *Method of and Apparatus for Locating Terrestrial Conducting Bodies* — 1931

US1890786 — *Radio Distance or Location Finder* — 1932

US2066135 — *Apparatus for Locating Bodies Having Anomalous Electrical Admittances* — 1936, 2-box locator, much like Fisher's

US2066561 — *Metalloscope* — 1937, Fisher 2-box

US2129058 — *Transformer for a Metal Locator* — 1936, describes an adjustable DD coil

US2139460 — *Means and Methods for Geophysical Prospecting* — 1938

US2160356 — *Geophysical Instrument* — 1939, another 2-box

US2167630 — *Electrical Prospecting Method and Apparatus* — 1938, bore hole unit

US2201256 — *Electrical Apparatus and Method for Locating Minerals* — 1940, 2-box variation

US2220070 — *Method and Apparatus for Magnetically Exploring Earth Strata* — 1940, bore hole instrument

US2268106 — *Radiowave Prospecting* — 1941

US2208029 — *Electrical Prospecting Apparatus* — 1946

US2447316 — *Variable Frequency Oscillatory System* — 1948

US2451596 — *Unitary Balanced Inductor System* — 1948, concentric coil

US2608602 — *Detecting Device* — 1952, bore hole unit, describes phase discrimination

US3015060 — *Method and Means of Prospecting for Electrically Conducting Bodies* — 1961

US3020470 — *Submerged Body Detection System* — 1962

US3105934 — *Method and Apparatus for the Remote Detection of Ore Bodies Utilizing Pulses of Short Duration to Induce Transient Polarization in the Ore Bodies* — 1963

US3471773 — *Metal Detecting Device with Inductively Coupled Coaxial Transmitter and Receiver Coils* — 1969